NOBEL PRIZES
Cancer, Vision and the Genetic Code

NOBEL PRIZES
Cancer, Vision and the Genetic Code

Erling Norrby
The Royal Swedish Academy of Sciences, Sweden

NEW JERSEY • LONDON • SINGAPORE • BEIJING • SHANGHAI • HONG KONG • TAIPEI • CHENNAI • TOKYO

Published by

World Scientific Publishing Co. Pte. Ltd.

5 Toh Tuck Link, Singapore 596224

USA office: 27 Warren Street, Suite 401-402, Hackensack, NJ 07601

UK office: 57 Shelton Street, Covent Garden, London WC2H 9HE

Library of Congress Control Number: 2019035635

British Library Cataloguing-in-Publication Data
A catalogue record for this book is available from the British Library.

NOBEL PRIZES
Cancer, Vision and the Genetic Code

Copyright © 2019 by Erling Norrby

All rights reserved.

ISBN 978-981-120-085-4
ISBN 978-981-120-086-1 (pbk)

For any available supplementary material, please visit
https://www.worldscientific.com/worldscibooks/10.1142/11421#t=suppl

Printed in Singapore by Mainland Press Pte Ltd.

Preface

This is my fourth book on Nobel Prizes in the natural sciences. The three previous books — *Nobel Prizes and Life Sciences, Nobel Prizes and Nature's Surprises, Nobel Prizes and Notable Discoveries* — have been published at three-year intervals. Now another three years have passed and it is time for the fourth book. It has been given a title that illustrates the main research areas covered — *Nobel Prizes — cancer, vision and the genetic code*. Hence three very different fields of advancing knowledge in the fields of physiology or medicine are discussed. Only rarely has the field of cancer been mentioned in the motivations for prizes in this category. However, obviously insights into the physiological conditions of cell metabolism and cell replication recognized by prizes are indirectly of importance for understanding the mechanisms underlying an altered cell behavior leading to an escape from physiological growth control mechanisms. There were two discoveries highlighted by the 1966 prize. These were the findings that viruses could possibly be involved in the emergence and development of abnormally dividing cells and that selected hormones could be used for suppressing the growth in patients of cells originating in particular organs like the prostate and the mammary gland.

The prize to Peyton Rous was unique because it recognized a discovery already made fifty years before it was awarded and the prize to Charles Huggins was special because it recognized a surgeon and a discovery of immediate clinical importance. In Chapters 1 and 2 the personalities and the professional scientific achievements of Rous and Huggins are presented. Chapter 3 represents an attempt to summarize the impressive developments after 1966 of the comprehensive field of research into cancer, a disease that remains a daunting challenge to biomedical research. However, major advances have been made and step by step new modalities of treatment have become available to control what were previously considered cases of the disease impossible to manage. The view on the role of viruses in the appearance of cancers has shifted between suggestions that they have no importance to their being the cause of all cancers. Since I have a background as a virologist I have put a particular emphasis on explaining the current perspective on this matter. Different viruses are involved in some 15-20 % of all cancers and in some cases this insight offers an opportunity to prevent certain forms of cancer by vaccinations.

Many colleagues have helped me to ensure that the extensive field of cancer is presented in a timely and appropriate way. In particular Klas Kärre at the Karolinska Institute read the text of the first three chapters and made valuable comments. He introduced me to the concept of the "hallmarks of cancer" and provided valuable references. Stephen Chanock at the National Cancer Institute, National Institutes of Health also kindly read the text. In my writing about Rous I initially had some help from Georg Klein. Rous archives are stored at the American Philosophical Society and via Annie Walcott I made contact with the librarian of the Society, Charles Greifenstein. He facilitated my review of the archives. Neeraja Sankaran shared with me the fairy tale by Christopher Andrewes. Finally Christer Sylvén and his wife reminisced about their responsibility way back in time for serving as diplomatic hosts of Rous and his wife during their Nobel week in Stockholm.

The wide-ranging chapter 3 was written by means of advice from many sources. Joe Goldstein provided valuable detailed information about the Lasker Prizes. Felix Mitelman gave advance on the developing field of cytogenetics. Robert C. Gallo, a pioneer in the field of cancer viruses, provided useful background material and in particular anecdotal information about fellow scientists in the field of cancer research. Roberto Cattaneo provided information on oncolytic viruses, Karen Nelson on microbioms, Harald zur Hausen on papilloma viruses and diets and cancer and Gösta Gahrton on treatment of childhood leukemia. Finally Shuguang Zhang kindly arranged meetings with cancer immunologists at MIT.

Chapters 4 and 5 were a challenge to write since the prize concerned the physiology of seeing and in addition a Finnish-Swedish scientist, Ragnar Granit, working at the Karolinska Institute was recognized together with two colleagues. I had already written about Granit in my preceding book, since he was heavily involved in the 1963 prize to the three neurophysiologists John C. Eccles, Alan L. Hodgkin and Andrew F. Huxley. The previous presentation of the large field of neurophysiology, the one field in parallel with immunology which has been recognized most frequently among prizes in physiology or medicine, was helpful when attempting to explain the neurophysiology of seeing both colors and in the penumbra. The key person in reviewing Chapters 4 and 5 was Sten Grillner. He knew the field exceptionally well since he was the successor of Ragnar Granit at the Karolinska Institute. I am also very grateful for contacts with the succeeding generations of the Granit family. Ragnar Granit's son Michael, visited me for a very pleasant and valuable conversation at the Center for the History of Science at the Royal Swedish Academy of Sciences. He spoke very warmly and informatively about his father. Sadly Michael died unexpectedly very soon after our encounter. I have also had important contacts with Granit's grandson Jacob and great-grandson Joakim. In chapter 5 I mention briefly a particular friend from many encounters in La Jolla, CA. It was the remarkable scientist Manfred Eigen, who received half the prize in chemistry in 1967. Sadly he also passed away during the later stages of writing this book. In this context I had contacts with his surviving wife Ruthilde.

Throughout my previous three books on Nobel Prizes I have returned to the continuously developing and impressively powerful field of molecular biology. Since the 1968 prize concerned the cracking of the genetic code I could summarize the enormous incremental knowledge in the field until that time and once again emphasize the almost limitless importance of being capable of both reading and writing the books of life. It was possible once again to return to the 1940-50s developments in the field and close my books on Oswald Avery and Erwin Chargaff. Jacek Hawiger kindly guided me to Avery's grave. The story of Marianne Grunberg-Manago crying in the kitchen was described to me by Robert Haselkorn.

Knowing the genetic language has already changed and will even more so in the future change the way we practice medicine. The Karolinska Institute could have decided to recognize only a single prize recipient in 1968, but chose to fill the maximum number of three. The central figure in the discovery of the code was Marshall Nirenberg. He had a unique humble but competitive personality and seems to have been liked by all colleagues in his immediate surroundings. Robert Gallo, Ham Smith, Michael Sela and others gave valuable insights into his warm but somewhat withdrawn personality. Joe Goldstein worked briefly with Nirenberg and has written about the golden age at NIH in the 1960s and the beginning of the 1970s.

The co-recipient of the 1968 prize Gobind Khorana was an impressive chemist. It was on his way to synthesize a complete gene that he made seminal contributions to the cracking of the code. The third man in the prize trio, Robert Holley, was a more withdrawn scientist. However, he and his group pioneered the elucidation of the structure of a critical so-called transfer RNA. After this important contribution he settled at the Salk Institute in La Jolla. I got some insight into his humble character and his talent in making small statues by contacts with Tony Hunter and Gerald (Jerry) Joyce at this Institute.

Peter Reichard was a very central figure as a reviewer for the candidates to the 1968 prize. As long as he lived he gave valuable information, but sadly he also died during the time of writing this book. However, his wife Vera Bianchi also provided valuable information on separate matters, in particular the Italian Luigi Villa who made an interesting nomination discussed at the beginning of Chapter 8. This concluding chapter widens the perspective and discusses how it has become possible for us to see the invisible. The development of the electron microscope, as in the case of Rous, recognized Ernst Ruska more than 50 years after the discovery. Since he was not awarded his shared prize in physics until 1986, there will be almost two decades until a full review can be made of the process that led to his prize. Interestingly from what can be seen until the present time the committee of physics retained a very lukewarm attitude to Ruska's candidacy. Hans Gelderblom provided a number of very valuable references describing the story of the development of the electron microscope. My class mate in medical school Nils Sjöstrand also gave helpful information.

In the second half of Chapter 8 I cannot refrain from discussing the growing insights into the origin of life. This development is the result of the remarkably expanded understanding of the various distinctive, in some cases very surprising functions carried by different forms of RNA. When telling this story I have had invaluable help by having access to the unpublished biography of James (Jim) Darnell. As I have described, the developments in this field have registered a number of very unanticipated discoveries and it is certain that there is more to come. Besides Darnell, Ulf Pettersson, Jerry Joyce and John Glass gave valuable information. It may seem that by writing about the origin of life I have now exhausted the field of molecular biology. However health permitting there might be more to come.

The 1969 Nobel Prize in physiology or medicine was awarded to Max Delbrück, Alfred D. Hershey and Salvador E. Luria for their pioneering work on the replication of bacterial viruses, bacteriophages. Research in this field has shown the way into a wider understanding of the complexity of genetic interactions. Thus again viruses are central in our development of an understanding of how evolution works. It is very tempting in some future writing to come back to a presentation of what has been referred to as "the virocentric view of life". The message is simple. Archaic primitive virus-like structures were decisive in the very early development of life on Earth. Their descendants can also be predicted to also have a critical role in the later evolution of life. They could supplement vertical gene transfer, from parent to offspring cells, by a horizontal gene transfer. Hence it is thanks to, not in spite of viruses that the development of a species with self-consciousness — us — has been possible on Earth. In following the amazing developments in the field of virology, in which I for many years now have no longer been active as an experimenter, I have had great help from many colleagues and friends from before. I would like to particularly mention Frederick (Fred) Murphy, who kindly let me stay with him during repeated visits to Washington. He is the author of a unique pictorial documentation of the history of virology and we could talk about this discipline all night long.

The writing of this book also depends on the generous involvement of a number of other people and also organizations. Following my yearly applications, Thomas Perlmann, the secretary of the Nobel Committee at the Karolinska Institute, and Karl (Kalle) Grandin, head of the Center for the History of Science at the Royal Swedish Academy of Sciences, have granted me access to the unique Nobel archives. I have previously emphasized that they are exceptional sources of information and will not elaborate further on this fact, except to note that they may even be considered to have a world heritage quality. The material presented in the book should speak for itself. When visiting the Nobel Forum at the Karolinska Institute I have had the pleasure of interacting with Ann-Mari Dumanski and Tatiana Goriatcheva, the latter of whom however regrettably retired about a year before the material for the book was finished. Since English is not my native tongue, all chapters in their final form have been read by Harry Watson. We have had the pleasure of collaborating on all my four books and I am glad to express my gratitude for the expedient and educative way in which he has improved my texts. The Sven and Dagmar Salén Foundation have generously defrayed the costs of this reviewing.

I am grateful for the full support given to me by World Scientific Publishing in the development of yet another book on the Nobel Prizes. In particular I would like to thank its Chairman Professor Kok Khoo Phua and the highly competent editor Kim Tan who with her colleagues have given support and advice in the processing of the material for the book. I have been provided with pleasant working conditions at the office in Singapore and my visits to that city have allowed re-establishing contacts with persons of historic importance, like Sydney Brenner. I had the privilege of interacting with him as late as only a few weeks before his death.

My professional office is at the Center for the History of Science. Once again I would like to emphasize the pleasure it is to have this environment as my second home, not least because my regular home milieu has become very quiet since my wife Margareta needed to be institutionalized because of her Alzheimer's disease. In November 2018 she finally died, a blessing in disguise. I have already mentioned the head of the Center, Karl Grandin, but would like to add thanks for all the hours he has patiently spent with me developing materials for the book, be they pictures or tables. The same generous support has been provided by him for preparations of slides I have used in my frequent lecturing about the rich Nobel material I have accumulated. The contributions by my other colleagues should also, with equal emphasis for creating a unique ambience, be highlighted. They are Maria Asp, Anne Miche de Malleray, Jonas Häggblom and Åse Frid. A very special thanks goes to my next-door office neighbor, Bengt Jangfeldt. I treasure our friendship and the continuous lessons he has given me in how to manage the written word. Our almost daily encounters provide a unique opportunity for inspiring exceptional meetings of the natural sciences and the humanistic disciplines.

Then it only remains to find a proper dedication for this fourth book. Previous books for obvious reasons have been dedicated to my wife, to our children and to our grandchildren. Since I do not as yet have any great grandchildren, there is a need to find someone else. I have decided to bestow this dedication on the rich collective of friends that my unique profession as a scientist has allowed me to gather. As a representative I will mention only one, the larger-than-life Craig Venter. He has placed remarkable confidence in me and I have for more than ten years been the Vice Chairman of the Board of Trustees of the J. Craig Venter Institute. One would think that with the dynamic pace of life that characterizes Craig's existence there would be little space left for the less spectacular but potentially humanly important considerations. I will give two examples to illustrate that this need not be the case.

A number of summers ago the charming sailing boat Sorcerer II, previously used for sampling of genes in the oceans around the world by Craig and his collaborators, had been brought to the Stockholm archipelago for later sampling of the Baltic Sea. The conditions were optimal for showing the remarkably rich number of rock islands of many different sizes. Finally we were going to dock the boat for a few days at the bridge at our summer house at the island Blidö. Craig then proposed finding his way to this place by use of a kayak and off he went with two crew members in their own kayaks. We then picked them up in the big boat one-third of the way along, on which occasion Craig said to me: "I have found the essence of life!" No wonder I became curious. "What is it?" "The essence of life is naked skin against a smooth rock." In order to fully appreciate the unique essence of the statement one needs to know that many of the rocks in the archipelago have been polished by the inland ice some 10,000 years ago to have a silk-like surface. All I could say was "Now you have come a long way!" A few years later I was visiting La Jolla for a board meeting and I was in a blue mood because of the progressive developments of my wife's disease. I tried to explain how I attempted to move forward in life by taking intellectual and physical challenges to which Craig wisely added "And then there are the emotional challenges!" At the board meeting of the J. Craig Venter Institute some ten days after my wife had died, Craig proposed to the Board that a permanent bench with memorial inscription of my wife should be established at the institute. This was unanimously approved and later on seeing this bench was a source of strong emotions.

In 2018 I received a cultural stipend to spend three weeks at Villa San Michele at Anacapri. It is a remarkable and mesmerizing place created by the enigmatic physician, author and animal lover Axel Munthe. It provided a very peaceful environment for writing and in addition I had the privilege of meeting a number of other scholars, each of them developing his or her particular cultural obsession. I thank Kristina Kappelin and her staff for providing a place of great ambience. In 2019 I was back for another two weeks. It was during this time that I wrote this preface, and hence the signature below.

April 2019, Anacapri

Contents

Preface v

Chapter 1

The Long Wait 1

Rous's Early Life 4
The Discovery and an Interlude of Other Science 7
Martin Arrowsmith 11
The First Evaluation of Rous. The Critical Role of
    Reproducibility of Scientific Data 17
The 1930s — Rabbit Warts and a New Start 21
A Foresighted Fairy Tale 25
New Nominations and New Conclusions 29
Evaluation of Rous's Continued Work in the 1940s 34
Tumor Viruses Come of Age in the 1950s 36
Klein's Insightful Review 41
Approaching a Nobel Prize at Last 44

Chapter 2

Hormone Treatment of Tumors and the Prize Events in 1966 47

Discovery Is Our Business 50
The Nobel Committee Reviews a Surgeon 54
A Nobel Committee in a Quandary 60
The Prize Events in 1966 67
The Aftermath and Nobel Medals 78
Coda 83

Chapter 3

Rous Virus and the Elucidation of the Genetic Nature of Cancer 85

A Premature General Attack on Cancer 86
The American "Nobel Prize" 90
A Critical Amendment of the Central Dogma 93
The Secrets of Rous Virus Finally Unraveled 98
Oncogenes Take the Stage 104
Tumor Suppressor Genes 106
Aging Cells and Cancer 111
Chromosomes and Genes in Cancer 114
Immune Defense and the Development of Cancer 121
The Human Microbiome and Cancer 125
Viruses and Cancer 128
The Hallmarks of Cancer 134

Chapter 4

The Rock Foundation of Nobel Prize Developments 139

A Harmonious and Challenging Upbringing 142
Training to Become a Neurobiologist 145
Fundamentals of the Process of Vision 148
The Rocky Road Towards a Stable Academic Position 151
A Change of Homeland and Post-War Developments 153
A Shift of Focus in Science 157
The Early Enthusiasm of the Nobel Committee 161
Turmoil in the Evaluation Process 166
Granit's Candidacy Back on Track 171
Finally an Expanded Basis for a Prize 177

Chapter 5

Visionary Contributions Gave a Happy Trio 181

A Student Born to Become a Scientist 184
Hartline's Career Takes Off 187
A Rich Personality 190

The Nobel Committee Reviews Hartline 193
An Exceptional and Narcissistic Eyewitness 198
Wald's Discoveries Catalyze the Prize Discussions 205
The Festivities and a Charming Mishap 214

Chapter 6
The Prime Author of the Saga of the Genetic Code 225
An Important Nobel Prize Given for Experiments
  Later Shown to be Flawed 228
The Lady Is a Trump 234
The Chagrin of Chargaff 236
Premature Discoveries Revisited 244
Crick and the Early Speculations on the Code 245
The Development of a Humble and Unassuming Scientist 248
NIH Provides an Important Home to Nirenberg 251
A Major Breakthrough 254
The Second Major Breakthrough 262
The Chemistry Committee Reviews Nirenberg 265
Evaluations of Nirenberg for a Nobel Prize in Physiology or Medicine 270

Chapter 7
The Formation of a Trio for the 1968 Prize 283
Todd's Perspective on the Biochemistry of Nucleic Acids 285
A Star Biochemist with an Exceptional Background 288
The Review of Khorana for a Prize in Chemistry 290
The Late Nomination of Khorana for a Prize at the Karolinska Institute 294
The Third Man and a Single Molecule 296
Holley as a Candidate for a Prize in Chemistry 298
An Exemplary Review of Holley for a Prize in Physiology or Medicine 300
Decisions on the Prize in 1968 306
Time for the 1968 Nobel Festivities 308
Life After Cracking the Genetic Code 318

Chapter 8

To See the Invisible and to Read the Unprinted 327

A Seemingly Analysis of the Wrong Nominee 329
Virus Particles Visualized for the First Time 331
Construction of an Electron Microscope 335
Possibilities to Examine the Structure of Cells 340
The Lukewarm Reception of Ernst Ruska by Physics Committees 342
The Three-Dimensional Structure of Complex and
   Aggregated Macromolecules 343
The Evolution of the Genetic Language Used Since the Dawn of Life 346
Insights Into an Unknown World of RNA 352
RNA and the Origin of Life on Earth 364
To the Greatest Benefit of Mankind 374

References 383

Index 391

Chapter 1

The Long Wait

> AN ULTRAFILTRATE
> CAUSED SARCOMAS IN HEN
> CANCER INFECTIOUS?

The precision and versatility of Nature is remarkable. Cell division is at the heart of all organic matter. It allows for replication in a highly faithful manner. This stability depends on the impressive fidelity of duplication of the central information-carrying molecule, the DNA. The error in insertion of a proper matching nucleotide in the double-stranded molecule, remarkably, is only one in a million. And still there are errors. These may be detrimental to the function of cells and potentially cause disease in multicellular organisms, but accidentally they may sometimes also be of value. The progress of evolution depends on such mistakes.

The evolution of ever more complex forms of life on Earth is a magnificent story. In the dawn of this process macromolecules with information-storage and operative capacity evolved, eventually to be enclosed in membrane structures. This allowed coordinated interaction between molecules in time and space. The first cell was formed some 3.8 billion years ago and it developed a progressively more intricate machinery for successful interactions with the environment. The cells of all present living organisms have their origin in the very first cell that came into existence. This simple fact was not identified until in the 19th century by the famous German pathologist Rudolf Virchow. In a popularized epigram *Omnis cellula e cellula* — all cells (come) from cells — he highlighted the simple truth that present day extensively diversified forms of life have arisen by billions and billions of cell divisions. They all have their origin in the last universal cellular ancestor (LUCA). But the existence of diversity emanates from the critical astrophysicist definition of life as "a self-replicating chemical system *with a capacity to make mistakes* (italics mine)." In an evolutionary perspective it is the rare mistakes which improve the survival value of the system that is critical. Evolutionary changes over eons of time eventually allowed the emergence of an amazingly complex multicellular system — modern man. Our brain has a relatively large size compared to our body, a fact correlated with our advanced form of self-consciousness, a remarkable capacity to memorize experienced events and plan for future actions. In spite of impressive scientific advances it remains for us to explain the molecular bases of these exceptional qualities. Suffice to note that with the emergence of modern man the possibilities for further evolutionary changes on Earth have changed dramatically and irreversibly since we — for good or for bad — have become masters of such changes in the future.

Multicellular organisms represent remarkable machines since they continuously renew almost all parts of themselves at different rates as they carry out their diversified daily tasks. In our bodies we have about 120 different kinds of tissues, generally a composite of cells of various qualities, specialized to fulfill certain selective functions. The cells of these tissues vary extensively in their turnover. Our nerve cells, heart muscle cells and fat cells replicate only rarely, whereas the cells in our skin/hair, intestinal membrane or the bone marrow derived white and red blood cells show a rapid turnover. Thus, for example, every minute about 100 million red blood cells are formed in our body! It is not an overstatement to note that the regulation of all the cells dividing at different rates represents an extremely complex system, the functions of which we have only a limited insight into. It can be conjectured that there must exist very delicate control mechanisms that aim at keeping our bodily functions in a coordinated order — to maintain *homeostasis* — under prevailing environmental conditions.

Particular mechanisms are required to take care of and eliminate the naturally emerging senescent cells. A special requirement for the control of the behavior of cells and of means for their possible destruction arises when an individual cell accidentally escapes normal cell division control mechanisms and start replicating at the cost of neighboring cells. This may lead to an abnormal local accumulation of like cells. An example is the excessive replication of superficial skin (epidermal) cells, as in plantar warts. However, in particular cases the independence of an abnormally replicating cell may become further accentuated and the local tumor evolves to threaten the whole individual in whom it originates. What is initially referred to as a *benign* amplification of cell replication evolves into a *malign* form of cell division. There is a host of mechanisms that aim at quickly identifying and destroy cells that happen to escape the rigorous cell division control mechanisms. And still on rare occasions one of these misbehaving cells may sneak through the fine network of controls and initiate an existence on its own. By an evolutionary process this so-called *transformed cell* in the worst case can develop to eventually destroy its host and hence unintentionally its own existence. Since it originally derives from a single malfunctioning cell it represents a *clonally* evolving disease. In this particular case the "survival of the fittest" may become suicidal. It is due to the aimless replication of freebooter cells. This potentially life-threatening process can be seen as the dark side of the powers of evolution in action. The developing cell clone dies together with its host, unless we manage to keep the tumor cells alive by cultivating them in the laboratory, which is a questionable form of immortality.

The term *karkinos*, the Greek word for crab, was introduced by Hippocrates around 400 years BCE to characterize visually identifiable life-threatening forms of cell transformation. The occasional tumor with its clutch of enlarged blood vessels gave him the image of a "crab" and the term with its vivid mental implications caught on. The term *cancer* has developed both to have a general and also more restricted use. It is applied to describe all kinds of serious tumor diseases but in addition to also specifically define tumors deriving from different kinds of epithelial cells, *carcinomas*. There is also another Greek work of importance in this context. This is the word *onkos*, a Greek term for "mass" and from this derives the modern term *oncology*, the study of tumors. As pointed out in an excellent book on the history of cancer, *The Emperor of All Maladies. A Biography of Cancer*[1] the etymological origin of the term *onkos* is an Indo-European word for *carry*, an equally appropriate metaphor. Because of its origin as a foe from within the body it is not surprising that the psychological existential management of the diagnosis cancer is often very difficult. It is hard to accept the self-destructive impact of cells originating in your own body.

The emergence of tumors and their impact on human lives has been recorded throughout the development of mankind. As civilization has progressed the relative importance of this group of diseases has increased. In the early days of average life spans of some 25–30 years mostly due to prevailing infectious diseases and accidents, cancers played a minor role as a cause of death. But in industrialized civilizations with dramatically increased longevity this kind of disease has become responsible for about a quarter of all deaths. To this should be added that most cancers occur after reproductive age has ceased and hence they cannot be genetically selected against. During the 20th century our insights into the processes of uncontrolled cell growth have increased dramatically, not least by the introduction of molecular genetics during the second half of the century.

The Nobel Prize to Peyton Rous is unique. He studied *sarcomas* — malignant cells originating in connective tissue cells — in domestic fowls. In 1911 he made the discovery that homogenized tissues of this kind of tumor passed through a filter which retained bacteria and other microorganisms (an ultra-filtrate) could transmit a tumor disease of the same kind to other domesticated fowls. The tumor could be transferred from one fowl to another retaining its tissue characteristics. This remarkable observation was not recognized for a prize in physiology or medicine until in 1966, when Rous, still an active researcher, had reached the age of 87. Using review material presented by various committee members at different times, pertinent primarily to Rous, it is possible to describe the development of scientific knowledge about cancer through time. In the case of Rous's work for more than half a century, three separate phases of accumulating knowledge can be distinguished. Before this is discussed let us look at Rous's background and his development into a dedicated scientist.

Rous's Early Life

The home environment in which Rous was brought up did not have any academic flavor. In the early 1800s his great-grandfather had come to America from Henham in Suffolk, U.K. His father Charles was a grain broker in Baltimore. On a visit to Texas he met and married Frances Anderson Wood of Huguenot origin from Virginia. Her ancestors, who like the Monod family[2] had left France after the Edict of Nantes had been revoked, had settled in Virginia. Prior to the eruption of the Civil War Frances's father bought land in Texas and eventually was able to move his family there. Rous's parents settled in Baltimore and three children were born; Peyton in 1879 was the oldest and two younger sisters. When Peyton was 11 years old his father sadly died. There are many examples of similar challenges of forthcoming Nobel Prize recipients; for example Anthony Hodgkin, Arne Tiselius and Sune Bergström[2]. The mother decided to stay in Baltimore and not to join her parents who had successfully established themselves in Texas. In spite of the rather mean resources she had access to her aim was to give her children the best education possible. By staying in Baltimore she decided Peyton's future because of the establishment of the Johns Hopkins University in that city. This institution came to pioneer, in the New World, a university training that went beyond a transmission of learning by formal lectures and had the added ambition to advance knowledge gathered at the bedside. But it was not only Peyton who profited from the environment, both his sisters made impressive careers, one as musician and musicologist and the other as a painter of considerable merit.

Rous received a scholarship to study at Johns Hopkins University and Medical School. The four founding fathers of this institution have been

The four founding fathers of the Johns Hopkins University: Henry Welch (1850–1834), William S. Halstead (1852–1922), William Osler (1849–1919) and Howard Kelly (1858–1943).

The Long Wait   5

described as follows — "pathologist William Henry Welch, a stout bachelor whose pastime was a week of swimming, carnival rides and five-dessert dinners in Atlantic City; surgeon William Steward Halstead, whose severity with students masked an almost debilitating shyness; internist William Osler, king of pranks; and gynecologist Howard Kelly, snake collector and evangelical saver of souls." It is not surprising that such a quartet managed to establish a unique school carrying its own values.

During his early studies Rous, being a naturalist, generated some financial support by writing articles for a Baltimore newspaper, "The Flower of the Month." He was one of the last students to experience Osler's teaching. Osler, a highly respected Canadian physician, was not only a renowned practical joker; he was also unique in his teaching by bringing medical students out of the lecture halls. Bedside clinical training was a new concept introduced by him. However, there is no evidence that Rous made any particular impressions on his early teachers. During his second year of Medical School he was infected with tuberculosis bacteria in a finger. The infection spread to his axillary glands, which were removed after which he had to interrupt his medical studies. He went to his grandparents in Texas where he got a job with an uncle on a ranch near Quanah. This was a different world to him. Rounding up cattle on horseback over a wide expanse was no small challenge. The camaraderie with uneducated men taught him new lessons, not provided at college, something he carried with him throughout life. After the loss of a school year he was back at Johns Hopkins, as a resident under Osler. Later in his studies he spent two years as an Instructor of pathology at the University of Michigan under the tutelage of Alfred Scott Warthin. This teacher came to have a major influence on Rous's future developments. Warthin offered to relieve him of his duty to teach summer school, and in addition shared some of his consultant fees with him. Furthermore, he encouraged him to learn German and hereafter sent him for advanced pathological training under C. Georg Schmorl in Dresden.

After his return to the U.S. it was discovered that Rous had developed pulmonary tuberculosis. He was sent to the Adirondacks in upstate New York to recover and fortunately this had the wished-for effect and the disease was not to return and trouble him later in life. It was under Warthin that Rous developed his first three publications. They presented studies of cerebrospinal fluid and lymphocytes. Encouraged by this early success he applied, on Warthin's suggestion, for a grant from the Rockefeller Institute for Medical Research. He was successful in receiving a grant and was able to pursue work leading to additional publications some of which appeared in the *Journal of*

Simon Flexner (1863–1946).

Experimental Medicine. These came to the attention of the Director of the Institute since its inception in 1901, Simon Flexner and he called upon Rous. He was first recruited to assist the pathologist Eugene Opie, but in 1909 there were major changes at the Institute. Flexner wanted to leave the responsibility for cancer research at the Institute to devote his energy to the increasingly important problem of poliomyelitis. Warthin was not overly enthusiastic about Rous's move to the Rockefeller Institute. He stated that working at that institution meant to accept tainted money and in addition he gave the subjective recommendation that whatever Rous did he should not work on the cancer problem. But that was to no avail. Rous came to spend the major part of his professional life studying this kind of disease.

In 1915 Rous married Marion Eckford de Kay who with her background contributed much to the humanistic interests of the family. They had three daughters and the oldest of them, also called Marion, we will meet again. She married Alan Hodgkin, who received the Nobel Prize in physiology or medicine three years prior to his father-in-law[2].

The Discovery and an Interlude of Other Science

In 1911 Rous made the critical experiment to analyze if an ultrafiltrable agent could transmit transforming properties between cells. One may ask why he tried this approach. He had acquired some knowledge about transplantation of tumor cells between experimental animals and he was of course aware of the discussions of the importance of contagious agents smaller than bacteria,

A young Peyton Rous in the laboratory. [From Ref. 19.]

The Long Wait 7

Vilhelm Ellerman (1871–1924) and Oluf Bang (1881–1937). [From Ref. 19.]

identified by the use of filters which could hold back such microorganisms. An additional incentive for Rous to examine the potential transfer of a transforming agent through fine filters, referred to in his original 1911 publication and also in his Nobel lecture, was a finding in 1908 by two Danes, Vilhelm Ellerman and Oluf Bang, that fowl leukemia could be transmitted between birds by use of material passed through the so-called Berkefeld filters. At the time it was not believed that *leukemia*, the uncontrolled replication of cells of bone marrow origin, in fact originally described by Virchow, was comparable to uncontrolled replication of cells forming solid tissue, like the aforementioned cancers involving fibroblasts, *sarcomas*, and epithelial cells, *carcinomas*. It took some time before it was understood that aggressive tumors may develop in essentially any kind of differentiated cells in the body and that the underlying mechanisms were similar — accumulated genetic changes as was discovered later. There is a plethora of terms used for cancers depending on their origin, like tumors originating in the kidney, nephromas, or in cells in the central nervous system, neuroblastomas etc.

Since Rous discovery concerned an ultrafiltrable agent causing a solid tumor, sarcoma, it was discussed potentially to have a relevance also to the problem of cancers in mammalians. This was not the case at the time of the different forms of leukemia, involving mobile white blood cells in the body, studied by Ellerman and Bang. However, as we shall see, a reviewer during the 1930s emphasized the contribution by the Danish scientists and credited that on par with Rous's discovery. So what did Rous find and what was the appreciation of his observations among cancer researchers at the time?

Rous shouldered the new responsibilities allocated to him by Flexner seriously and it did not take long before he made the aforementioned discovery of a filterable agent of potential importance in solid tumor formation. In 1909 a farmer had brought a Plymouth Rock hen carrying an irregular tumor mass at the breast to the Institute. It caught the attention of Rous who examined the tumor in the light microscope and decided to transfer minced tissue from the tumor to another similar kind of hen. Somewhat to his surprise he was successful. He recognized that the transferred tumor appeared like the original tumor in the light microscope. He then did the critical experiment and transferred not whole cells but ground material which he passed through filters known to retain bacteria. Again tumors developed in the fowls. In the microscope the tumors had the same appearance as the original tumor. Thus it seemed like the ultra-filtrate, at the time a loose definition of a virus, could induce a tumor with microscopic characteristics similar to the original tumor. These observations were heretical. The scientific community argued that certainly some cells had managed to pass through the filters and unknowingly been transferred to the recipient hen. In order to disprove this possibility Rous freeze-dried his material to disrupt cells, but many colleagues still remained skeptical. It was too early to put virus and cancer on the same radar screen.

A Plymouth Rock hen with tumor. [From Ref. 19.]

The dominant German school of cancer research vehemently denied the possibility of any importance of infectious agents in cancer, and even Flexner himself was skeptical, to say the least. However, Rous was not discouraged. There was a flurry of publications from his pen in high-quality journals like *Journal of Experimental Medicine* and *Journal of the American Medical Association*. He identified not only chicken tumor No. 1 virus, rapidly named by other scientists *Rous sarcoma virus*, but two additional bird tumor viruses. One of them caused an osteo-chondrosarcoma tumor and the other a "sarcoma showing an intracanalicular pattern of blood sinuses." He documented that his virus-induced tumors were capable of invading neighboring tissues and forming metastases. Hence they behaved like true cancers.

Interestingly Rous also pioneered the use of the embryonated chicken egg for studies of his viruses and the tumor cells. Many years later this came to be appreciated as a very valuable substrate for propagation of many viruses and for producing them in larger quantities, as described earlier[3]. It was not until the early 1950s that this technique for growing viruses to a large extent was substituted by monolayer tissue culture techniques. Treatment with trypsin became a critical component in the development of the latter technique and incidentally Rous published as early as 1916 that this kind of treatment allowed the creation of a suspension of individual cells from solid tissues. Furthermore Rous in his early studies also examined immunological resistance. He registered that it was of two different kinds, either directed against the tumor cells themselves or against the transmissible agent, which Rous at an early stage referred to as a virus. His colleague Murphy preferred the term "transmissible mutagen." It would take time before the scientific community caught wind of the significance of Rous's findings. In fact from the beginning of World War I he put his tumor studies to rest for almost 20 years and changed to a field of particular relevance to the time of war. It concerned the problem of storage of blood. He did not return to the virus-cancer problem until the mid-1930s.

It would be going too far to review in more detail the work during this interlude in Rous's research pursuits. It was a productive period and he continued to publish high-quality scientific papers. He and his collaborators developed a solution that could be used to store blood for transfusions. This Rous-Turner solution has been used into modern times. He also contributed important studies of the physiology of the liver and gall-bladder functions. Among other findings can be mentioned the demonstration of the important active role of the gall bladder. It was an efficient organ for concentrating bile and the mechanisms for the pathological formation of gallstones became elucidated by these studies. Rous also demonstrated the use of vital dyes in the living animal and examined vascular permeability in other experiments. Finally in an approach to isolate the Kupffer cells in the liver, which have a capacity to engulf larger particles — to be "big eaters," macrophages — he cleverly made a monocellular suspension of liver tissue, let the cells that could do this engulf magnetic iron particles and finally collected the cells searched for by an electro-magnet. A colleague referred to this as "high peak laboratory virtuosity." No doubt Rous was a born experimenter.

Martin Arrowsmith

One summer day at the end of the 1970s I and my family had docked our sailing boat Karolina in the northern part of Stora Nassa, a remarkable gathering of 365 islands in the uniquely island-rich Stockholm archipelago, well protected from the prevailing winds. After a few hours we were joined by another sailing boat of the Laurin 32 kind which is "spetsgattad," meaning that it is wedge-shaped both front and aft, like the seaworthy Colin Archer boats. I recognized the skipper as Hans G. Boman, an experienced sailor and in addition a high-profile and idiosyncratic scientific colleague. He was one of the pioneers in the field of molecular biology in Sweden. On the initiative of Arne Tiselius he was appointed to the first research position in this field established in the country in 1963 by the Swedish Natural Science Council. After a few years he moved from Uppsala University to the newly established University in Umeå in the northern part of Sweden. In a short time he managed by his own contributions and by those of other promising young scientists that he recruited to put the young university on the world map of science. Its department of microbiology has remained a very strong institution up to the present time.

Hans G. Boman (1924–2008).

Half-way through his career Boman moved on to Stockholm University and initiated a completely new line of research. It pioneered the development of insights into the emerging field of natural immunity. At the Massachusetts Institute of Technology (MIT) he had learnt about the qualities of the silk butterfly *Hyalophora cecropia*. Using this experimental system he was able to identify previously unknown antibacterial peptides. He developed this work in collaboration with Swedish colleagues and coined the term *cecropines* for this new kind of peptide antibiotics. This was a Nobel-class discovery, but when the new field of what came to be called *innate* immunity, in 2011, a few years after Boman's death, was recognized by a prize in physiology or medicine, the recipients were instead Bruce A. Beutler, Jules A. Hoffman and Ralph M. Steinman.

Sinclair Lewis (1885–1951) and his book *Arrowsmith*.

The reason for mentioning this encounter is that, while sitting on the rock, coincidentally I was reading *Martin Arrowsmith* (also entitled only *Arrowsmith*)[4] by the 1930 Nobel laureate in literature Sinclair Lewis. For several reasons this book has caught the interest of scientists and scientific historians and Hans mentioned that it was required reading for his students. We had a long discussion about the book during the white summer night.

When Lewis had finished his successful books *Main Street* (1920) and *Babbit* (1922) he was contemplating writing about the Christian labor movement but then he met a young scientist, Paul de Kruif who had just been fired from the Rockefeller Institute. He was able to provide extensive personal impressions of operations at this private American research institute. This made Lewis switch to writing about scientists at the McGurk Institute as it came to be called in the *roman à clef* he embarked upon. One of the figures depicted in the novel is supposed to be modeled on Rous.

Originally the book was intended to be a joint project, but soon it was taken over by the already well-established writer Lewis. However, in a foreword to the text he wrote:

"To Dr. Paul H. de Kruif I am indebted not only for most of the bacteriological and medical material in this tale but equally for his help in the planning of the fable itself — for his realization of the characters as living people, for his philosophy as a scientist. With this

Paul de Kruif (1890–1971) and his book *Microbe Hunters*.

acknowledgement I want to record our months of companionship while working on the book, in the United States, in the West Indies, in Panama, in London and Fontainebleau. I wish I could reproduce our talks along the way, and the laboratory afternoons, the restaurants at night, and the deck at dawn as we steamed into tropic ports."

De Kruif was a career scientist at the Rockefeller Institute who got into a situation of conflict with the management. The reason was that he had written a critique in *The Century Magazine*, at the time a well-known publication, of the way the Institute was managed, in particular the way in which supposedly independent research and commercial interests and public prestige of the institute were mixed. These polemic articles came to the knowledge of Flexner. He interpreted the presentations to be in conflict with his ambition to create an atmosphere of team spirit at the Institute and hence forced de Kruif to resign[5]. Since de Kruif had already developed ambitions to shift from a research career into one of writing he accepted his fate. Although he did not get more than introductory credits for his fundamental contributions to Lewis's successful book, which in fact was a genuine co-production, he later, on his own, managed to create a very influential book furthering the career of many successful scientists, as already referred to[6]. In 1926 he published the book *Microbe Hunters*[7] illustrating the self-sacrifice and success of a number of pioneering scientists in the field of microbiology and immunology. This book has inspired many promising young students to choose science, not least in the field of microbiology, as their future career.

The main character in Lewis's book, Arrowsmith, gets an offer to work at the McGurk Institute after he has tried, unsuccessfully, a career as a country doctor. He was very pleased by this opportunity to advance and become a scientist, since he had learnt that it appeared to be impossible to combine the two careers. After some time he discovers viruses that can destroy bacteria and comes to appreciate the possibility of using this kind of agent — viruses referred to as bacteriophages — to manage a severe epidemic of cholera in a remote island in the Caribbean. In his pursuit of a careful case control study, which leads to the death of both his wife and a colorful missionary Swedish health professional, Gustav Sondelius, another important role model for Arrowsmith, he returns to his home institution. He then learns that the new principle he has discovered has already been described by Frederick Twort and Felix d'Herelle, two near-Nobel Prize candidates introduced earlier[6] and referred to below, and secondly that the Institute wants to use his discovery for future commercial purposes, which is against his high ideals. He himself wants to secure more data and pursue his reductionist science before this step is taken. In the end he leaves the Institute to join a smaller, isolated research enterprise together with a similarly-minded, unworldly colleague.

Clearly the close interaction between de Kruif and Lewis provided a very special background to the development of one of the latter's most appreciated novels, cited at length in the 1930 presentation speech at the Nobel Prizes ceremony by the renowned Swedish author Erik Axel Karlfeldt[8], himself later a recipient (posthumously!) of the 1931 prize in literature. Karlfeldt highlighted this famous satire on medicine and science by referring to "…medical schools with their quarrelling and intriguing professors," by contrasting "the unpretentious country doctor (modeled on Lewis father?)" with "the shrewd organizer of public health and general welfare, who works himself into popular favour and political power," and "large institutes with their apparently royally independent investigators, under a management which to a certain extent must take into consideration the commercial interests of the donors and drive the staff to forced work for the honor of the institutes." Karlfeldt also referred to the fact that "Martin Arrowsmith develops into one of the idealists of science. The tragedy of his life as research worker is that, after making an important discovery, he delays its announcement for constantly renewed tests until he is anticipated by a Frenchman in the Pasteur institute." Finally Karlfeldt also referred to "Arrowsmith's teacher, the exiled German Jew, Max Gottlieb, who is drawn with warmth and admiration that seems to suggest a living model. He is an incorruptibly honest servant of science, but at the same time a resentful anarchist and a stand-offish misanthrope, who doubts whether the humanity, whose benefactor he is, amounts to as much as the animals he kills with his experiments." Gottlieb was modeled on Jacques Loeb, who during the first 24 years of awarding Nobel Prizes had received 78 nominations for a prize in physiology or medicine. Loeb had been successful in bringing unfertilized frog's eggs to develop into larvae, but the "examiners remained unconvinced as to the general significance of parthenogenesis."[9] Loeb was de Kruif's mentor but he could not save him in his conflict with Flexner.

Jacques Loeb (1849–1924).

The development of medical sciences went through progressive metamorphosis in the late 18th century. Under the influence of developments in Europe, in particular in Germany and in France, it was appreciated that descriptive morphologically-oriented research needed to be supplemented by experimental research. Developments in bacteriology led the way. These new approaches rapidly caught on in the U.S. and the pioneering institutions in this country were Harvard University and, as already mentioned, Johns Hopkins University. At this time the success of business in the U.S. had led to some individual industrialists becoming immensely rich. New trust laws had to be introduced to prevent upsetting the balance of democracy in the country. President Theodore Roosevelt took the initiative to review the situation and in 1911 the Supreme Court ruled that Standard Oil was an illegal monopoly. At the time Standard Oil was the dominating competitor on the international scene of The Oil Production Company Nobel Brothers — abbreviated Branobel — in Baku, the center of the oil boom at the time. Baku is located on the Caspian Sea, in present-day Azerbaijan, but at the time was a part of Russia. This company was managed by Alfred Nobel's brothers Robert and Ludvig. About one-seventh of the legacy serving as a basis for the Nobel Prizes derived from Alfred's financial involvement in this company. The consequence of the 1911 decision by the Supreme Court was that Standard Oil was split into 34 separate companies, paradoxically resulting in a major increase in John D. Rockefeller's

The Long Wait 15

wealth. On the positive side this immense wealth allowed the establishment of a very influential philanthropic institution, the Rockefeller Institute. It was critical how the goals of this private institution were to be defined. A balance had to be struck between furthering free and independent research and the exploitation of the data generated for successful business ventures. This is well highlighted by the story presented in *Arrowsmith*. However, it should be added that this conflict of interest has been progressively reduced as the Rockefeller Institute with time has established itself as a uniquely successful institution for basic science.

So what about Rous in Arrowsmith? The McGurk Institute is presented in the book as a temple of science providing a unique opportunity for talented scientists to develop their work uninhibited by financial constraints. It also introduced the way the institute tried to stimulate interactions between the groups of researchers by an attractive joint luncheon milieu. But then there is the critique of despotic leaders — DeWitt Tubbs modeled on the Director, Flexner, — and the group leaders — like Rippleton Holabird, modeled on Rous and the first director of the Rockefeller Institute Hospital Rufus Cole — accused of being more interested in the form for pursuit of research rather than its content and also in public relations and commercial exploitation of the findings in the science conducted. The commercial exploitation does not seem to have attracted Rous. Instead he himself presented his long life in science as professionally obsessive and even at periods as halcyonic. There were, however, some negative traits, like his bulldoggish, persuasive and inflexible behavior, referred to by others in their description of him. In his biographical memoir[10] Christopher H. Andrewes used the following concluding paragraph:

> "In his youth Rous had red hair, as befits his name, but he sometimes referred to himself as "the last of the red Rouses." He had not, however, the brisk temper sometimes associated with red hair. He did not always suffer fools gladly, but was invariably kind and courteous to his colleagues and co-workers."

Rous retained throughout his career the monolithic view that viruses were the only explanation of the origin of cancer cells. Thus he may have been reluctant to enter new fields of science when such had been made available. It also remains a puzzle why he left the field of cancer research for almost 20 years relatively soon after his original discovery. Rous's performance at the Rockefeller Institute is a good example of the tireless pursuit in the important field of cancer research progressively leading to major new insights and eventually also to recognition by a Nobel Prize highlighting the original discovery. However, as we shall see, Rous completely misjudged the potential future development of the field that he had made available for research by his original discovery.

The luncheon meetings in the famous tea room at the Rockefeller Institute are depicted in the book as an environment for self-aggrandizement. Clearly the book needs to be viewed as polemical and as overemphasizing potential problems, but the developments after 1925 at the Rockefeller Institute and later Rockefeller University have come to highlight other much more positive qualities of this unique institution and its scientists. Its emphasis on independent research groups led by top-notch scientists has made it a beacon in the advance of science and also led to recognition by many Nobel Prizes in chemistry and physiology or medicine. In 2017 the 25th prize was awarded to the institution. It has developed to become one of the most successful research institutions in the world. There is a unique atmosphere in the Rockefeller University campus at the foot of 66th Street, York Avenue in New York. I can testify to this since over decades this has been my home staying at the Abby Aldrich Rockefeller Hall when I visited New York, having my breakfasts overlooking East River and the tramway from Roosevelt Island against the background of the impressive Queensboro Bridge.

The First Evaluation of Rous. The Critical Role of Reproducibility of Scientific Data

The first nomination of Rous was in 1926. It was submitted by none other than Karl Landsteiner. This giant of science has been described to some extent previously[3, 6]. He discovered the human blood groups, for which he received a Nobel Prize in physiology or medicine in 1930. He also demonstrated that the infectious agent causing polio was a virus in 1909 and in the 1940s he laid the grounds for studies of cellular immunology. His homeland was Austria, but after the First World War the country was in disarray and Landsteiner was forced to move abroad to be able to continue his research. He first established himself in the Netherlands but upon an invitation from Flexner he moved with his family to New York in 1923 and took up a position at the Rockefeller Institute, where he remained till his death in 1943. Thus he and Rous worked at the same institution for two decades.

In his nomination of Rous, Landsteiner emphasized the pioneering discovery of a filterable agent as a cause of tumor formation. The agent not only caused the formation of malignant cells but also decided their characteristic histological type. Reflecting the time of nomination when the nature of viruses was still poorly defined, he introduced a comparison to the discovery of bacteriophages by Twort and d'Herelle. He also referred to some recent confirmatory data by William E. Gye and F.I. Barnard. Landsteiner's nomination was subjected to a preliminary review by Folke Henschen, professor of pathology, introduced in my previous books [3,6], including Rous together with three other candidates, Bernard, Gye and Katsusaburo Yamagiwa. He noted that Rous and in unspecified confirmatory work, Gye and Bernard had investigated the pathogenesis of tumors. The follow-up studies made by Gye and collaborators were essentially a repeat of Rous's original work. The observations made were phenomenological and no significant data were added. However, Gye later developed to become an authority in the field. He was for 16 years, starting in 1934, the Director of the Imperial Cancer Research Fund laboratories at Mill Hill in the U.K. Henschen concluded that at the present time it was not motivated to subject these studies to a thorough investigation. He had doubts that they had any relevance to the development of tumors in mammals.

Henschen's evaluation of the nomination of Yamagiwa was different. He proposed a more extensive analysis of this nominee together with a renewed analysis of the researcher indicated to have given impulses to his work, namely Johannes Fibiger. A combination of the two had been proposed by the famous German pathologist Ludwig Aschoff. Thus Henschen made a thorough follow-up evaluation of Fibiger and Yamagiwa. His appraisal of the contributions by the former had been documented in his earlier evaluations, but now he gave full credit also to Yamagiwa's discovery. He wrote (underlined; translated from Swedish):

"Through the discovery made by Yamagiwa the experimental tumor research has been provided with an excellent method. Hereby has been created a firmer foundation for the research and possibilities to examine

Katsusaburo Yamagiwa (1863–1930).

18   *Nobel Prizes: Cancer, Vision and the Genetic Code*

in more detail the biological processes underlying the development of a tumor cell from a normal cell."

And in a later comparison of the different methods used by the two candidates he commented "First when Fibiger in 1913 discovered a secure, although cumbersome method and Yamagiwa in 1915 described a simple, easily accessible and secure method for induction of tumors, possibilities were made available to study their etiology."

As a conclusion Henschen made a very strong plea for a combined prize. This was also the proposal that the committee handed to the College of Teachers, but apparently the college was not satisfied and asked for more advice. The committee then came back on October 22, after having made additional reviews of Fibiger and Yamagiwa and the discoverer of bacteriophages, d'Herelle, and recommended that no prize should be awarded. The prize money was to be reserved for the coming year. Fibiger had to wait for another year for his prize, later judged to be controversial, not to say wrong. Let us therefore first give a background to his assumed discovery and then judge what, in an historical perspective, went wrong when the committee took its decision.

Fibiger had a solid background in bacteriological research studying passive immunization against diphtheria with the 1905 and 1901 laureates Robert Koch and Emil Adolf von Behring in Berlin in the 1880s. After his return to Copenhagen in 1900 he became professor and director of the Institute of Pathological Anatomy of the University of Copenhagen. Just prior to that, he had pioneered modern clinical trials using random allocation of diphtheria patients to be treated with a hyperimmune serum or a control serum. His interest in cancer formation emerged in 1907. Using a strain of rats from Tartu, Estonia, to study tuberculosis he had found the development of "tumors" in the stomach which were interpreted to be capable of spreading in the animals. This assumed tumor formation was associated with the presence of roundworms, nematodes. After several years of studies he concluded that the nematode, which represented a new species, was the cause of the cancer. The new species was originally named *Spiroptera carcinoma*, a name Fibiger continued to use, although it was later renamed *Spiroptera neoplastica*.

Fibiger was nominated 18 times between 1920 and 1927 and the two main reviewers were the professors of pathology, Folke Henschen and Hilding Bergstrand, the latter being 14 years younger. In 1926 there was also a nomination of Yamagiwa, as already mentioned. The two reviewers throughout their repeated analysis of Fibiger held different opinions with Henschen arguing for a prize in 1926 to both of the nominees and Bergstrand against. In 1927 there were 7 proposals for a prize to Fibiger, but none for Yamagiwa, who the previous year had been recommended to share the prize. This absence of a nomination for Yamagiwa easily could have been remedied by a nomination from a committee member or from the secretary in due time. No such proposal was submitted. Two new strong candidates for a prize had now entered the scene, the physiological chemist Otto Warburg also working in the field of cancer research studying the oxidation in cells, and Julius Wagner-Jauregg, a physician who had introduced the use of malaria treatment in the management of dementia paralytica, caused by a general meningo-encephalitis at the end stage of syphilis. An additional reviewer was added to evaluate Warburg. It was the professor of chemistry Einar Hammarsten, who was introduced at length earlier[2]. Henschen and Hammarsten were positive regarding a prize to all three candidates and Bergstrand against Fibiger as before. The committee then proposed to combine Fibiger with Warburg for the reserved 1926 prize and to give the 1927 prize to Wagner-Jauregg. One of the four committee members questioned whether Fibiger was worthy of a prize and mentioned instead another candidate, Charles S. Sherrington, who was to receive his shared prize in 1932. In the end the College of Teachers decided not to include Warburg with Fibinger. The latter received the reserved 1926 prize "for his discovery of the Spiroptera carcinoma" and Wagner-Jauregg the 1927 prize "for his discovery of the therapeutic value of malaria inoculation in the treatment of dementia paralytica." Yamagiwa was left out since he was not nominated and Warburg was saved for the 1931 prize. This he received after a major intervention by Hammarsten.

The background to this, in a historic perspective, highly questionable choice of Fibiger in 1927 has been subjected to several analyses[11, 12, 13]. The bases for Fibiger's claim to have discovered a mechanism for appearance of cancer was briefly presented above. Two major mistakes were made by the committees in managing the Fibiger case. One was that Henschen was used as a reviewer. He had himself nominated Fibiger for a prize in 1923. Thus he could not possibly be an objective referee. Today it is carefully scrutinized what kind of relationships a reviewer might have with a proposed candidate. The other mistake was that the committee made a recommendation in the absence of experiments by other scientists *confirming* the proposed discovery. This is a serious error, since validity of, not least, dramatic new discoveries, relies on the fact that they can be demonstrated to be reproducible in other laboratories. This is the reason why a full scientific paper should always contain a detailed material and methods section, so that anyone who so wishes can reproduce and expand the data. It is the confirmatory nature of scientific discoveries that gives the whole enterprise of seeking new knowledge by experimental research such a unique solidity. One argument for the establishment of Nobel institutes, discussed previously[2], was that they could be used to confirm data considered to be recognized by a prize. However, for natural reasons they rarely came to serve this function. During the time they were in operation until the mid 1960s they instead came to serve as centers for building Swedish scientific competence of great value to the work by the Nobel Prize committees.

Fibiger died of colon cancer a month and a half after he had received his Nobel Prize. During his lifetime the only discussion concerning his findings were if they truly represented cancerous tumors. After his death it was found that the strain of mice he had used had been fed a diet deficient in vitamin A. This was interpreted to lead to certain tissues possibly becoming more vulnerable to mechanical irritation, like the one caused by the nematodes. Finally three research groups published data in 1935, 1937 and in 1952 showing that Fibiger's findings could not be reproduced[11]. Thus although another review[12] of Fibiger's findings gave a major emphasis to the need for data presented being judged in the proper historical context, it cannot be accepted that unconfirmed results were awarded a Nobel Prize. By way of contrast the data published by Yamagiwa were readily confirmed. Tar painting of the ears of rabbits induced true tumors and this could be readily repeated in other laboratories. The tar-painting technique was in fact used by Rous to study the progression of the papilloma virus-induced tumors some ten years later, as we shall see. It is thus regrettable that this milestone in the development of cancer research was not recognized by a prize[13]. Finally one may ask if the debacle in recognizing new important cancer research embodied by the prize to Fibiger caused a certain restrain in forthcoming committees in recognizing scientific advances in the field of cancer. It would last almost 40 years until cancer research was again recognized by the 1966 prize to Rous and Huggins. During this time there had been a number of important developments in the field.

The 1930s — Rabbit Warts and a New Start

Rous returned to studies of the virus-cancer problem in 1934 after twenty years of involvement in completely different kinds of high quality research as mentioned. In order to pursue his concept of virus-induced cancers he needed to study tumors in mammals. He also wanted to shift his focus from tumors in connective tissue, the sarcomas, to those emanating from epithelial cells, the carcinomas. The critical stimulus to do this was Richard Shope's discovery of viruses causing skin warts in rabbits. Shope was a respected veterinary virologist with an Iowa farming background. Farmers in this state had experienced two outbreaks of highly infectious influenza-like diseases in swine in 1918 and 1929. The disease looked very much like influenza in man. The nature of the infectious agent was unknown at the time. In studies at the Rockefeller Institute the young Shope and his mentor Paul Lewis attempted to further characterize it. When Lewis died of a yellow fever infection contracted in the laboratory Shope continued this work on his own.

Richard Shope (1901–1966). [From Ref. 19.]

A bacterium later named *Hemophilus influenzae* was isolated, but when transferred to healthy pigs no disease developed. Shope then passed material from diseased animals through ultrafilters and found that an agent in the filtrate could cause disease. However, this experimental disease appeared milder than the one seen in the natural epidemics. When the two agents were combined in a mixed infection full-blown disease was encountered. Although it is true that influenza, like many other mucosal virus infections, paves the way for secondary bacterial infections, it later turned out that the virus found represented the single critical etiological agent. Still *H. influenza* remains an important pathogen in its own right. Second to pneumococci it is the most important cause of bacterial pneumonia. Shope's finding was important because it paved the way for the identification of a virus causing influenza in man. Furthermore, it allowed the scientists on both sides of the Atlantic to jointly document that the swine virus discovered by Shope was a surviving form of the virus causing the pandemic in humans that killed more than 50 million people in 1918. This story has already been told in one of my previous books[6] and we will return to it in the further discussions of Christopher Andrewes below.

In 1928 the Rockefeller Foundation established a separate virus laboratory to develop its international program. It was the Rockefeller Institute Department of Animal Pathology at Princeton. Shope became one of its original members and this led to him developing a friendship with Rous. In the early 1930s Shope made two observations of relevance for future studies of viruses and tumors, one in 1931 on fibromas/myxomas in rabbits and another in 1933 on papillomas in the same kind of animals. The early studies of the ultra filtrates of fibromas showed development of transient tumor tissue in inoculated animals and the appearance of a specific immune response. A possible relationship to a previously described rabbit myxoma agent was found and subsequently immunization with the newly discovered virus was used to protect rabbits against myxomatosis. However, this work was left unfinished and instead Shope discovered two years later that warts in wild cottontail rabbits could also be transmitted after ultrafiltration. Hence the disease was caused by a virus. The infectious agent in addition could be transmitted to domestic rabbits but it was not readily passed from one domestic rabbit to the next. Importantly, Shope observed that warts in one domestic rabbit had acquired a malignant character. It had become a cancer. These findings were published in 1933 in the renowned house publication of the Rockefeller Institute, *Journal of Experimental Medicine*. Since Shope was more of a field virologist he offered his papilloma material to Rous, the dedicated pathologist, for further examination. Rous was very stimulated by this new opportunity to be back in cancer research and, as he expressed it, "again to be sailing free on the broad ocean."[10] Studies of the Shope papillomas and cancers derived from them came to occupy Rous and collaborators for the coming three decades. More than 50 publications were presented and almost all of them in the above-mentioned house journal.

The results from these studies seemingly were not as dramatic as the original finding of a virus causing sarcomas, but they were important in the formulation of some critical parts of the perspective we have on cancer development today. It should be remembered that the conclusions drawn were based on the use of animal experimentation and application of morphological techniques. In addition the presence or absence of infectious virus or virus products serving as antigen could be examined during different stages of cancer development. It turned out to be difficult to study a line of this kind of tumor cell by passage from one animal to the other, by the use of out-bred animals, the only kind available at the time. We will return to the importance in experimental work on cancer of using inbred experimental animals later, when discussing the forthcoming discoveries of a host of new tumor viruses in addition to Rous virus and Shope papilloma virus. However, Rous and collaborators made the important observation that tumors could be passed from one animal to the other if suckling rabbits were used. In 1940 they published data showing that a tumor had been passed through fourteen generations of rabbits. It was originally called V2, but when the same designation was used during the war for German rockets it was changed to VX2. Out of all these studies grew an important general concept of *progression*. The establishment of a malignant tumor was interpreted to include many steps with the virus contributing mainly to the early steps. It could then remain as a passenger or disappear from the scene completely. The papilloma virus by itself could only cause benign tumors like plantar warts in humans. However, when additional factors come into play, as in the environment of genital warts, a progression into a malignant tumor may occur. We will return to the family of human papilloma viruses in Chapter 3 when discussing the possibilities of preventing certain cancers in humans by immunization.

Rous took advantage of Yamagiwa's findings that tar could contribute to the process of carcinogenesis. Interesting similarities were found when comparing tumors induced by the virus and by painting with tar. Using a combination of the two led to malignancy being "telescoped into a few days instead of taking months or years." Different publications described "Conditional neoplasms and sub-threshold neoplastic states" and "The initiating and promoting elements in tumor formation." These were important new concepts in the understanding of tumor formation and development.

Nominations of Rous continued through the coming decades. In 1934 he was again nominated by Landsteiner. Bergstrand made a preliminary evaluation which concluded that it was justified to make a full investigation. He had noticed that the perspective had changed since recent discoveries had shown that both "sarcomas and cancers in rats and mice could be transferred by cell-free filtrates." However, in this year no full follow-up investigation was made. In 1936 Landsteiner repeated his nomination and this year a separate nomination was given by another colleague at Rockefeller Institute, Alexis Carrel, the controversial Nobel Prize recipient mentioned in the beginning of the next chapter and also discussed previously[6] and and in addition by E. Libman. This year it became Henschen's turn to make a preliminary review. Being generally of the opposite opinion to Bergstrand he did not recommend a full review. He referred back to his 1926 evaluation and he reiterated (translated from Swedish) "Although Rous has demonstrated that certain sarcoma-like tissue changes can be induced by microbes in fowl, there is no evidence that allows the conclusion that this has any relevance to malignant tumors in general and especially in man (indicating that they) should be caused by the influence of microorganisms on tissues." But he also noted that interesting new things had happened. The discovery of phages and their replication in bacteria had given more substance to the term "ultrafiltrable." He also referred to the more recent studies of the rabbit papillomas and the important new findings made. These were summarized by the statement "In these studies Rous follows an independent and original way (of research), which he himself a long time ago has selected and it may be that this can lead to the goal concerning (the understanding of) certain forms of tumors."

A Foresighted Fairy Tale

During the 1930s new information led to intensified discussions of the nature of viruses. This also included their possible role in tumor formation. This problem was discussed at length in letters exchanged between a small group of virologists and tumor biologists. They included, besides Rous, the already mentioned Andrewes and Gye. In their correspondence the members of the group referred to themselves as the "three musketeers." Andrewes who had a particular poetic talent inspired the discussions in this group by writing an allegorical Christmas fairy tale which was sent to Rous in 1935. This particular essay was never published but has been carefully analyzed by historians of science[14]. It remains in the archives of the American Philosophical Society, which is the home of Rous's legacy which we will return to in the next chapter. The fairy tale reads as follows:

A Christmas Fairy-Story for Oncologists

Once upon a time there was a family of viruses, much like other viruses. Some of them liked mutton, others poultry. Some of them multiplied wantonly in the cells they inhabited, and smashed up their homes; others, more restrained, practiced a magic which caused their homes to proliferate. But after a while most of the homes got broken up and the viruses had to seek new ones in new hosts. These new homes they penetrated by means of powerful tusks in their upper jaws.

Now the hosts didn't like this business and they defended

themselves in two ways. They provided themselves with weapons called antibodies which smashed the tusks of the viruses so that they couldn't enter new cells and died of inanition and exposure. Or else they made terms with the enemy and said "Come right in and live in our cells. But you must practice birth-control and leave the furniture alone. If you do that you may come even into the inner sanctum of our germ plasm and be tolerated as feudal retainers from generation to generation.*

In this feudal existence most of the viruses lost all ambition and ceased to clean their teeth, which became subject to dental caries and fell out. And after a few generations the race was quite tusk-less. Some of them retained their teeth and their self-respect (like the rabbit papilloma virus) but even they were apt to become edentulous in the fat luscious cells of the domestic rabbit. Others — the poultry-loving race — didn't bother as a rule, but were readily able to sprout tusks again whenever it seemed worth the trouble.

Now this was all very well, except for one complication. The hosts which had admitted viruses upon terms as feudal retainers had left nothing to chance. They had always built their cells with virus-proof cement. Occasionally, however, dry rot got into the walls after exposure to hydrocarbons with a phenanthrene nucleus, or a fire-work party of X-rays would cause cracks to appear. Then the young bloods would see chance for revolt. Perhaps they would run riot and smash up the home, as their ancestors did. If they did, — well it was only one cell amongst millions, and the young bloods when they found themselves homeless, perished miserably for they had no tusks to help them enter new cells. But perhaps, they would practice their ancestral proliferation — magic, and then no one could stop them. Some were cleverer at it than others, some improved with practice; and the results were neoplasms of all degrees of malignancy. Ultimately the host died and the viruses with them, — for still they didn't know how to grow tusks. And pathologists, who thought that all viruses

---

* This is Hans Andersen's version. Grimm says they were only allowed to live a saprophytic existence in the Servant's Hall and weren't allowed in the cells at all. I don't know what is right.

had tusks, studied the neoplasm and grunted into their beards "The parasitic hypothesis of cancer is dead."

The hosts became alarmed at the increasing incidence of cancer and called upon the Goddess Evolution at a great religious revival. But, alas, the goddess had no interest in diseases which attack creatures *after* they have successfully propagated themselves, and turned a deaf ear to their prayers.

The text was signed with the initials C.H.A., which readily could be reshuffled to H.C.A. — Hans Christian Andersen.

I had the privilege of meeting Andrewes in the 1960s during my post-doctorate period. He was a very charming and amicable elderly colleague, caring in particular for us younger virologists. His warm voice would fit very well for a reading of his fairy-story at the fireplace. Gye was the central actor in this story, since it was he who kept the virus theory of cancer alive when Rous had left the field for some two decades. Andrewes was one of the pioneering virologists in the U.K. and had a solid training at St. Bartholomew's Hospital Medical School in London. He then spent two years in the mid-1920s at the Rockefeller Institute, where he got to know Rous. After this he returned to England and joined Gye at the National Institute of Medical Research at Hampstead Heath, London. Together they repeated and attempted to extend Rous's early work consolidating the theory of a possible virus etiology of cancer. After this he made a number of important contributions to the field of virology to be discussed further below.

A critical part of the fairy tale is the proposal that the virus might stay in cells without destroying them. This was a truly heterodox thinking, which had received support from the interaction Andrewes had with another early giant in the young field of virology. This was Frank Macfarlane Burnet. This remarkable biological thinker was introduced at length in one of my previous books[3] in connection with his shared Nobel Prize in physiology or medicine in 1960. However, this prize was not awarded within the field of virology, where he was a dominant figure, but in immunology, which he advanced by his intuitive conceptual thinking. During the early 1930s Burnet was examining viruses of bacteria, the bacteriophages. He made many important contributions in this field, but conceptually the largest leap was probably his proposal that the viruses could remain in cells which continued to replicate happily. At the time this was referred to by the German word "anlage," English "predisposition." This was the origin of the phenomenon later called lysogeny, for which Lwoff received his shared Nobel Prize in 1965[3]. The idea picked up in Andrewes's fairy tale was truly ahead of its time. It would take development of central knowledge about the biochemical character of the genetic material and techniques to demonstrate interaction between virus nucleic acids and host cell genetic material before the nature of lysogeny — and still later the interaction of virus genetic material and host cell nucleic acid in tumor cells — could be demonstrated. It appears that Andrewes himself thought he might be skating on thin ice as evidenced by his ambivalence in the asterisk reservation.

It has been speculated about who Andrewes had in mind when he addressed the oncologists in the title of his story. One of them probably was a collaborator of Rous in the early publications on the hen virus and sarcomas, the already mentioned Murphy. By his own formulation he never believed in the virus theory. He preferred to use the vague term "transmissible mutagens." It was never clarified what the meaning of this term was and how it differed in its nature from a virus.

As mentioned Andrewes was a central figure in British virology for many decades and his early contribution to the field of cancer was one of many. He was involved in the identification of the first strains of influenza virus. Together with Richard E. Shope, whose work on influenza has already been described above and whose work on tumor viruses will also be discussed later in this chapter, and Wilson Smith, he was nominated in 1958 for a Nobel Prize in physiology or medicine. He was reviewed by Gard as discussed in my first book on Nobel Prizes[6], but the discovery was not considered worthy of a prize, in part because Patrick P. Laidlaw, who had made important early contributions to this discovery, was no longer alive. In 1967 and again in 1968 Andrewes was nominated for a Nobel Prize in physiology or medicine by a British colleague David Tyrrell. In the latter year Gard made one more evaluation of his candidacy. At this time a considerable amount of knowledge about viruses causing respiratory infections had accumulated. Besides influenza virus, it had been found that parainfluenza viruses, adenoviruses and in particular a large group of more than 100 rhinoviruses could infect the respiratory tract. The latter group of viruses had received their name on a suggestion by Andrewes. Rhino means nose and is of Greek origin. Tyrrell for some 30 years was the head of The Common Cold Research unit at Salisbury, U.K., which had a central role in research on this kind of viruses. A large part of the work was based on the involvement of volunteers and the slogan used was "Free 10 day autumn or winter break: You may not win the Nobel Prize, but you could help find a cure for the common cold."

In 1968 Gard made a supplementary review of Andrewes. He revisited the field of influenza and elaborated on all the new knowledge that had been gained in that field. Then he addressed Andrewes's, involvement in researching the etiology of the common cold and the importance of the work at the Salisbury unit of which he was one of the initiators. By his argumentation regarding the different important contributions by Andrewes including the work on rhinoviruses Gard proposed that he was prize-worthy, which was accepted by the committee. This may be one example of an excessive use of the term prize-worthy, and may also reflect Gard's authority in the committee. As we shall see in Chapter 7 there were as many as 52 candidates considered worthy of a prize in 1968. Hence given the prevailing competition there was no possibility for Andrewes to claim the prize. In fact it was Tyrrell himself who carried the main responsibility for the advances in understanding of common cold viruses and their infections. Not surprisingly he called his memoirs "Cold War." It could be added that Tyrrell, like Andrewes was a very warm and pleasant colleague. I got to know him in particular in the 1980s when he and I and Luc Montagnier, the co-discoverer of the AIDS virus, who received a shared Nobel Prize in physiology or medicine in 2008, formed an advisory group ("the three wise men"!?) to give guidance on the use of chimpanzees at a colony in The Netherlands for research purposes supported by funds from the European Union.

It is now high time to return to Rous.

New Nominations and New Conclusions

In 1937 new nominations were submitted, again by Landsteiner and also by the 1922 Nobel Prize recipient (awarded in 1923) Otto Meyerhof. Now, again, it was Bergstrand's turn to make a review, but this time it was comprehensive and covered 12 pages. In the beginning he referred back to his 1934 review in which he had taken the position of hesitation, although he had a vague feeling that Rous's discovery might have a wider biological importance. Thus he returned to the history of tumor virus research and also to the advances in the field of virus research in general. He mentioned the original finding by Ellerman and Bang of transmission of leukemia with cell-free filtrates and also cited a number of follow-up studies. In these, different forms of leukemia had been encountered and there was vague discussion of speculative "mutational" events caused by the infectious agent. An immunological protection against the enigmatic virus agent had been observed but not against the tumor cells caused by the infection. Infection with certain agents causing leukemia also led to the development of sarcomas. A parallel was drawn to cancers induced by different tar products where the place of application may decide whether a sarcoma or leukemia developed. In the discussions of mutations a reference was made to the plant virus TMV, representing an important model system for studies of the nature of viruses. Bergstrand drew the critical conclusion that leukemias should be considered as just another kind of cancer formation, something which was not appreciated at the time of Ellerman's and Bang's original observation, as already mentioned.

He then returned to Rous's discovery and by comparative reasoning he now argued that it had a very great value, although he noted that as regards what priority it had it was presented three years after Ellerman's and Bang's original observation in studies of leukemia in birds. Parallels were drawn with several studies of different kinds of tumors induced by tar products. Citing Carrel it was vaguely speculated that the tar chemicals could change the metabolism of cells and that products might be released which encouraged the cells to proliferate. He then reviewed the recent studies of rabbit papillomas. Shope's discovery of 1933 was presented and an interesting reference was given to one of the first animal virus infections ever identified. This was the appearance of myxomas —represented by mucus-containing benign tumor cells — in rabbits after infection by an ultrafiltrable agent which was first observed by Giuseppi Sanarelli as early as 1898. Later on it was found to represent the first identified member of the extensive group of poxviruses, which also includes smallpox virus.

Already in 1887 John Buist published data from the examination by a special form of light microscope, so called dark field microscopy, of stained smears and vaccine lymph from rabbits infected with the Jenner vaccine against smallpox, later named vaccinia virus. Particles referred to as "elementary bodies" were identified, but it would take more than 40 years before they were shown to represent the infectious particle. The definitive identification of the particulate nature of the virus had to wait another decade when the introduction of the electron microscope allowed a visualization of the particles carrying infectivity as we shall see in Chapter 8.

The question of which infectious agent was the first one to be convincingly demonstrated to be ultrafiltrable — to be a virus — has been debated[15]. It has been strongly argued by Marc van Regenmortel[16] to be the agent causing foot-and-mouth disease in cattle which was described as ultrafiltrable by Friedrich A. J. Loeffler and Paul Frosch in work supported by Robert Koch. This happened in 1898, the same year that Sanarelli's findings were published, but in its conceptual distinction the former work was the most convincing and had the larger impact. A few years later it was demonstrated that the agent responsible for yellow fever could cause the disease in humans after having passed through a filter holding back bacteria. Sanarelli believed that this disease was caused by a bacterium. In 1908 bacterially sterile material was collected from a 9-year-old boy who had died of poliomyelitis and used to transmit the infection to two kinds of non-human primates. In experiments a year later it was demonstrated that the agent causing paralytic disease could pass through a filter holding back bacteria. The central figure in these studies was the already mentioned Landsteiner[6], a frequent nominator of Rous.

The development of filters and calibration of their pore size developed to be a science on its own. With time it had become possible to produce filters with graded pore sizes and demonstrate that infections with different characteristics were caused by viruses of seemingly variable sizes. In history books the first agent to be mentioned as ultrafiltrable is the plant infectious agent, tobacco mosaic virus (TMV). The Russian Dmitry Ivanovsky made the first filtration experiments with this agent, but he was himself not convinced that what he had found was not some conventional microorganisms that had sneaked through his filter. Similarly there have been doubts about the conceptual impact of the findings in follow-up studies of TMV in 1898 by the Dutchman Martinus W. Beijerinck, who believed that the agent existed in a "liquid" form. The story of the early "virus" studies illustrates that it may be very hard to draw solid conceptual conclusions from the first tentative findings. It may be too tempting to rewrite history, sometimes even emphasizing national preferences.

Although Bergstrand's review should focus on Rous it devoted attention, to a considerable extent, also to Shope's early work on the fibroma-causing agent. It also compared Shope's studies to Sanarelli's much earlier work on the myxoma virus mentioning the already presented finding by Shope of the existence of a possible cross-immunity. Bergstrand also commented on some early suggestions of the occurrence of "changed virus" or "mutants" of the fibroma agent and the fact that it could be "infectious" but not "contagious." By way of contrast the myxoma virus was highly contagious. Thus the tumor-like formations occurring with these different agents needed to be further compared, but this was not to be done since Shope's interest in 1933 was caught by another virus, the papilloma virus in rabbits. This, as mentioned by Bergstrand, was the virus-tumor system gratefully taken over by Rous. He mentioned Rous's finding that papillomas in domestic rabbits could develop into cancers. This took a long time and by comparison to cancers developing after prolonged exposure to tar the question was raised about the role of the infectious agent in the process. Could it be "a secondary phenomenon" or the result of some "unspecific" stimulation?

The review then went on to discuss various forms of papillomas in humans. Four forms of papillomas in man were listed; verruca vulgaris — the common warts on hand of feet; condyloma acuminatum — venereal warts; papilloma laryngis — warts in the area of the vocal cords and papilloma vesicae — warts in the urinary bladder. Bergstrand mentioned that already in 1907 Giuseppe Ciuffo had published data in an Italian journal that the skin papilloma agent was ultrafiltrable and that later similar findings had been made with material from condylomas. Regarding larynx papillomas he referred to a Swedish colleague Ullman who in 1923/24 had published new data in *Acta Oto-laryngologica*, which Bergstrand judged "to be highly reliable." Ullman had inoculated himself and some assistants intracutaneously with mashed skin papilloma material from a 6-year-old boy. After a considerable time, local papillomas developed in two cases. The possibility that bladder papillomas were also caused by an infectious agent was left as conjectural. Bergstrand finally noted that apparently the three latter forms of papillomas, but not the common skin warts, could develop into cancer. He mentioned in particular penile cancer and cited that it occurred frequently in the Orient mentioning that "it is for example common among Chinese "coolies" because of their defective cleanliness." After his extensive review, which in fact dealt to a very limited extent with Rous's contributions, Bergstrand came to the following conclusion (translated from Swedish):

> "It seems apparent that certain aspects of the problem of formation of tumors are intimately related to questions regarding virus diseases and if one considers the discoveries which have been a necessary precondition for the developments observed, Rous's demonstration that sarcomas in hens, transmissible by a cell-free filtrate, is one of these. It is therefore my view that this discovery is worthy of a Nobel Prize in medicine and physiology."

This is the first time that Rous is judged to be worthy of a prize by a reviewer. It was confirmed by the eight-member committee at its meeting on September

27, 1937. However, it would take 29 years until a prize would be awarded.

Two years later, in 1939, Rous was again nominated and this time by Howard W. Florey, the forthcoming 1945 laureate for contributions to the development of penicillin. In 1939 there were many nominations of candidates involved in work on viruses. Besides Rous, Thomas M. Rivers, Wendell M. Stanley, Frederick C. Bawden and Norman W. Pirie were nominated. The three latter candidates have been extensively discussed previously[6]. Like Rous, Rivers was employed at the Rockefeller Institute.

Thomas Rivers (1888–1962). [From Ref. 19.]

He had been briefly reviewed the previous year and was subjected to two additional evaluations, by Henschen and by the physicist The Svedberg, this year. The conclusions of the reviews differed. Although Rivers was a central figure in the powerfully developing field of virology he had not contributed any specific discovery worth considering for a Nobel Prize, according to Henschen. Svedberg was more generous in his evaluation as we shall see.

In addition to his brief review of Rivers, Henschen also made a thorough evaluation of Rous over 14 pages. It provided a repeat presentation of Rous's early work, with an emphasis on the facts that the tumors induced did not contain any demonstrable virus and that attempts to repeat Rous findings using tumors from mammals had been unsuccessful. In these early experiments no virus had been identified. Like Bergstrand, Henschen also discussed the rabbit fibroma although these studies had been exclusively performed by Shope. Rous's important development of the rabbit papilloma model system was described in some detail, noting again that no virus had been found in the established tumor, which in this sense could be compared to tar-induced tumors. Henschen then discussed papillomas observed in other species including man. Summarizing the state of the art of the research field Henschen noted that there are endogenous — hereditary — and exogenous factors — carcinogenic chemicals and viruses — which may have importance in tumor developments. He commented that at the time almost no knowledge had been retrieved about how the virus might change the regulation of cellular growth. His conclusion read (translated from Swedish):

The Long Wait 33

> "...that Rous's discovery of 1911, seen in the light of current virus research, should be seen as belonging among the most important in tumor research, since Fibiger was awarded a Nobel Prize for his discovery of the spiroptera carcinoma (sic, my remark). The follow-up conclusion was that he had found reasons to waive his previous hesitant attitude towards worthiness of a prize and that Rous *without doubt* (my italics) had made discoveries that deserved to be awarded a Nobel Prize."

The same year the committee as mentioned had also called upon Svedberg to make a review of the five virologists. It is an impressive evaluation of a wide-ranging field of biology by a physico-chemical scientist with broad perspectives developed in connection with work on the Nobel Committee for Chemistry for many years. Of course there was an emphasis in this review on the accumulating knowledge about high molecular biological structures, like for example virus particles. He summarized that the work by Rous and Rivers were more directly focused on solving problems of relevance to medicine, and that Rous was the most original among these two scientists. His discussion of the three other scientists has been reviewed before[6]. In a generous conclusion he proposed to combine Rivers and Rous in one prize and the three other nominees for another prize. In the end Rous was the only one to receive a prize in physiology or medicine whereas Stanley received a shared prize in chemistry in 1946. Somewhat surprisingly the committee in 1939 recommended that none of the five candidates in the field of virology, including Rous, at the time should be considered for a prize.

Evaluation of Rous's Continued Work in the 1940s

In total Rous was to spend 60 years of research at the Rockefeller Institute, later University. In 1945 he had become an emeritus professor, but he carried on his laboratory work and twenty percent of his 300 scientific publications derive from this period of "retirement." Rous had a particular friend in Henry R. Dean, professor of pathology and Master of Trinity Hall, Cambridge. In 1926–27 he was a visiting scientist in Dean's laboratory at Cambridge and he returned to this laboratory on a number of occasions hereafter. It was Dean who nominated Rous for the 1946 Nobel Prize in physiology or medicine, after nine intervening years including the Second World War without any proposal for a prize. At this time there were some new professors on the committee. One of them was the qualified cell biologist Torbjörn Caspersson, who had made pioneering studies of nucleic acid turnover in cells. He was introduced at length in my first book on Nobel Prizes[6]. Caspersson was selected to evaluate Rous together with the earlier reviewer Henschen, who was asked to supplement his previous evaluations.

Caspersson's evaluation is impressively comprehensive and covers 17 pages. He intuitively started his presentation reflecting on virus functions required to allow its replication and possible other effects of its interaction with cells which might cause a disturbance of cell division. It should be recollected that at the time the term "biochemistry" had just started to be used and molecular biology and the establishment of the central dogma was almost two decennia away. Other examples of virus-induced tumors, like different forms of warts, were cited as not representing true tumors — cancers — since they were benign. Caspersson appropriately in parallel discussed the growth of insights into mechanisms for development of Rous sarcoma and of leukemia in fowl. Although at the time essentially nothing was known about virus replication — the growth of bacteriophages in their hosts had just started to become a hot topic in experimental virology — Caspersson wondered if one might separate the production of new particles and possible other influences on cell functions. He concluded that Rous's justification for a prize was defined not only by his early discovery of a virus causing tumors but in addition by his observations of the tumorigenic consequences of the infection by the two viruses he studied. In his work Rous had emphasized that viruses only represent one particular form of cancer-inducing agent. He had introduced the terms "actuating" and "provocative" carcinogens. Caspersson's review then appropriately compared tumors induced by a virus and by selected chemicals, in particular different hydrocarbons, a topic of rapidly growing interest at the time. Interestingly he also mentioned that there were tumors with a growth rate that was influenced by the presence of hormones. A reference was given to Charles Huggins' early work, which will be discussed in the following chapter.

In addition Caspersson's review discussed the recent discovery of other viruses, like the Bittner factor, that might lead to development of tumors. We will return to this below when discussing the development of the field in the 1950s. These findings appropriately highlighted an expansion of the speculative relations between virus infections and tumors. He also discussed "endogenous viruses" and their possible role in general genetic phenomena, a remark reflecting the state of ignorance at the time of the molecular basis for such central phenomena. The conclusion of the extensive review was that Rous was worthy of a prize.

In 1946 Henschen had his final year as chairman of the committee. He retired from his professorship the following year at the age of 65. In his review he emphasized the important renewed and expanded high-quality work on tumors caused by viruses that Rous and his collaborators had performed using Shope's papilloma virus in rabbits. He emphasized that Rous himself in his writings had expressed the view that viruses only represent one among several other factors that alone or in combination with other pathological stimuli could lead to the emergence of a cancer. Henschen maintained his previously-expressed opinion that Rous's discovery deserved to be recognized by a Nobel Prize. This was supported by the conclusion of the committee at its meeting on September 20. However, the "permanent" secretary and member of the committee, Göran Liljestrand added to the protocol his view "that the discovery, because it had been made a long time ago, should not be selected for a prize"!

Tumor Viruses Come of Age in the 1950s

Nomination of Rous continued during the 1950s, but no new evaluations were made. The case would appear to be closed. In 1951 he was nominated by Edmund V. Cowdry, a qualified organizer of cancer research in St. Louis. He finished his nomination in the following way "I believe that this (Rous's contributions) has had a far greater influence on medical research than the discovery by Johannes Fibiger of spirochetal carcinoma for which he received the Nobel Prize in 1926. Since that time no Nobel Prize has been given in the field of cancer research." In 1954 there were three more nominations one of which was submitted by Shope, the discoverer of the virus to which Rous devoted his later studies. It was very comprehensive and highlighted the importance of 21 selected articles by Rous, predominantly in the *Journal of Experimental Medicine*, with specific comments to each publication. The committee took no action and Rous was not mentioned in the protocol. Another nomination was made in 1957 in which Rous was mentioned together with six other candidates by Selman A. Waksman, the 1952 recipient of the prize in physiology or medicine. In 1961 both Waksman and Shope, again in an extensive presentation by the latter, reiterated their nominations of Rous. This time the committee finally took action and let Georg Klein, a young professor at the Institute and a recent member of the committee make a review. This is an impressive and commendable review over 36 pages. It also included two additional candidates besides Rous, Ludwig Gross and Sarah E. Stewart. Let us first introduce Klein and then discuss the rapidly expanding field of newly identified tumor viruses.

Klein is probably the person who next to Gard has had the greatest influence on my development as a scientist. His Department of Tumor Biology at the charming Solna Campus of the Karolinska Institute was a place I returned to essentially every week during my pre- and post-doctoral years, although I had to travel a few kilometers since the Department of Virus Research was located at the State (National) Bacteriological Laboratory in Huvudsta. In Klein's department there were remarkable arrangements of seminars in cramped quarters, involving the

Georg Klein (1925–2016). [By permission from the G. Klein family.]

best scientists of the time in their respective field, tumor biology, molecular biology, genetics, immunology etc. Discussions were intense and revelations representing heterodox thinking were abundant. In 1972 I became Klein's colleague as professor of virology which stimulated many contacts, not least in the Nobel Committee for physiology or medicine in which simultaneously we were members or adjunct members for some 15 years. When I started in 2007 to write books about Nobel Prizes in physiology or medicine Klein became an invaluable source. His memory was impeccable. In October 2016 I visited him to give him a signed copy of my third book on Nobel Prizes[2], expressing from my heart my deep appreciation of our friendly professional acquaintance. On this occasion we discussed Rous and I received some anecdotal insights. Regrettably this was our last encounter since Klein died soon afterwards on Nobel day (December 10) at the age of 91 years. It should be added that he was a Renaissance man, writing books on humanities, philosophy and popular science. He was equally acquainted with Rilke's poems as with Crick's latest theories on mechanisms of genetics. Because of his influence on the development of science at the Karolinska Institute and also his major importance in the work to select Nobel Prize recipients it is worth providing a broader background to his life.

Georg Klein was born into a Hungarian-Jewish family in 1925. He escaped miraculously from a Nazi labor camp and managed to hide during the remainder of the Second World War. In 1947 he had an opportunity to visit the Karolinska

Institute and was offered a working place in Caspersson's laboratory. Klein managed to return to Hungary to marry a fellow medical student, Eva, and bring her back to Sweden. From then on they were inseparable in work and in private life. They rapidly developed their Ph.D. theses and in 1957 a personal chair, unique for its time, was arranged for Georg Klein. The early work by himself, Eva and their group was performed in Caspersson's Laboratory of Experimental Cell Research at the Medical Nobel institute, described earlier [3]. In the early 1960s a separate new laboratory for Tumor Biology, which became the Mecca of science already alluded to, was built. An important part of the laboratory resources was used as animal quarters hosting a large number of colonies of different strains of inbred mice. This kind of experimental animal, with each line representing a unique lineage of identical twins, allowed controlled studies of the role of genetics for cancer development, of the functions of potentially oncogenic viruses under different genetic conditions and of the potential role of immune reactions in the control of the development of tumors. Many different kinds of oncogenic viruses were studied at the department illustrating the rapidly expanding field of tumor virus research at the time.

William E. Castle at Harvard University was the first American to search for genetic inheritance in mammals. He had a student, Clarence C. Little, who in 1929 established the Jackson Laboratory at Bar Harbor, Maine. By brother-sister inbreeding of mice a variety of genetically homogenous lines of mice were established. These inbred animals were to become of enormous importance in experimental genetic research. The unique immunological complex at the surface of cells that determine species and individual specificities was identified and characterized by use of these lines of mice. This work brought George D. Snell, studying throughout his career the major histocompatibility complex (MHC) at the Jackson Laboratory, a shared Nobel Prize in physiology or medicine in 1980[3]. In the early 1930s it was observed at these laboratories that females of certain lines of mice developed breast cancer at an unusually young age. One of the researchers at the laboratory, John Bittner, cross-bred lines of mice with a high and a low frequency of breast cancer, respectively. The results obtained were unexpected and it was found that some agent was spread by the milk from mothers representing the high frequency tumor line. It turned out to be — an ultrafiltrable agent — a virus referred originally to as the Bittner factor. This virus had been mentioned in Caspersson's 1946 review, as presented earlier. However, these findings did not suffice to convince cancer researchers in general of the importance of tumor viruses. It took the contribution of another, very stubborn scientist working in the early 1950s.

His name was Ludwig Gross, of Jewish origin from Poland, who arrived in the U.S. during the Second World War.

Gross harbored an unsophisticated view that all cancers were caused by viruses. In his idiosyncratic experimentation he took a new approach. He inoculated recently-born mice instead of using adult animals. The logic was that the expected immunological immaturity of these very young animals might allow the replication of potential tumor viruses. Gross prepared cell-free filtrates from leukemia tissues of an inbred strain of mice with an increased frequency of leukemia. He was lucky and a virus, later given his name, Gross virus, was identified. But there was more to come. Many scientists were very skeptical of Gross's findings. Several attempts to repeat his observations failed.

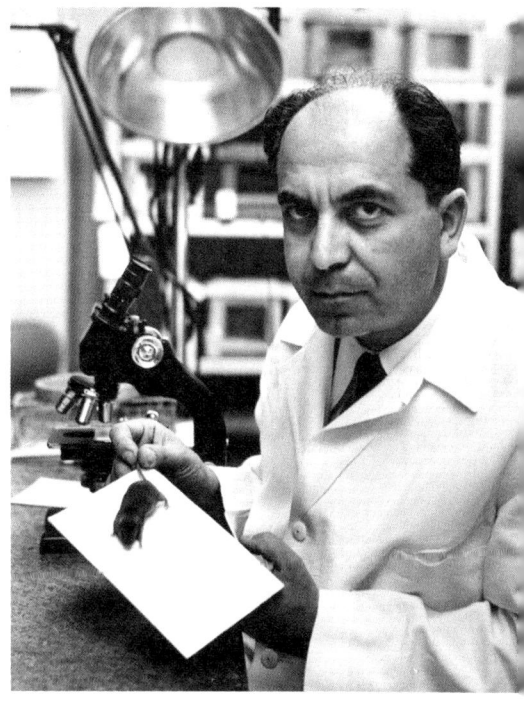

Ludwig Gross (1904–1999).

Among those who failed were two middle-aged women scientists at the National Institutes of Health (NIH) in Bethesda, MD. Their names were Sarah Stewart and Bernice Eddy. However, in spite of their failure they believed that Gross was right and tried a different approach. At this time, efficient techniques had been developed to grow cells in tissue cultures. This enormously increased the possibilities for the development of virological research. A number of new viruses of importance to man were identified and could be propagated in the laboratory. Stewart and Eddy prepared cell-free filtrates of tissues from mice of strains showing an increased frequency of leukemia and inoculated them into tissue cultures of fetal mouse cells. Material from these cultures was then inoculated into mice of the selected strain. However, this did not allow them to repeat Gross's findings, but instead a much unexpected result was obtained. The inoculated animals developed a number of different kinds of solid tumors, in the breast glands, in hair follicles, in salivary glands, in the skeleton, etc. A new kind of tumor virus had been discovered later called *polyoma* — the cause of many (*poly*) tumors, giving the *oma* ending. It was later found that the virus they had discovered was related to the kind of agent that is the cause of papillomas studied by Shope and Rous. These different findings led to waves of speculation that viruses might play a role in *all* forms of tumors, a view which grew during the

Bernice Eddy (1903–1989) and Sarah Stewart (1905–1976). [From Ref. 19.]

1960s and led to the development of the Federal Conquest of Cancer Program, which we will return to in Chapter 3. Both Stewart and not least Eddy were very special research characters and we will meet Eddy again in a discussion of polio vaccines which were developed in the 1950s[6].

Eddy had worked at NIH since 1937. Because of her familiarity with scrutinizing the safety and efficacy of vaccines, she was given responsibility for controlling the first formalin-inactivated polio vaccines produced under the scientific supervision by Jonas Salk in the early 1950s. She isolated live poliovirus from some of the batches of vaccines produced by the California-based Cutter Company. The virus she isolated caused disease in monkeys. For unexplained reasons this important information was disregarded and when the vaccine was used in sizable cohorts of as many as 192 people, many of them children, developed polio. This so-called Cutter episode had major political consequences, with the resignations of the Director of NIH and the Secretary of Health. Eddy was moved to another kind of work, but managed to stay at the Division of Biological Standards. Her discovery of polyoma virus together with Stewart prompted her to reflect on the conditions used for production of polio vaccines in which cell cultures of monkeys were used. Could it possibly be that they contained endogenous, potentially oncogenic, viruses that might survive the techniques of inactivation of polio virus? This was a horrifying thought, since more than 69 million Americans were to be vaccinated during the 1950s after the introduction of the formalin-inactivated vaccine. Logically this independent scientist took material from the kind of tissues used to produce the cell cultures in which the vaccine was prepared, mashed them and passed them through a fine filter and injected them into newborn hamsters. Of the 154 animals injected about a hundred developed tumors!

This was crisis at its worst since at the end of the 1960s a change from the inactivated Salk vaccine to the live Sabin vaccine was imminent. In a live vaccine the risk for survival of possible tumor viruses was even greater than in a formalin-inactivated product. At the same time Ben Sweet and Maurice Hilleman, at the Merck, Sharp and Dohme Co. , found a previously unknown tumor-causing monkey virus, SV (simian virus) 40, in a kind of monkeys , however, not those generally used for vaccine production. Furthermore, the infection by this virus, which turned out to be related to Steward's and Eddy's polyoma virus, seemed to be destroyed by the formalin treatment used to produce the inactivated vaccine. But still the knowledge that potentially tumor-igenic viruses might have occurred in polio vaccines caused a considerable stir.

Fortunately follow-up studies over many decades have not demonstrated any increased risk of the appearance of any particular forms of tumors in the early large cohorts of people who had received the inactivated or live polio vaccine. It has therefore been possible to globally expand the use of vaccines to eliminate polio and at the time of writing the World Health Organization has initiated the end game for a final eradication of the disease and the virus causing it throughout the whole world. The story of SV40 and its role in polio vaccines has been told in its full length[17] and the dilemmas in choosing cells in which to produce virus vaccines has also been excellently reviewed in a newly published book[18]. The SV40 virus was identified because it caused vacuolization — appearance of vesicles inside cells — in cell cultures and hence it was referred to as the vacuolating agent. It has turned out that Shope's original *pa*pilloma virus, Steward's and Eddy's *po*lyoma virus and Sweet's and Hilleman's *va*cuolating agent (SV40) belong to the same family which by using an acronym was named papova viruses.

Klein's Insightful Review

It is now time to return to Klein's impressive and extensive 1961 evaluation of Rous, Gross and Steward. The candidates were discussed separately in the order mentioned. Since Rous's work until 1946 had been reviewed in some detail by Caspersson, Klein focused on more recent developments by Rous himself and in particular by others using Rous virus for their experiments. This discussion covered almost half of the 36-page-long review. Work by many other cancer virus biologists had highlighted and amplified Rous's early work on the role of the dose of virus used, age and strain characteristics of the animals inoculated and the presence or absence of virus or virus antigens in the tumors generated. A vague concept of "slumbering" viruses started to emerge. It was also remarked that Rous virus might be of importance in mammalians.

Harry Rubin.

Klein paid particular attention to the significant work by Harry Rubin and Howard Temin, whom we will meet again in Chapter 3. They used tissue cultures of single cell layers to examine the transforming effect of Rous virus. In 1954 Renato Dulbecco and Marguerite Vogt had shown that if virus of a dose infecting a small fraction of the cells in a culture was trapped by a layer of agar gel on top of the monolayer of cells at the bottom of the vessel, the local replication originating from a single infectious particle led to formation of a visible hole in the cell layer. These holes were referred to as *plaques* and counting the number of plaques became a means of determining the quantity of infectious virus in a sample. Rubin and Temin took a similar approach, but instead of recording local cell destruction they identified local atypical growth of cells with an altered morphology, referred to as *transformed cells*. It is a characteristic of normal cells that as they replicate on a glass or plastic surface they stop dividing when they get in contact with other neighboring cells growing on the bottom of the cultivation vessel. This is referred to as "contact inhibition (of cell division)." However, transformed cells do not react to contact inhibition in the same way but pile up on each other and a *transformed focus* is formed instead of the plaque caused by replication of a virus destroying cells, a cytolytic agent. This was a way to identify and quantify the transforming potential of Rous virus and to study this in more detail. However, in the absence of the forthcoming techniques of molecular biology, examination of the virus-cell relationships were limited to changes identified by histological or immunological tools. It should be mentioned that in 1975 Dulbecco and Temin together with David Baltimore were awarded the Nobel Prize in physiology or medicine "for their discoveries concerning the interaction between tumour viruses and the genetic material of the cell." We will return to this discovery in Chapter 3. Klein's review of Rous ended by noting that he had continued to make pioneering contributions, not least in the context of combined effects of viral and chemical carcinogens.

Gross's discovery of virus-induced leukemia in mice was also discussed at length by Klein. In his work Gross was inspired by the discovery of the Bittner factor which, spread early after birth by milk, could cause breast cancer in certain strains of inbred mice. The cleverness of Gross's approach, as mentioned, was to infect animals less than 24 hours old, but he was also somewhat lucky, as a good scientist should be, to choose a strain of inbred mice that allowed the identification of the tumor-causing effect of the inoculums. When other scientists wanted to repeat his finding later they often failed. This turned out to be due to minor differences even between closely related strains of inbred mice. In later studies other scientists managed to detect additional kinds of leukemogenic viruses in inbred mice strains, such as for example Friend virus. Charlotte Friend was a highly qualified American virologist. In 1957 she discovered another virus causing leukemia in experimental animals. In studies of Ehrlich ascites (the name of the fluid in the abdominal cavity) carcinomas, a very popular material for work on cancer at the time, she discovered virus particles by use of the electron microscope. This led to the identification of what came to be called Friend's leukemia virus. She developed into one of the key scientists in studies of viruses and cancer. When she turned 60 she was diagnosed with lymphoma. She continued to work in her laboratory and died from her tumor five years later.

Let us now return to Klein's analysis of Gross. Besides the virus causing leukemia he also discovered other viruses causing tumors associated with the parotid gland, the suprarenal glands and also involving connective tissue, different kinds of sarcomas, in his experiments. The viruses responsible for these tumors were left for others to identify. Klein's conclusion was that although Gross was the first to identify a leukemia virus in mammals, this discovery was not of the magnitude that it motivated a Nobel Prize. It was Steward who together with Eddy identified the agent which was the cause of the range of different tumors observed by Gross in his extended experiments. The discovery of polyoma virus, already mentioned above, was very important but as Klein noted it would not be possible to recognize Steward by a Nobel Prize without at the same time considering Eddy, who was not nominated. He made an interesting note of the fact that polyoma virus contains DNA as its hereditary material, compared with Rous virus which had been demonstrated to contain RNA. At this time it had been realized that viruses contain either one of the two kinds of nucleic acids, never both of them simultaneously. Klein mentioned in the text that this difference in nucleic acid characteristics "opens the first door for a tumor virus to enter the fascinating field which presently is developing in the border zone between virology and nucleic acid chemistry." This is a true visionary statement since, as we shall see, the introduction of molecular biological techniques revolutionized the studies of mechanisms by which viruses could change the growth characteristics of cells.

The final two pages of Klein's extensive review again placed Rous center stage. Still at the mature age of 82, which he had reached at this time, he continued to make critical contributions to the field. Klein noted that the importance of Rous's original discovery had increased for each year that had passed, recently in an accelerated way. Tumor virology had come to take an increasingly more central place in the wide field of cancer research. He also noted that it would only be a matter of time before viruses of importance for tumors in man would be discovered. Whether they would be of importance for all kinds of tumors in man or only for certain forms remained a subject for forthcoming research. Klein did not discuss the theoretical possibility that development of cancer cells might lead to activation of dormant viruses in the cells, a possibility which potentially could be conjectured at the time. In the final paragraph he noted that "The only Nobel Prize which earlier has been awarded in the field of cancer research sadly has not withstood the test of time." And further on (translated from Swedish) "he (Rous) has been declared worthy of a prize in all previous evaluations (only from 1937 and onward, my remark). When I now agree with these earlier evaluations, I do this convinced that Rous is the hitherto strongest candidate in the field of cancer research and that it will be very difficult in the future to find any other candidate in this field who can compare with Rous when it comes to the originality, importance and scope of the contributions made." The committee agreed with Klein and reaffirmed that Rous was worthy of a prize, but the prize went, somewhat unexpectedly, to a hearing physiologist, George von Bekesy[3].

Approaching a Nobel Prize at Last

Nominations of Rous continued. In 1962 he was nominated by William H. Stein, a colleague at Rockefeller Institute, who in 1972 received a shared Nobel Prize in chemistry. In parallel with Rous he nominated Karl von Frisch, who in his turn received a shared prize in physiology or medicine in 1973 for his discoveries in the very special field of animal behavior. Fritz Lipmann also nominated Rous as one of three possible candidates in 1962. In 1964 Shope renewed his nomination of Rous and he repeated this the following year. In addition in 1965 there were two joint nominations of Rous and Shope, one by a renowned virologist Anthony C. Allison from the Mill Hill Laboratory outside London and one by Edward A. Doisy from St. Louis, the shared recipient of the 1943 Nobel Prize in physiology or medicine. Allison could not refrain from noting in his nomination that "Shope now has a fatal illness (a cancer, my remark) and Rous is an old man…." "These joint nominations prompted the committee to ask Klein to make a joint evaluation of the two candidates Rous and Shope. The emphasis in the evaluation was on Shope and concerning Rous, only a summary of his most important recent contributions was given. It was noted that during the latter years he had focused his work on chemical carcinogenesis. Shope's original observations in the 1930s of papilloma virus and its tumor-causing effects in wild and tame rabbits were revisited once more. It was emphasized that the relevance of these early findings had been accentuated by the more recent observations of tumor-causing effects of the related RNA-containing viruses and also a number of DNA-containing viruses. Among the latter were the different papova viruses and also another kind of DNA virus, adenoviruses. During this time I was involved in structural studies of adenoviruses and because they were tumor-causing I received a 6-year postdoctoral position from the Swedish Cancer Society. We learnt a lot about the structural components of this kind of virus but we never studied its transforming effect directly.

Shope had returned to his work on tumors caused by papilloma viruses in the late 1950s. At this time he was able to draw the conclusion that the interaction between the virus and cells could be of two different kinds. One form was the complete replication of virus leading to the formation of virus particles and the other an immature, non-infectious form. Other workers had made the interesting observation that a full maturation of virus with formation of virus particles only occurred in the most superficial layer of cells in the skin, the epidermis, but not in its deeper layer. This allowed for both a persistence of the infection and for release of infectious particles at the surface of the body that may carry the infection to another individual. The same situation applies to warts on hands and feet in humans. They are categorically benign and can develop to occur in a high frequency. They can also disappear over a short period of time presumably due to some targeted immunological reaction.

During my late teenage years I had a larger part of the sole of one foot covered with warts, an infection I had contracted in sports halls playing handball. To get help I visited the professor of dermatology at the Karolinska Institute, Sven Hellerström. He prescribed magnesium sulfate to be taken orally! Once I did this the warts disappeared in a short time. So much for hypnotic influences on the immunological defense mechanisms!

In his review Klein also mentioned the technical developments that allowed the preparation of purified virus particles that could be examined in the electron microscope and also that the genetic material of the virus, the DNA, which could be extracted, had been shown to be infectious on its own. When Klein weighed together Shope's original identification and characterization of papilloma and myxoma viruses and their potential role in formation of tumors he came to the conclusion that he too was worthy of a Nobel Prize. As a consequence the committee this year listed both Rous and Shope as qualified to receive the prize. However, in this year the prize in physiology or medicine recognized the three French intellectuals and front-line early molecular biologists, François Jacob, André Lwoff and Jacques Monod "for their discoveries concerning genetic control of enzyme and virus synthesis."[2]

Fortunately both Allison and Doisy repeated their nomination of Rous and Shope in 1966. No more reviews were made this year, but in the end the committee decided to recommend that the prize should be given to Huggins and the 87-year-old Rous. Before we speculate on how this could come about it is time to introduce the former surgeon and scientist.

Chapter 2

Hormone Treatment of Tumors and the Prize Events in 1966

HORMONE TREATMENT CAN
REDUCE GROWTH OF TUMOR CELLS
NOT A PANACEA

At the beginning of the twentieth century cancer was a serious and frequently lethal disease among many other equally life threatening diseases. The relative importance of cancer has increased not least in industrialized countries because of improved longevity and major advances in treatment of other medical illnesses. The introduction of antibiotics and prophylactic use of vaccines to prevent infectious diseases as well as major improvements in management of coronary heart diseases have been critical. In some countries death from cancer today even supersedes that caused by the latter kind of afflictions. In the course of the previous century there have been a number of discoveries in biology and medicine that have progressively, and sometimes dramatically, changed our understanding of the mechanisms of cancer development. Many of these advances have been recognized by Nobel Prizes after the award to Rous and Huggins in 1966. To a large extent the revolutionary insights have become possible because of the introduction of techniques of molecular biology. But let us first see how it all started. The selected samples given will refer to various relevant Nobel Prizes awarded or in a few cases discoveries potentially worthy of a Nobel Prize in physiology or medicine. To get the full story it is worth consulting leading books on this topic, be they the more popular, already mentioned, *The Emperor of All Maladies. A Biography of Cancer*[1] or a modern textbook like *The Biology of Cancer*[2]. The early means of treatment were surgery and irradiation with X-rays. At a somewhat later stage chemotherapy and immunotherapy were introduced.

The use of surgery had already been initiated in the nineteenth century and in the absence of other means of treatment its application initially increased with time. Sometimes it was effective provided that the tumor was strictly localized and that no complications, like untreatable infections, became associated with the surgical procedure. As the techniques advanced, in particular concerning antisepsis, anesthesia and possibilities for blood transfusions, the surgical interventions became increasingly bold but the success rate varied extensively depending upon the nature and location of the tumor. There are examples when surgery, such as for example removal of colorectal carcinomas at an early stage, has a very high success rate, equivalent to a cure. Many attempts have been made to combine surgery with local irradiation. After the Second World War techniques to provide a focused radiation improved markedly. Mammary tumors were a preferred target, and when assumed to have been successful the intervention was referred to as radical. This term has its origin in the Latin word *radicalis*, meaning root. Thus in its use by surgeons it implies a complete extirpation of all cancer cells. But radical in other contexts has another meaning, namely someone who is in opposition to the group, which also may apply, since it is the only one of few cancer cells that unknowingly were left behind, which may lead to detrimental future developments.

With the discovery of X-rays, recognized by the first Nobel Prize in physics in 1901 to W. Conrad Röntgen, a new means of treatment became available. Four years later, the prize in physics recognized the discovery by Pierre and Marie Curie of a powerful — and hence a useful — source of X-ray irradiation. However, radiation treatment like surgery was limited to only having a focal efficacy. Furthermore, it became apparent that irradiation by itself could be a cause of cancer. In fact Marie Curie had to sacrifice her life because of the large exposure she had had to irradiation during her active years in science. She died from leukemia in her mid 60s. In 1946 a Nobel Prize in physiology or medicine was awarded to Hermann J. Muller "for his discovery of the production of mutations by means of X-ray irradiation." The connection between damage to the genetic material, at the time of undefined chemical nature, and the uncontrolled growth of cells, became increasingly significant as science progressed. In the 1970s it became apparent that the origin of all cancers was due to damage to the genetic material, as we shall see. Still the treatment of certain tumors, not least of mammary origin, with both surgery and irradiation, meant an improved situation for many patients. But there was a need for more specific and effective means of treatment. This was offered by the introduction of chemotherapy.

Sidney Farber at Harvard Medical School pioneered the successful treatment of childhood leukemia by chemotherapy. He is often referred to as the father of this kind of treatment and is remembered by his name having been loaned to the Dana-Farber Cancer Institute in Boston. Surprisingly until 1968 he had not been nominated for a Nobel Prize. As already mentioned in the previous chapter it took a long time before the uncontrolled growth of white blood cells was considered to represent, in principle, the equivalence of cancerous formations of epithelial cells. In addition it should be mentioned that there are several different forms of leukemia. There are in principle two forms of white

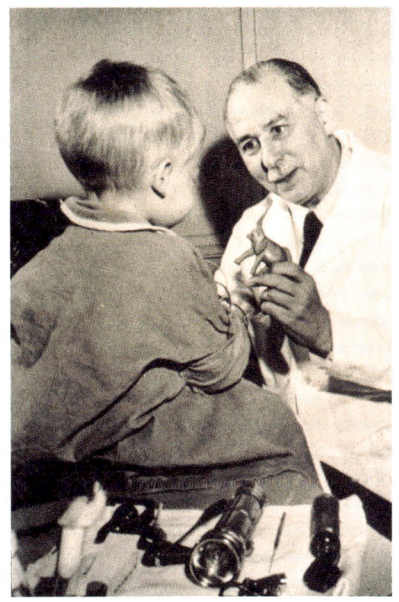

Sidney Farber (1903–1973).

blood cells circulating in the blood. They both have their origin in the bone marrow, but then differentiate into, myeloid cells, remaining associated with the bone marrow and lymphoid cells associated with lymphatic tissue throughout the body. Tumors deriving from them have distinguishable names, myeloid leukemia (acute and chronic forms) and similarly acute (lymphoblastic) leukemia, and lymphomas deriving from more mature cells. Farber's target was the acute lymphoblastic leukemia. The substances he tried were antifolates. A considerable transitory effect was recorded.

A similar kind of effect, albeit weaker, had been noticed previously in bold treatments with mustard gas. This gas was used as a horrible chemical weapon in World War I and its destructive effects on the bone marrow were noted by two American pathologists Edward and Helen Krumbhaar. An incident in the Second World War in the Italian city of Bari confirmed that survivors of exposure to a large dose of the gas essentially had no white cells in their blood and their bone marrow looked "scorched." However, not unexpectedly, the different forms of this kind of chemical treatments were found to represent uses of a two-edged sword, since the substances by themselves were also mutagenic and hence carcinogenic.

There was a need for more effective drugs and such were developed in the George H. Hitching's laboratory at the Burroughs Wellcome Co. in New York, by him and his critical collaborator Gertrude Elion, described earlier[3]. By a lucky choice they focused on compounds interfering with the synthesis of DNA and RNA. A number of compounds effective in treatment of both leukemias and lymphomas in children were discovered and it later turned out that some of them could be used to induce immunosuppression of use in transplantation surgery. Hitchings and Elion, together with James W. Black, the father of beta blockers used to treat heart disease, received the 1988 Nobel Prize in physiology or medicine "for their discoveries of important principles for drug treatment." The situation for children afflicted by childhood leukemia has markedly improved thanks to effective chemotherapy. Similarly chemotherapy has become an essential adjunct form of treatment in many cancers and not infrequently remains the only alternative available for treatment of disseminating cancer. Over the years a number of additional so-called anti-metabolites have developed and found their use in treatment of many forms of cancer. We will see some examples of this in the next chapter.

Discovery Is Our Business

Surgeons have rarely been recognized by Nobel Prizes in physiology or medicine. If the criterion of being an active surgeon throughout one's career is applied the list of "cutting-edge heroes" is limited to only four. The discipline of surgery has hitherto only been mentioned once in a prize motivation. This was in 1909 when the prize recognized Emil T. Kocher, specifying that he was being awarded "for his work on the physiology, pathology and surgery of the thyroid gland." The other three laureates who remained active in surgery throughout their career were Werner T. O. Forssmann, Joseph E. Murray and the central figure in this chapter, Charles Huggins. Forssmann developed a technique for catheterization of the heart and received his prize together with André F. Cournand and Dickinson W. Richards in 1956. Murray was the father of organ transplantation, and shared a prize in 1990 with E. Donnall Thomas, a pioneer of bone marrow transplantation, as described earlier[3]. Antonio Moniz was recognized in 1949 for his introduction of the technique of leucotomy, but as discussed[4] it was his assistants who performed the operations. Other names sometimes cited are scientists who used surgery in their studies of experimental animals; Allvar Gullstrand (1911), dioptrics of the eye; Alexis Carrel (1912) vascular sutures and organ transplants in animals discussed previously[5], Frederick Banting (1923, shared), the discovery of insulin, also described earlier[5] and the recipient of the shared 1949 prize, Walter Hess. The professor of surgery at the Institute at the time, Jules H. Åkerman, was the one to introduce Carrel at the prize ceremony. He had in fact introduced a new technique for joining blood vessels using experiences from the well-known French embroidery tradition at the time. However, as pointed out earlier, his so-called triangulation technique cannot be considered a discovery, but possibly a simple form of invention. Further it remains debatable if his other contributions in development of tissue culture techniques and attempts at transplantation sufficed to motivate a prize.

Huggins was not only a full-fledged surgeon; he was also an impressive experimentalist. Above his desk in the laboratory was a plaque carrying the motto of the subtitle given above. He was of Canadian origin, born in Halifax, Nova Scotia in 1901. After three years of studies he earned his bachelor's degree at the local Acadia University at the age of nineteen. He demonstrated a broad interest not only in natural sciences, spending his summers at Columbia University in the U.S. to improve his insight into physical and organic chemistry, but he was also deeply involved in the excellent classical studies offered at the college he attended. Apparently already early on he managed to build a solid self-confidence. When he learnt that Harvard was the best university in North America he applied to its medical school. He was one of the few foreigners admitted. The studies stimulated his intellect and he managed them excellently. He moved on to an internship in internal medicine at the University of Michigan. Inspired by acquaintances at the surgical faculty he changed the intended focus of his specialization and instead trained in this discipline under Fredrick A. Coller. When he had reached the level of instructor at Michigan he was recruited to the newly established University of Chicago Medical School.

Huggins in the laboratory. [From Ref. 11.]

Dallas Phemister, the chairman of surgery decided that Huggins should take responsibility for urological surgery, about which he knew essentially nothing. He was employed as one of the eight original full-time staff members at the Medical School and since he had a long life, dying at the age of 95, he became the last survivor of this group. Because of the culture of dividing the clinical responsibilities among the Chicago medical community the number of patients initially was small. This gave him time not only to develop his insights into the theory of urological surgery but also to extend his skills in the laboratory. He soon became an enthusiastic spokesperson for this kind of surgical specialty. His early laboratory work concerned the physiology of bone formation stimulated by a finding by others that such a formation could be induced by contact with the dense connective tissues surrounding muscles (fascia) used to repair the urinary bladder. However, there was another topic that was soon to engage his full attention. This was prostate physiology.

Premister arranged for Huggins to get acquainted with high quality European medical science. In 1931 he visited Otto Warburg's laboratory in Berlin. At this time Warburg, introduced already in the previous chapter, was at the peak of his career, as described earlier[4]. He received his Nobel Prize this same year. Huggins nourished a hope to work in Warburg's laboratory but this did not materialize. However, as we shall see Warburg remembered their contact and followed Huggins's work closely. Instead of working in Germany, Huggins developed a broad acquaintance with biochemistry under the guidance of Robert Robinson, a specialist in studies of the enzyme phosphatase in bones, at the Lister Institute in London, to whom he was introduced by chance.

Back in the U.S. he returned to his studies of the prostate. Using dogs he developed a technique to separate the secretions of the gland from urine and was able to analyze their composition. Methods were developed to conveniently measure the content of acid phosphatase and other essential substances present in the samples. Critically, he observed that the prostate activity was increased by testosterone, but decreased by estrogens. Using experimental animals of different ages it was observed that older dogs developed hypertrophy of the prostate. This provided an interesting model for studies of similar phenomena in man. A seminal — from Latin *seminalis*, of seed origin — discovery was made. Extrapolating from studies of dogs it was demonstrated in patients that prostate hypertrophy and also prostate cancer could be suppressed by reducing the male hormone androgenic activity achieved either by treatment with the female sex hormone estrogen or by castration. This was the first time that the development of a cancer could be demonstrated to be influenced by a systemic treatment although as we shall see it had been observed earlier that mammary cancer could be influenced by estrogen treatment. It had the important conceptual implication that cancers were not as anarchic and autonomous as was believed at the time. These revolutionary findings were published together with Clarence Hodges in 1941. Previously all forms of treatment of cancer had been based on a direct attack on the tumor by surgery or by irradiation. Clinicians around the world rapidly adopted this new form of cancer treatment and they have remained in use into the present time.

Huggins found that some patients did not respond well to the treatment because of a continued secretion of androgenic hormones from their adrenal cortexes. It was then demonstrated that surgical removal of the adrenal glands could lead to regression of the prostate cancer in these special cases. However, this dramatic intervention required supplementary hormone treatment since the adrenal glands are the source of many different critical hormones. Such hormones became progressively available for substitution for medical use in the 1940s. Departing from the limited knowledge at the time of the different adrenal-derived hormones, various forms of replacement therapy were tried. Due to the inherent complexity of the surgical approach the technique only became of limited practical use. The encouraging results of studies of systemic treatment of prostate cancer led to corresponding studies of another tumor in the human reproductive system, mammary cancer. It was already known ever since the end of the nineteenth century that removal of the ovaries was beneficial in some cases of this form of tumor. In 1952 Huggins found that there was hormone dependence also in many cases of mammary cancer. Estrogens stimulated breast cancer growth and androgens retarded its progression. Various hormone treatments were tested and in certain cases the drastic intervention of simultaneous removal of the adrenal glands was also tried. However, the relative effects observed were not fully comparable with those seen in prostate cancer.

For many years Huggins's research group received support from a philanthropist, Ben May, a highly successful businessman from Mobile, Alabama. A growing support led to the establishment in 1951 of the Ben May Laboratory for Cancer Research. Huggins became its first Director, a position he held until 1969, but he continued to work there until his death 28 years later. In 1962 Huggins reduced his involvement in clinical work and devoted most of his time to research as the Wilam B. Ogden Distinguished Service Professor. Huggins's wife Margaret, an operating room nurse when they met and married in 1927, played a major role in supporting his deep commitments to science, by taking full control of their home and providing support to junior professional colleagues. This allowed her husband to spend so many hours in the laboratory. They had a long life together, although she died 14 years earlier than her husband. Of their two children, a son became a physician and there is also a third generation of Hugginses in medicine.

It seems that Huggins was proud of and enthusiastic about his science. He carried the torch of his mentor Phemister, viewing science as a warm and enriching human endeavor. He was always curious and supportive of his junior colleagues, frequently inquiring about their most recent discoveries.

The Nobel Committee Reviews a Surgeon

The first nomination of Huggins was submitted in February 1949 by none other than Warburg, the 1931 recipient of the prize in physiology or medicine, who has already been mentioned. He simply stated that he viewed the introduction of hormone treatment of prostate cancer as the most important contribution to the cancer field. There was also a second nomination by Kellogg Speed, a professor of clinical surgery at the University of Illinois. He emphasized "this monumental advance in relief of human suffering in cancer along with its greatly expanded future possibilities." The professor of surgery at the Karolinska Institute since 1939, John Hellström, was selected to make a review. He had himself published articles on urological surgery. The review was thorough and started with some remarks on the role of prostate cancer in society. The relative importance of this form of cancer has progressively increased as a consequence of the prolonged average life span developing during the 20th century. The only means of treatment available at the time was surgery or irradiation, often with many side effects. He then noted that castration had been tried as a means of influencing prostate cancer during the 1930s but with unclear results.

The breakthrough came in 1941 with Huggins's publication of orchiedectomy and hormone treatment of 21 cases of prostate cancer, 14 of which had metastases. The use of this new method of treatment spread rapidly, even though it was acknowledged that it did not provide a cure. Hellström emphasized that Huggins's surgical/hormonal intervention was based on careful experimental studies determining the production of seminal fluid in dogs under different regimes of treatment. He also mentioned in this context the importance of using acid phosphatases as a measure of activities of the prostate. Determining the concentration of this enzyme was a means to monitor the activity of the gland. It was later substituted for by determination of prostate specific antigen (PSA). Hellström's conclusion was that Huggins introduction of castration and treatment with estrogenic substances hitherto represented the largest step forward in treatment of prostate cancer. They were pioneering. He paid tribute to the fact that they were based on experimental studies in animals and that the systemic approach introduced was something conceptually new in the treatment of certain (hormone-dependent) cancers. In summary he clearly stated that Huggins's discovery should be recognized by a Nobel Prize. Surprisingly the committee concluded that Huggins should not "at the present time" be considered for a prize.

The following year Huggins was nominated again, this time by R. Geissendörfer, Frankfurt am Main, Germany. Hellström made a brief supplementary review reiterating that he considered Huggins worthy of a prize. He also provided some additional information on recent results of removal of the adrenal glands in prostate cancer patients not responding to estrogen hormone treatment. By use of corticosteroids, which had recently become available, it had been possible to critically compensate for the comprehensive loss of hormones caused by the operation. Hellström referred to one of his younger colleagues, Curt Franksson, who during a prolonged visit to Huggins's clinic had became an eyewitness to the success. Franksson later became professor of surgery at the Karolinska Institute and became the surgeon who was to introduce kidney transplantation in Sweden. The committee still hesitated and wanted to delay a decision of a prize to Huggins. Three years later (1954) Warburg repeated his earlier nomination. This time the committee decided to let the rising star in endocrinology, Rolf Luft, make a review. A few years later, in 1958, Luft became professor of endocrinology, a position he retained until 1980. We worked together in the Nobel Committee for many years and became good friends. In the late 1950s he pioneered the discovery of a mitochondrial disease, later on demonstrated to be due to a defect in a gene in the separate, extra-nuclear genome of this particular cellular organelle. In addition to Huggins, Luft had one more candidate to review. This was Edward C. Dodds. Dodds had developed a means for production of synthetic substances with estrogenic effects. A substance named stilbestrol came to have many different uses in patients including treatment of prostate cancer. Luft considered this contribution worthy of a Nobel Prize, but the committee, not surprisingly, did not agree with this and declared Dodds not at the time worthy of a prize.

Rolf Luft (1914–2007).

As for Huggins, Luft was more critical. He accepted that the methods, castration and estrogen treatment, strongly advocated by him were effective, although they did not provide a cure for the disease. Searching earlier literature Luft found a number of examples of previous indications of a role of hormones in the development of certain cancers. Already in the early part of the twentieth century experimental evidence of a relationship between estrogen activity and breast cancer in animals had been presented. Corresponding studies of prostate cancer could not be performed in experimental animals due to the absence of any useful model. Luft also ventured a certain hesitancy in the interpretation of the effects of surgical removal of the adrenal glands as a means of improving the situation in advanced prostate cancers. His interpretation was that it might be more useful in mammary cancer. Thus although Luft accepted that it was thanks to Huggins that the new surgical and hormone treatment interventions in the case of prostate cancer and, to some extent, mammary cancer was introduced into medicine, he questioned his prioritizing the discovery of the concept of hormonal dependence of tumors. He interpreted that the pioneering conceptual discoveries had been made by others and declared his view that Huggins was not worthy of a prize. The committee upheld its position that he was "not at the time" worthy of a prize.

In 1956 and resubmitted in 1957, Huggins was nominated by the 1943 (awarded in 1944) recipient of the Nobel Prize in physiology or medicine Edward D. Doisy, the discoverer of the chemical nature of vitamin D. After a description of the essence of Huggins' work he stated "Moreover, it has given both relief and hope to those suffering from cancer and encouragement to scientists engaged in seeking methods of control and cure of cancer." In 1957 there was also another extensive nomination of Huggins by Choh Hao Li, an exiled Chinese scientist at University of California, Berkeley, who pioneered identification and synthesis of protein hormones during the 1950s and 1960s. Interestingly Li had requested Rous to provide support to his nomination, which he happily did in a long attached letter. He eagerly supported Li's nomination and he did not mince his words in the conclusion, stating: "…one of the most effective biological scientists living to-day, and he is all the more effective because his work is done primarily for the living human creature."

This time a third professor at the Karolinska Institute, Axel Westman, was requested to make a review. He had been professor of obstetrics and gynecology since 1943 and in 1957 he was one of the three ordinary members of the committee. He first referred back to the earlier reviews by Hellström and Luft. Like Luft he went back to the history of the early findings of hormone dependence of tumors as identified in experimental animals. As a first example he cited

Loeb's — introduced in the previous chapter — demonstration in 1919 that the ovaries were of importance in the development of breast cancer in mice. He also gave other examples, but then praised Huggins's experimental approach to the cancer problem. The critical question Huggins raised was if in certain forms of tumors the cells retain some characteristics of the cells of origin and if this can be exploited to suppress their growth, for example by hormone treatment. Huggins's experiments demonstrated in dogs that a functioning prostate tissue was dependent on intact testicular functions. Already in the original publications 20 patients with advanced prostate cancer with metastases were treated by removal of the testicles. Four of the patients in this original group survived more than 12 years. Westman revisited this field and noted that in 1956 there was a publication of studies involving 1,818 cases of advanced prostate cancer consolidating the conclusion that castration and estrogen therapy were highly beneficial.

The effects of the treatments introduced by Huggins varied among patients and attempts were made to find markers on tumor cells in different patients which would allow the prediction of possible beneficial effects of their application. In the 1950s Huggins stimulated another scientist at the Ben May Laboratory for Cancer Research, Dr Elwood Jensen to shed some light on this problem. Using radioactively labeled estrogen Jensen surprisingly found that the hormone was not metabolized but remained attached to the cells. This led to the formulation of a theory that cells contained certain molecular structures to which the hormone could bind — a *receptor*. Some decades later it was possible to isolate the receptor and to demonstrate that the hormone, by attaching to this structure, could elicit a production of specific proteins via a number of intermediary metabolic steps leading to induction of synthesis of specific messenger RNA. The possibility of demonstrating the presence or absence of estrogen receptors was later found to have an important practical application. It could be quantified if the cells of a breast cancer in a particular patient did or did not have the receptor. Only patients with cancer cells carrying the receptor responded to hormone treatment. Another important consequence of the identification of the estrogen and other similar receptors was the possibility to develop drugs that interacted with the receptor. The foundation of this important work, which can be predicted to have been discussed by Nobel Committees during the coming decades, albeit not resulting in a prize, was laid at the Ben May Laboratory, which Huggins headed until 1969, when Jensen took over as Director.

Let us now return to Westman's evaluation of Huggins. In relation to

The aging Huggins and Elwood Jensen (1920–2012), his successor as director of the Ben May Laboratory for Cancer Research.

Huggins's increasing involvement in hormone treatment of mammary cancer in the 1950s, Westman discussed a number of ongoing studies at the Karolinska Hospital in particular by the professor of surgery Hellström, who has already been introduced. Following Huggins's recommendations he had performed removal of adrenal glands and ovaries on both sides. Cortisone, the discovery of which had been recognized by the 1950 Nobel Prize in physiology or medicine[5], had now become available for treatment of patients. It was used in the clinic. This markedly improved their survival rate. In many cases it was observed that there was a major regress of the tumors and also of metastases in different parts of the body. By reference to these fresh Swedish data Westman, without reservation, concluded that Huggins work deserved to be recognized by a Nobel Prize. In spite of this the committee still hesitated. It wrote in its concluding protocol "… that the continued developments (of the field) should be followed, before a definitive position is taken to his work concerning the importance of hormones in certain forms of cancer and (the value of) hormone therapy of these."

Warburg did not give up and submitted a new nomination in 1958. In addition there were three nominations from Boston, one by the Clinical Professor of Genito-Urinary Surgery, Harvard Medical School, J. Hartwell Harrison, with an attachment of seven pages, and two shorter ones from William P. Murphy and Joe V. Meige. In addition there was a nomination by F. H. Bauer, Heidelberg. Westman made another supplementary evaluation. It departed from the statement that "it is apparent to me that Huggins' work deserves to be recognized by a Nobel Prize." However, having stated that, he introduced some doubts as to the interpretation of the effect of hormones on the development of certain cancers. He stated "It is therefore uncertain to what extent estrogen and androgen principles, respectively, on their own, have importance for the development of cancer or if one has to account for a hitherto unknown factor, possibly attached to the hormone substances." However, these vague critical remarks did not hold the committee back. For the first time it concluded that Huggins was worthy of a prize.

In 1959 the insistent Warburg nominated Huggins again, but this time he was joined by another towering figure in the history of medicine, Albert Szent-Györgyi, the recipient of the 1937 Nobel Prize in physiology or medicine for his studies of vitamin C and its role in biological combustion. Westman made some supplementary remarks over two and a half pages. He concluded that after the introduction of advanced surgical treatment and irradiation therapy the therapeutic use of hormones represented the major advance in the treatment of cancer. It can be noted that chemotherapy was not mentioned, but it was only at this time that the first encouraging results of chemical treatment of certain forms of cancers, like leukemia; were obtained as we shall see. In the final protocol of 1959 there was one dissenting voice among the committee, which otherwise unanimously supported a prize to Severo Ochoa and Arthur Kornberg. It was Westman who apparently was so impressed by his reviewing of Huggins that he instead proposed him alone for the prize.

Huggins was nominated again for the 1961 and 1962 prizes. In the latter year the earlier 1954 critical reviewer, Rolf Luft, was requested to make a follow-up evaluation. He made a thorough analysis of Huggins's work, not least the frontline contributions over many years by him and his collaborators at the Ben May Institute. The review was divided into three sections; the introduction of hormone treatment of prostate cancer in man; other contributions evaluating endocrine treatment of malign tumors in man; and experimental studies in animals of relevance to cancer research. Only the contributions in the first section were considered to be of importance in discussions of a Nobel Prize. Even though hormonal treatment of certain cancers had been considered before Huggins, he was clearly the pioneer in introducing the practical use of this form of treatment of disseminating prostate cancers in man. Luft now considered this achievement as the first milestone in modern cancer treatment. Huggins's and his collaborators' studies of animals to understand the mechanism of the hormone treatment on tumors were described as laudable, but they had not as yet presented data of fundamental and generalizing importance. The early 1960s still only represented the dawn of what came to be rapidly expanding molecular biological studies of cancer. After a two-page summary of all arguments by many different nominators over almost 20 years Luft presented his conclusions that Huggins was the pioneer in introducing hormone treatment of generalized prostate cancer and that he had also made many other important contributions to the field of experimental cancer research. Luft's conclusion read (translated from Swedish) "However, I consider Huggins's total contributions to cancer research to be of such importance that they motivate a Nobel Prize. He should be one of the foremost candidates for a Nobel Prize today within clinical and experimental medicine. He is also the central figure in an important field of research, which as yet has not been recognized by a Nobel Prize." It would seem quite acceptable that Luft disregarded the prize to Fibiger, but his reference to "total contributions" is more questionable since the prize in physiology or medicine is given only for a single discovery.

There was only one, unspecified, nomination of Huggins in 1965, by Janet S. F. Niven at Mill Hill, London. In a two-page review Luft recapitulated his comments from the previous year and added in a final sentence: "Huggins to an exceptional degree fulfills the requirements for a Nobel Prize in physiology or medicine as specified in the statutes." In spite of this strong endorsement the prize this year as mentioned at the end of the previous chapter recognized the three French intellectuals, Jacob, Lwoff and Monod[4]. In the critical year of 1966 there were three nominations of Huggins, a brief one by E. C. Dodds, London, another brief one by a Finnish-Swedish physiologist from Lund University, Georg Kahlson, and, encouraged by the latter nominator, one more extensive proposal from a professor of physiology at Chicago University, Dwight J. Ingle. Since this was the year when the prize finally recognized Huggins and Rous, one would expect that, as in essentially all similar previous situations, a concluding summarizing review of the two strong candidates would have been commissioned. However, this was not the case! Hence one needs to reflect on why the two candidates were brought forward to finally receive their prize in this particular year of intense competition.

A Nobel Committee in a Quandary

At this time in the mid-1960s an impressive number of strong contestants for the prize were listed. In fact in 1966 there were no fewer than 35 candidates considered alone or in combinations with one or two others to be worthy of being recognized by a prize. It seems that this may have caused an "embarrassment of riches" because the committee could not agree on a single prize proposal at its concluding meeting of September 26. Another meeting was needed. It presented its final recommendation to the Faculty of Medicine first on October 13.

Already when the first nominations for a Nobel Prize in physiology or medicine were received in 1901 they were divided into different disciplines and this procedure, with certain modifications, was used until the 1970s. In 1901 the wide field of physiology or medicine was divided into six sections;

I: anatomy, histology, II: general biology, physiology, physiological chemistry, III: pathology, IV: medicine, surgery, ophthalmology, V: hygiene, etiology and VI: immunology. This division into separate fields was maintained for many decades but then, because of the changing profile of the medical sciences, it became progressively modified with time. In 1966 the nominations were sorted into five sections; I: cell biology, cell structure, II: physiology, pharmacology, III: physiological chemistry (surprisingly the more modern term of biochemistry had not as yet been adopted), IV: medicine, surgery, therapy, and hygiene, and V: microbiology, immunology. Twelve years later when Jan Lindsten took over as secretary of the committee the subdivision of nominations into different fields was finally eliminated. They no longer served any purpose. There is not and has never been any rotation between different disciplines in the choice of candidates for a certain year. Not surprisingly prizes are never given in the same or neighboring fields two years in a row, but that is easy to keep track of without dividing nominations into different fields. The most important factor in bringing the discussions to a conclusion is the access to insightful reviewers with the capacity to present the strength of a certain candidate in a lucid and convincing way. I know from my own experience that the final discussions in early September represent a true chamber play, an intense giving and taking of arguments. Let us conceptualize how the discussions may have played out this particular year.

In the absence of the world-renowned cell biologist Torbjörn Caspersson, who was introduced in Chapter 1, and who most likely was not a member of the committee since he himself had been considered for a prize just a few years earlier[3], Klein probably presented arguments for the candidates in group I. This group included, for example, candidates who had made discoveries in the field of chromosomal structures. Important new information on the structure of chromosomes and abnormalities correlating to specific diseases should have been revealed and reviewed. The discoverer of the Barr body, an inactivated X chromosome identified in female cells by the Canadian geneticist Murray L. Barr and the discoverers of chromosomal changes specific for certain genetic diseases, Jérome Lejeune and Raymond Turpin must have been discussed. None of these prize-worthy candidates were ever to become recognized by a prize. In fact as we shall see, new insights into the important cytogenetic chromosomal structures, which originally had been highlighted by the 1933 Nobel Prize in physiology or medicine to Thomas H. Morgan "for his discoveries concerning the role played by the chromosome in heredity" were never to be recognized again by a prize. This may seem somewhat astonishing considering the importance of identification of specific chromosomal aberrations in certain genetic diseases such as, for example, the occurrence of three instead of two copies of chromosome number 21 in children with Down's syndrome. Chromosomal aberrations with particular characteristics are also of importance in studies of cancer cells as we shall see in the next chapter. Thus it may seem somewhat surprising that cytogenetic changes specifically associated with certain diseases were never recognized by any prize. Klein probably also initiated the discussion of the cell biologists Palade and Keith R. Porter. They also had been reviewed by Caspersson. Palade was to receive the Nobel Prize in physiology or medicine in 1974 together with Claude and de Duve as will be seen in Chapter 8.

Under the heading of physiology, pharmacology, group II, there were several strong candidates reviewed during the year. The forthcoming prize recipients in physiology or medicine Ragnar Granit, Haldan K. Hartline and George Wald were highly praised in the review made by Yngve Zotterman, but they had to wait until the following year as discussed in Chapters four and five. Other praised and prize-worthy candidates in the field of neurophysiology such as Ulf von Euler and Bernhard Katz had to wait for four more years. Another qualified candidate in the same field, Stephen W. Kuffler, was never to receive the prize, as discussed earlier[4]. Finally one more strong candidate also mentioned earlier, Hugh Huxley[4], who together with the prize recipient in 1963 Andrew Huxley — no relative — had formulated the fundamental "sliding filament theory" of muscle contraction, was considered worthy of a prize based on a review by Arne Engström. However, this important discovery was never to be recognized by a prize.

There were four candidates, Francis B. Colton, Herman Knaus, Gregory Pincus and John Rock, nominated for the development of oral contraceptives in the early 1950s. This field was reviewed by the professor of obstetrics and gynecology, Ulf Borell, who was a member of the committee. He considered Pincus, but not the other three candidates, worthy of a prize, but pointed out that an important candidate Carl Djerassi had not been nominated. There seems to be no question of the fact that the introduction of chemical contraceptives might be considered appropriate to be recognized by a Nobel Prize in physiology or medicine as being of benefit to mankind. But it appears that the Nobel Committees over the years have shunned recognizing the importance of candidates representing behavioral sciences. Sigmund Freud was never recognized in spite of having received 33 nominations for a prize in physiology or medicine[6] and also one in literature! Similarly as discussed earlier[4] the important discovery of the role of smoking in the development of lung cancers was never recognized by a prize. This hesitation to recognize the significance of behavioral sciences deserves to be reflected on since modern medicine to a large extent relies on advices in the choice of lifestyles.

There were eight candidates in the field of biochemistry considered worthy of a prize. They had been reviewed by Sune Bergström and Hugo Theorell. Bergström was not a member of the committee because he was considered worthy of a prize in 1965, as mentioned in my previous book[4]. Adopting a new principle introduced by the first year secretary Bengt Gustafsson he nominated Bergström in the absence of any external nomination. Gustafsson was a very efficient secretary, as mentioned briefly earlier[3]. Thus he began as secretary in 1966 and was to carry this important responsibility for a total of 12 years. Hence I had the pleasure to be a part of the committee for six years under his aegis. Bergström remained listed as a candidate worthy of a prize although no further evaluation of him had been made. It would take 16 years before he was to receive his shared prize. Another of the candidates in chemistry, Luis F. Leloir, was reviewed and praised by Peter Reichard. LeLoir was to receive the prize in Chemistry in 1970 for his studies of the synthesis of carbohydrates. One of Theorell's favorite candidates, Britton Chance, considered worthy of a prize for many years never made it as discussed earlier[4]. Giuseppe Bertani, a first time member of the committee introduced earlier[4], most likely argued for candidates in the field of molecular biology, also in group III. Among those were the molecular geneticists Seymor Benzer, Sydney Brenner and Marshall Nirenberg. Since Nirenberg was to become one of the three prize recipients in physiology or medicine in 1968 we will return to him in Chapters 6 and 7. Out of the two other strong candidates Benzer was discussed in one of my previous books[4]. He received the Crafoord Prize, but no Nobel Prize. Brenner had to wait for his Nobel Prize until 2002, almost forty years after he was declared worthy of a prize for the first time in 1966.

In group number V for candidates representing microbiology and immunology there was a host of highly qualified candidates. There were the scientists who had pioneered the use of bacteriophages in exploring the fundamentals of genetics. Among those Max Delbrück, Alfred D. Hershey and Salvador Luria were to receive the 1969 prize. However, Seymor Cohen in the related field of genetics of plant and bacterial viruses, was never recognized by a prize. Other highly qualified candidates in the same field, Gerhard Schramm, Alfred Gierer and Heinz Fraenkel-Conrat, discussed in my previous books[3,5] as having discovered the full infectious properties of the RNA of a virus, were very close to a prize but never received it. Then there must have been major discussions about possibly recognizing the fathers of polio vaccines, Jonas Salk, Hilary Koprowski and Albert Sabin. Sven Gard, one of the most influential members of the committee for two decades was also discussed in this context. He had been nominated because of his seminal description of the kinetics of inactivation of poliovirus by formalin, allowing the production of a safe inactivated polio vaccine. His vaccine was used to eliminate polio from Sweden, one of the few countries which managed to do this by use of only inactivated vaccine and which therefore never had the problem of vaccine-associated polio occasionally seen following the use of live vaccines. Gard declined to be a candidate and in the end it was considered that the 1954 prize to John Enders, Frederick Robbins and Thomas Weller had recognized the critical discovery in this field[5].

There were also very strong candidates in the field of immunology, not least researchers who had established the field of cellular immunology discovered to be of matching importance to humoral (antibody-mediated) immunology. As discussed previously[3] Bruce Glick, Robert Good, James L. Gowans and Jacques F. A. P. Miller pioneered the field. Astrid Fagraeus wrote a generally positive review and recommended these candidates for a prize. However, there seems to have been some resistance to her proposal and in the end Gowans was considered not worthy of a prize and the other three were put on a backburner, listed as not at present worthy of a prize. It remains somewhat surprising that in the field of immunology which has been recognized by a relatively large number of prizes[3], the original identification of the existence of T cells responsible for cell-bound immunity was never identified by a prize.

Finally there was a nomination in a particular clinically relevant field of immunology. The red blood cells carry specific antigens of different kinds. They specify the blood group characteristics, the so-called ABO system. The discovery of this system earned Karl Landsteiner a Nobel Prize in physiology or medicine in 1930. Later it was discovered that there is an additional blood group system, called the Rh (from the name of *rh*esus monkeys) system. A woman who has repeated pregnancies with a Rh-positive fetus and who herself is Rh-negative can become immunized which may cause harm to the fetus. This important insight into a possible development of a disease caused by harmful immunological reactions in connection with a pregnancy was recommended for a prize to Philip Levine and Alexander S. Wiener. In this case an exceptional choice of an external reviewer was made. This was Rune Grubb from Lund University. He proposed that the discovery should be declared worthy of a prize, but in the end the important contribution by Levine and Wiener was not recognized by a Nobel Prize in physiology or medicine. The discovery did, however, lead to ways of passively immunizing Rh negative women to prevent development of a disease in the fetus.

With such a host of strong candidates it is not surprising that the committee hesitated. The chairman, Sten Friberg, the Vice-Chancellor of the Karolinska Institute at the time, must have had a challenging task to bring his fragmented large committee to a final decision. The runners-up among this host of strong candidates in this year were Huggins, the only prize-worthy candidate listed in group IV and Rous, a member of group V. It can only be speculated what may have had the major influence on their final deliberations. It seems not unlikely that the five professors of clinical medicine, Ulf Borell, Curt Franksson, Rolf Luft, Carl-Axel Hamberger and Rolf Zetterström, were attracted to prioritize a prize in their applied specialties. They were probably supported by Georg Klein, Sven Gard and Giuseppe Bertani, all three with insights into virology, who would have thought that

Sten Friberg (1902–1977).

it was time for a prize focusing on cancer and in part the role of viruses in this context. Since Klein had recommended that Richard Shope should be considered worthy of a prize, which was accepted by the committee, it needed to be considered if he should be accepted as a candidate in parallel with Rous. The committee decided not to include him. Therefore the final choice included two candidates who had *not been* subjected to a final review in 1966, namely Rous and Huggins. Their worthiness for the prize was based on reviews made in 1965. The combination of Rous as a pioneering tumor virologist and Huggins, representing in a commendable way surgery, a clinical discipline rarely represented among prizes, and experimental studies, should have been appealing to the Faculty when time came for the final decision. The prizes in physiology or medicine over the years had to a large extent favored discoveries in physiology. To award a prize relating to cancer would mean to highlight a disease that had an obvious major medical impact. The importance could be readily explained to the public. In regard to Shope, who was excluded by the committee, it can be added that in between its two concluding meetings, on October 2, he had died — of cancer.

The committees throughout the years have come back to discussions of two particular matters. One is the definition of the important concept "discovery" used by Nobel in his will and the other is where to set the bar in comparing those that are strong contenders for a prize. During my twenty years on the committee at the Karolinska Institute we frequently returned to discussions of these issues. We learned that if we were too generous in defining important discoveries as worthy of a prize the usefulness of the term was reduced. Very strict and demanding criteria needed to be applied. Of course each one of us had a particular familiarity with the field in which we developed our own research. Naturally we also had an increased understanding of developments in this field and might have a preference for emphasizing its importance. However, the charm of the committee work was that it was a true give and take. For once our collegial interactions were not defined by territorial protectionism, to gain resources that furthered the particular science each one of us represented. Instead it was a matter of presenting the subject of a review for which you might be responsible in as clear and convincing a way as possible, but then to listen carefully to your colleagues presenting their candidates and the discoveries they had made. Once brought around to acknowledge the superior strength of a particular proposal presented by a colleague the interactive distillation process could progress. The work on the committee of course provided an exceptional education in the wide range of disciplines represented in the large fields of physiology or medicine. It was a friendly give and take, which sadly may not be that prevalent in academic interactions in general.

We appreciated the huge impact of our final recommendation and had a deep respect for the work performed by preceding committees over many decennia. It was the high quality of this work that had made the Nobel Prize unique in the world. Our responsibility was huge! In fact looking in the rear mirror I would say that of all the work I have done on numerous committees at the Karolinska Institute, there is nothing that compares to being an ordinary or associate member of the Nobel Committee. The spirit at the joint dinner following the meeting at which the final decision had been taken was high. Lasting friendships were built. But for a selected number of the committee members there was more work ahead. One of them was to formulate a brief and succinct prize motivation. Additional involvements for selected committee members were to work out press releases and to plan arrangements for events at the Karolinska Institute during the Nobel week. But all this was an enjoyable, although demanding task. We felt the responsibility to bring our insights and enthusiasm to the public at large. Announcement of Nobel Prizes provide a golden opportunity for the exposure of high quality science in the many different forms of mass media.

In the case of 1966 two separate motivations were formulated although the two prize recipients shared a single prize. Rous was praised "for his discovery of tumour (committee members had been schooled to English spelling)-inducing viruses." The plural form was used to emphasize that at this time a number of tumor-inducing viruses had been discovered. The motivation for Huggins's half of the prize was straightforward "for his discovery of hormone treatment of prostate cancer." It was probably out of veneration for Rous's mature age that the two prize recipients were not presented in alphabetic order, as is normally the case.

The Prize Events in 1966

The Nobel Prize recipients of vintage 1966 were a group of seniors. The youngest was the French physicist Alfred Kastler who was "only" 64 years old. Huggins had just turned 65 and had recently retired from his professorship in Chicago. The prize in literature was divided this year, recognizing Nelly Sachs, the younger at 65, and Joseph Agnon, the older at 77. The prize recipient in chemistry, Robert S. Mulliken, who like Huggins came from the University

All Nobel Prize recipients in 1966 with Rous in the middle and Huggins at the far left. [From Ref. 11.]

of Chicago, was 70 years old. Rous at 87 was the true grand old man of this selected group, and in fact throughout all time when it comes to prizes in physiology or medicine.

Rous's arrival in Stockholm caused some confusion. When scientists come to Stockholm to receive their Nobel Prize in physiology or medicine they are met at the airport by the secretary of the Nobel Committee at the Karolinska Institute often together with one or more professors at the Institute representing the relevant discipline. Similarly the Permanent Secretary of the Royal Swedish Academy of Sciences welcomes scientists who will receive prizes in physics or chemistry at the airport. This is the beginning of what is often referred to as the "fairy tale" week in Stockholm. In 1966 the secretary Gustafsson together with Klein was waiting at Bromma Airport, still used at the time, to welcome Rous and his family. They became concerned when no Rous family appeared among those who disembarked through the front door. However, they then checked the passengers gently disembarking through the back door and there were Rous and his wife. This sadly was the last anecdote Klein could share with me before he died on the Nobel day 2016.

Each Nobel laureate has a personal attaché during their week in Stockholm as previously described[3]. In Rous's case it was the career diplomat Christer Sylvén. He has a legal background and became associated with the Department of Foreign Affairs in 1960. After different trainee employments abroad he was back at the Department in Stockholm during 1965–1971. Thereafter he served as Ambassador of Sweden in many countries. I got to know him at the Royal Swedish Court at which he was Master of Ceremonies between 2000 and 2005, which overlapped with my first years as Lord Chamberlain-in-Waiting. Christer Sylvén and his wife took good care of the Rouses. Together they paid a visit to the outdoor museum Skansen and shared a Swedish Christmas dinner — the famous smörgåsbord — at the restaurant Solliden. The Sylvéns described the Rouses as "extremely interested and gracious." After the encounter in Stockholm contact was maintained by Christmas greetings during the few years that remained of Rous's life. The Rouses also sent a charming book for children, possibly by their daughter Marni, to honor the son that had been born into the Sylvén family in October of 1966. Finally Mrs Sylvén remembers that Rous was very insistent that her husband should stop smoking. He finally did so ten years later.

Rous was well prepared for the prize events in the Stockholm Concert Hall. The oldest of his three daughters, the already mentioned Marni, married Alan Hodgkin in the midst of the Second World War. The proceedings of the

1963 prize ceremony, when Hodgkin received his shared prize, were described by her in a vivacious and very charming way in a letter to her parents in New York. It was cited in full in my previous Nobel Book[4]. The Hodgkin family hence had the pleasure to participate in the ceremony again only three years after Hodgkin himself had received his prize. It all went well and Rous and Huggins were finally able to receive the insignia of their shared prize from the hands of His Majesty the King, the very popular Gustaf VI Adolf. Klein gave the

Rous and Huggins receive their Nobel Prizes from the hands of His Majesty the King.

introduction speech to the two laureates[7]. He emphasized all the reservations made regarding Rous's early findings. It was hard to replicate his results and they seemed to apply to birds but not to mammals. He then discussed the work on Shope papilloma virus in the 1930s which led to the important appreciation that tumor development was a step-wise process for which the term "progression" was introduced by Rous. Klein hereafter emphasized the radically changed situation in the 1950s with the discovery of a host of new tumor viruses, some of which could also transform human cells. He added to this the unanticipated observation that even Rous's avian virus could transform cells of different mammalian species, referring to findings by Swedish scientists in Lund and Uppsala. Before discussing Huggins's contributions he stated the important fact (translated from Swedish): "It is not yet clear in which way viruses induce cancer but there is much to indicate that the virus does not behave like a little boy setting fire to a hayrick and running away; part of the virus's own genetic material seems to be directly responsible for the malignant behaviour of the virus-transformed tumour cell." This was a visionary statement at the time and as we shall see way beyond Rous's own summarizing of his life-long conceptual thinking. Klein then gave an overview of Huggins's work and stated "This was a completely new type of cancer therapy, capable of helping a previously inaccessible category of patients, by the administration of non-toxic, naturally occurring hormones rather than by toxic or radioactive agents, and with few side effects." By way of conclusion Klein graciously emphasized the scope of Huggins's discovery by use of a citation from Rous. It read "...the importance of this discovery far transcends its practical implications; for it means that thought and endeavor in cancer research have been *misdirected* (my italics) in consequence of the belief that tumor cells are anarchic." As we will see the developments through time have not verified this critical statement. The tumor cell and its descendents are truly anarchistic. The progressive development of the tumor is based on accumulated genetic changes and adaptations. This allows the cells to multiply in an uncontrolled way and the tumor to acquire a capacity to increase in size and also to potentially spread from its place of origin to other parts of the body.

Inga Fischer-Hjalmars (1918–2008) addressing the prize recipient in Chemistry. On the left side Rous is listening. [From Ref. 11.]

Prior to Klein's address to Rous the Nobel Prize recipient in chemistry Robert S. Mulliken had been addressed. For the first time since Nobel Prizes started to be awarded in 1901 it was not a man but a woman who gave the laudation. It was Inga Fischer-Hjalmars. She became the first female professor of mathematical physics at Stockholm University. In this post she succeeded Oskar Klein, who for many years had a major influence on the work of the Nobel Committee for physics. In 1978, 12 years after her presentation at the Nobel Prize ceremony Fischer-Hjalmars became a member of the Royal Swedish Academy of Sciences — the fourth woman in Sweden to be recognized in this way. Thus she was selected to give this presentation more than ten years prior to being elected to the Academy. Together with Per-Olov Löwdin, she was the foremost authority on quantum chemistry at the time and she had served as a reviewer to the current Nobel Committee for chemistry. She was an impressive scientist, with deep involvements not only in science, but also the humanities. For a number of years she was the Chair of the Standing committee for the Free Circulation of Scientists at the International Council for Scientific Unions. From this and also from a much more general perspective I cannot refrain from commenting on the original caption of the picture which is shown. The last sentence reads translated from Swedish "In front of and behind the professor who speaks — an unusually sympathetic appearance for that matter! (sic, my remark) — sit representatives…." How condescending and paternalistic! What male chauvinism! The situation for women in academia has improved with time, but we still have a long way to go before we will see a reasonable balance between representation of women and men, both as prize recipients and as presenters at the podium of the Concert Hall in Stockholm in connection with the Nobel Prize ceremony. It is in fact referred to as the "penguin mountain" since the formal dress tails are required of male participants.

There had earlier been an opportunity for the involvement of a female scientist as an introductory speaker. This concerned the prize in physiology or medicine in 1939. As previously described[5] the College of Teachers at the Karolinska Institute selected Gerhard Domagk as the recipient of the prize "for the discovery of the antibacterial effects of prontosil." However, Domagk was forced by Hitler to decline the prize and no prize ceremony was arranged. Still, as can be seen from the official web site of the Nobel Foundation, an "Award Ceremony Speech" or more properly an account of Domagk's work was presented by Nanna C. Svartz, who was the pioneer female professor of medicine at the Karolinska Institute between 1938 and 1957. Thus if Domagk could have accepted his prize and furthermore if there had been a prize ceremony in Stockholm, there would have been a female presenter already at this relatively early time. It should be added that Domagk eventually did receive his prize. He visited Stockholm after the war in 1947, when he received his gold medal and diploma, but no prize money. He also gave his Nobel lecture, probably introduced by Svartz. After this detour let us now return to the laureates in focus.

Rous and Huggins enjoyed each other's company and both of them gave brief speeches at the banquet. Although their names in the announcement of the 1966 Nobel Prize in physiology or medicine were not in alphabetical order, let us present Huggins's involvement in the festivities first. He was a true laboratory scientist, in spite of his clinical background. His presentation at the Nobel banquet and also his lecture emphasized the importance of the research team and the conditions that allow it to flourish. In his banquet speech Huggins also referred to literature citing the second book of the *Iliad*, which is a catalogue of ships. His catalogue of thanks included his "Science-Widow," his wonderful colleagues "with satchel and shining morning face" and finally the fact that he had been blessed by becoming a part of the medical profession.

Three days after the banquet, on what is called Lucia day in Sweden, Huggins and Rous presented their Nobel lectures. Rous's Nobel lecture[8] provided a good historical background

Huggins and Rous in Stockholm. [From Ref. 11.]

to the early history of cancer virus research. It appropriately cited Ellenman's and Bang's discovery that preceded Rous's identification of the virus named after him. It did not reveal why Rous left his original field of research for some twenty years before Shope's studies of papilloma viruses lured him back into studies of tumor viruses. Interestingly Rous preferred not to use the generally accepted term "carcinogens," but instead referred to them as *oncogens*. This certainly did not improve the clarity of use of terms and in the following chapter we will return at length to a closely related concept, oncogenes, which eventually took on a major importance. He appropriately introduced the concepts of progression and promotion, which he himself had been instrumental in developing. The way the tumor mass may expand and the demands this puts on development of infrastructure of a large amount of independent tissue consuming energy and a separate newly formed blood supply were also remarked on. The developments of the work with papilloma viruses was well attended to, but a reference to the several tumor viruses found later in different animals, surprisingly, was not given, except that he mentioned Gross virus. Furthermore, a generalized statement was made that no viruses have been found in human tissues. In the end Rous mentioned approaches to chemotherapy, stating that "The leukemias of children have also been overcome by chemotherapy in some instances and also the singular, highly malignant lymphomas in African children known as Burkitt's disease." At this time it had been documented that the latter disease was associated with an infection with a herpes virus, originally called Epstein–Barr virus and later herpes virus type 4. Rous's remarks in general remain in stark contrast to the up-to-date information given in Klein's presentation speech. But the most serious mistake in the lecture was the statement that Rous did not accept the emerging dominating theory of cancer at the time, the somatic mutation theory. The essence of this theory was that cancer originated due to damage of the genetic material of cells and that progressively more damage accumulated as the cancer developed in the body. We will return to this serious lack of insight in the beginning of the following chapter.

In the beginning of Huggins's lecture[9] various hormone-dependent cancers in man and in experimental animals were listed. It was naturally emphasized that their particular dependence reflected their tissue origin and opened possibilities for a systemic control. Huggins was not the first one to recognize this, but he pioneered the practical use of hormone-dependence in the care for patients. He emphasized that the "key to the puzzle of the steroid hormones in cancer was the isolation of crystalline estrone by Doisy et al." The use of estrogen was in fact the first example of a rational chemotherapy of cancer. Doisy was recognized by a shared Nobel Prize in physiology or medicine in 1944 (the prize for 1943) for another major contribution, his clarification of the chemical nature of vitamin K. Huggins, prior to introducing his important studies of hormone control of prostate cancer, presented his in-depth studies of the phosphorous metabolism in the genital tract. He discussed at length attempts to also use hormone treatment in cases of breast cancer, both experimental and in humans. In reflecting on the many difficulties encountered he cited Emerson who said "The ambitious soul sits down before each refractory fact."

The Nobel week brought no rest for the aged Nobel Prize recipient Rous. The medical students traditionally invited him and his wife as well as the Huggins couple to festivities celebrating Lucia in the evening of December 13. There is a long tradition of making this a relaxed and enjoyable evening. It is full of pranks and not least singing. In student academia it has for a long time been popular to make new texts to an already well-established melody. In the 1950s Povel Ramel was a very much appreciated entertainer. In 1956 when I was a first-year medical student we used one of his melodies, the song "Ratata" and wrote four new verses. The name of the song became "Prostatá" (with emphasis on the last syllable), like the organ Huggins selected for his attempts to develop hormone treatment of cancer. The first verse of this song was created when we did our military service (split up into three summers so as not to interfere with our studies) at a regiment T1 (T stands for Trängen meaning "crowd") in

Linköping. Three additional verses were written in my parents' home, the rectory in Sundbyberg outside Stockholm. My co-authors in creating the texts were Ulf Broberger, Kjell Mattson and Thomas Brundin, of whom the latter should be credited especially. This song has up to the present time remained one of the most popular among medical students. However, at the end of the 1990s some feminist students objected to the text and recommended that the song should be removed from the official song book of the medical students. This led to the appearance of the most impressive headline I have ever been honored with. It was printed in the daily evening newspaper Expressen and it read (translated from Swedish) "Nobel Professor wrote a gender offensive song." My employees at the time at the Academy were somewhat impressed and I think I ranked a bit higher among them after they got to know about this "original sin" of mine. The day after this publication the chairman of the students' association called me saying "Don't worry about the article. We will continue to sing Prostatá!"

I told this story at an accidental meeting in the subway with an earlier colleague at the Karolinska Institute, the retired professor of endocrinology Sigbritt Werner. She then told me that she had a friend by the name of Sven Westman, a retired associate professor of inorganic chemistry at Stockholm University, who since 1999 has provided the medical students with texts to another popular Swedish song for their Lucia celebration of Nobel Prize recipients. Until the present time two new verses have been created referring to the particular prize recipients in a certain year. These verses are intercalated between three verses, originally created in 1999, sung in the beginning, and then the first verse repeated towards the end. The tune is one all students at Swedish universities are familiar with. It has its origin in a German song — O alte Burschenherligkeit — from the early 19th century. In Swedish it is known as "Du gamla klang- och jubeltid" loosely translated as "O, you old clang and cheerful time." The three verses, sung together with the two newly created verses each year, read as follows:

Oh, splendours of the century,
Ye Laureates astounding!
We sing to you in harmony.
with awe and praise abounding,
who bade farewell to mundane strife
to probe the very fount of Life
and raised in bold defiance
the banner brave of science.

We hail von Behring, Ross and Koch
as mighty men of healing
and Pavlov, Golgi, Hess and Bloch
our body's gears revealing.
And Kendall, Krebs and Theorell
explained our chemistry, as well
as Messrs Crick and Watson
whom we're depending lots on.

Tinbergen, Lorentz and von Frisch
dispelled our dear delusions;
yet, like McClintock still we wish
to jump to fine conclusions.
But knowledge grew by sweat and strain,
and Sperry split in two the brain,
and Levi-Montalcini
made us grow tall and skinny.

Ye laureates astounding!
Etcetera…

Let us now move from singing about Nobel laureates to more intellectual challenges.

Rous was also invited to participate in a television program. The very first TV recording in Sweden was made — of course! — of the Nobel Prize ceremony in 1950. It was managed by technicians from the Radio Corporation of America, who also brought along the technical equipment. The recording could be viewed on a monitor and film screen in associated localities at the Concert Hall in Stockholm. However, regular programs available to the Swedish public did not become available until 1957. They were managed by the agency Sveriges Radio using fees paid by the owners of television sets. There was no advertising and until 1969 there was only one channel. One of the programs regularly shown since the dawn of Swedish television was the Nobel Prize ceremony and since 1959 an associated program originally called "Science and man" and later "Geniuses speculate" was aired. This program survived for 55 years until it was taken over by BBC World in 2004. In the beginning the leader of the program was a very popular TV personality Bengt Feldreich. Rous was invited personally by Olof Rydbeck, the Director of Swedish Radio, to join in

Hormone Treatment of Tumors and the Prize Events in 1966   75

this program in 1966. The program was carefully structured by formulation of a series of questions, which the participant had received beforehand. In Rous's copy of this text, his handwritten comments in the margin indicate his possible answers. The program grew to become very popular and one of the questions Feldreich regularly returned to was, "Do you believe in intuition?" Rous enjoyed joining in the program and did well. In a letter Georg Klein praised his contribution. By way of thanks Rous received a vase from the Director that he proudly exhibited in his New York home.

The Nobel week allowed deepening of contacts between Rous and Georg Klein. Together with his wife Eva a private dinner was arranged for Rous and his family in their home. In those days the Nobel week was not as overloaded with official commitments as it is today. Private dinners in the home of the presenter at the prize ceremony were not uncommon. I still remember the dinner I and my wife hosted in 1976 for Carleton Gajdusek and Baruch Blumberg. It was a very lively event in which we offered live chamber music created by my uncle Gunnar who was a cellist in the Stockholm Philharmonic orchestra and Tore Wiberg accompanying him on our Bechstein grand piano. Rous's contacts with Georg Klein deepened after the prize, but it should be noted that there was no correspondence *prior to* the prize. It seems that they developed a pleasant friendship and some matters dealt with plans by Klein and collaborators to submit manuscripts for possible publication in the *Journal of Experimental Medicine*, to which Rous devoted some five decades of his professional life. They also discussed the occurrence of possible spontaneous regression of cancers, which led to the following response by Rous "Everson's paper showing the rarity of spontaneous regression of human cancer made me conclude that viruses cannot be the cause of the generality of these growths." It is not clear how Rous reasoned in this case but as we shall see towards the end of Chapter 3 his premonitions were correct.

After the announcement of the prize, there had of course been a correspondence concerning the practical arrangements in Stockholm. Letters were exchanged between the award recipients and the different hosts, the Nobel Foundation and the Karolinska Institute. The Director of the Nobel Foundation Nils Ståhle gave helpful instructions on the arrangements for the prize ceremony in the Stockholm Concert Hall and on the banquet in the City Hall. In his reply, expressing his thanks, Rous used the formulation "I mention this here not in self-esteem but to exemplify in a small way the far-reaching power of the Nobel Foundation," a power, to be clarified, that only concerns all financial and legal matters as well as the practical arrangements of festivities. However, the prize was awarded by the Karolinska Institute and other correspondence therefore was with its Vice-Chancellor Friberg, who together with the secretary of the Nobel Committee Bengt Gustafsson, signed the diploma. During the time I was the Permanent Secretary of the Royal Swedish Academy of Sciences, the President of the Academy and I traditionally signed the diplomas. It was also my task to instruct the Nobel Foundation to pay the predetermined prize amount to the laureate. As already mentioned the Foundation besides being responsible for the prize ceremony and the banquet also manages the money of the Foundation and supervise legal matters. The Foundation is not mentioned in Alfred Nobel's will. It was created in 1900 and became a legal body that was able to become the recipient of Nobel's legacy[5].

As Permanent Secretary I was also a member of the Board of the Nobel Foundation. I believed it important to make it clear that it is the prize-giving institutions that are individually responsible for all the meticulous work in their Nobel Committees. The final decision on the recipient(s) of the prize a certain year is taken *in pleno* at the Royal Swedish Academy of Sciences or the Swedish Academy, whereas at the Karolinska Institute it is taken by the Nobel Assembly since the end of the 1970s. Hence I argued that, instead of me ordering the Nobel Foundation to pay out the prize money, this money should be transferred to the prize-giving institutions which then could pay the proper amount to the Laureate. This may seem like a formality, but I thought it was important to emphasize that the Nobel Foundation does not award Nobel Prizes, which is a common misconception. My proposal was not well received. For the sake of clarity it should be mentioned that members representing the prize-giving institutions are in majority in the Board of the Foundation, emphasizing the relative responsibility of the different actors in the Nobel system. As the prestige of the Nobel Prize has grown there has developed an increased need to make it clear what different roles each one of these central actors have. The risk of an emerging power game otherwise is imminent.

Finally there was also a correspondence between Rous and Ragnar Granit, the recipient of the Nobel Prize in physiology or medicine the following year, who we therefore will soon meet again. Granit was the editor-in-chief of *Les Prix Nobel*, in which Rous's Nobel lecture was to be published. This annual booklet has been published by the Nobel Foundation since the first Nobel Prize award year 1901 until 2010. It was then decided to modernize this annual publication and regrettably no book therefore is available for the year 2011, although the appropriate information is available on the web. Since 2012 the annual book now called *The Nobel Prizes*, with the suffix *Les Prix Nobel*, has

Karl Grandin handing over the insignia to the King.

been published. It was printed on behalf of the Nobel Foundation by Science History Publications/USA division, Watson Publishing International LLC. In 2019 the book about the 2018 Nobel Prizes will be printed by the publisher of the present book, World Scientific. The present editor is Professor Karl Grandin at the Center for History of Science at the Royal Swedish Academy of Sciences, where I have my office. It could be added that Professor Grandin has a very special responsibility at the prize ceremony. He is the person who hands His Majesty the King the proper insignia for the King to then give to the prize recipient, who a moment later approaches him at the center stage to receive them.

The Aftermath and Nobel Medals

Rous also had an invitation from the Dean of the Medical Faculty at Uppsala University to lecture there, but he politely declined this. Instead he and his wife took the opportunity to make a stop-over in Cambridge on their way home. They stayed with their daughter's family, the Hodgkins to celebrate Christmas and to see their four grandchildren.

On his arrival back home Rous was overwhelmed by a huge congratulatory correspondence. He counted that there were about 600 telegrams, letters, hand written and typed, and other personal messages. He wanted to respond to them all and he repeatedly apologized for the fact that it took so long and that he was "under great strain" or "tired." All this correspondence can be found at the Archives of the American Philosophical Society in Philadelphia. It is an impressive read. There are congratulation from President Lyndon Johnson, Vice-president Hubert Humphrey, John Lindsay, Major of New York, from a host of previous Nobel laureates, from a huge number of colleagues in the field of tumor virology and many, many others. In a letter from one of Rous's virology colleagues, Harry Rubin, mentioned in Chapter 1, it was stated that the prize was "the most richly deserved award in the history of the prize." In the same vein Sidney Farber wrote an exuberant letter. In a response to Sven Gard, Rous wrote that he enjoyed their "walks and talks." Some colleagues sent their congratulations from Japan because the 9th International Cancer Congress was held in Tokyo in the month of October, 1966. I had the privilege of participating in this congress, my first trip around the world, since I continued on to a vaccine congress arranged at the Pan American Health Organization in Washington. Questions of viruses and cancer were of course extensively discussed at the former congress.

Finally it can be added that the archives associated with Rous's prize also contain an interesting correspondence from a friend trying to trace the origin of the inscription on the Nobel medal, which as mentioned earlier[5] reads *Inventas vitam iuvat excoluisse per artes*, literally translatable to "Inventions enhance life which is beautiful through art." Sometimes the loose translation "Let us improve life through science and art" is used.

The letter reads:

"I have been trying all along to find the source for the inscription on the Nobel medal. Finally this Thanksgiving (the letter was written November 20, 1967, my remark) weekend in Boston brought the desired answer thanks to my friend and former schoolmate, the widow of Werner Jaeger (foremost Grecist and Latinist first in Germany, later at Harvard). Mrs Jaeger has taught Latin at Milton Academy for over 20 years. She writes — and I quote verbatim:

VERGIL, Aeneid, Book VI, lines 660–664:

In the Elysian Fields, there are

Those who suffer wounds fighting for their country

And those who were pure priests, while alive,

And those who were pious poets, and spoke words worthy of Apollo,

Or those who adorned life through the invention of special skills,

And those who made others remember them by their good deeds.

Line four reads in Latin:

inventas aut qui vitam excoluere per artes.

This is clearly what has been changed to

inventas iuvat vitam excoluisse per artes.

Since the sentence here is a dependent clause, "or those who adorned," it is very logical that an educated man like Nobel or a friend of his changed it to "It is good to have adorned" etc. There is also the possibility that some later writer (Vergil was imitated constantly) used those famous Vergil words and changed them to the form in which it is given in the Nobel citation.

I asked a professor of classics whom I had to consult about some other matter, and he came up with this after some research with the Vergil dictionary. Neither one of us had at first thought of this particular passage, (which we both know extremely well and which is one of the most beautiful and famous ones), because of the slight change from the perfect "excoluere" to the perfect infinitive "excoluisse." I am rather ashamed I did not think of it, but he did not either,"

Sadly the next page of the letter is missing, so it is not possible to track the sender. And there is more to be told about the design of Nobel Medallions.

The Swedish King Oscar II in Council on June 29, 1900, settled the Statutes of the newly established Nobel Foundation. It specified that each prize-recipient should receive the prize, a diploma and a gold medal. The prize-awarding institutions therefore were called upon to decide the design of the diploma and the medal. Separate initiatives were taken for the Swedish prizes for the natural sciences together with the prize for literature and the Norwegian Prize for peace. In Sweden a joint committee was established with representatives from the three institutions, the Royal Swedish Academy of Sciences, the

Karolinska Institute and the Swedish Academy. The committee was chaired by the permanent secretary of the latter Academy C. David af Wirsén. There were three other members of the committee; Ragnar Törnebladh representing the Royal Swedish Academy of Sciences, Karl Mörner, the Vice-Chancellor of the Karolinska Institute, and a second representative of the Swedish Academy, Hans Hildebrand, a numismatist and preserver of the national heritage. The committee recommended that the front of the medal should carry Alfred Nobel's picture and specify his day of birth and of death. The reverse gave rise to more discussions.

It was agreed that there would be a joint text in Latin as described above. However, it was not for Nobel or someone he knew to propose this, it came from one of the learned members of the committee. It could possibly have been from af Wirsén himself, who was well acquainted with Latin. However it is most likely that it came from Törnebladh. He had just become a member of the Royal Swedish Academy of Sciences and one of his first tasks was to be a member of the committee led by af Wirsén. Törnebladh had written about and given lectures on Vergil. The two changes he introduced in the original text was to delete "aut qui" and insert "iuvat," meaning "it is a pleasure" after "vitam." Finally the tense was changed as already mentioned. In addition the reverse was to give the name of the prize recipient and the prize-awarding institution. Depending upon the category of the prize one of three different motifs was to be used on the reverse.

In physics and chemistry Nature is allegorically represented by a goddess resembling Isis (the foremost God in Egyptian mythology) who in her arms holds a cornucopia, the horn of plenty. The Genius of Science unveils her serious face. In medals awarded by the Karolinska Institute the Genius of Medicine simultaneously carries an opened book in

Medallions awarded to recipients of Nobel Prizes in physics/chemistry, in physiology or medicine, in literature and in peace.

Hormone Treatment of Tumors and the Prize Events in 1966 81

her lap and collects water from a well to quench the thirst of a sick girl. The medal handed out by the Swedish Academy finally depicts a picture of a young man with a pen in his right hand sitting under a laurel tree. Enchanted he listens to and writes down the song of the muse. Had the sexes been reversed it could be said that possibly the prize for literature to Bob Dylan in 2016 was not too far off the mark! It may be noted that there are female figures dominating the engravings; Nature is depicted as a goddess, the Genius of medicine is a woman and the Muse of course is a woman. Regrettably realities differ. At the time of writing the percentage of women is about 1% in physics, 2% in chemistry, 6% in physiology or medicine and 12% in literature.

A very central figure in designing the medal, in particular the reverse sides, was the artist Erik Lindberg, but the proposals he gave to the three academies caused a considerable debate. Originally af Wirsén had contacted his father, Johan A. Lindberg, the most famous medal engraver in Sweden at the time. However he was too busy with other commissions and therefore recommended his son, Erik, who was only 25 years old. He was temporarily living in Paris, where he received important inspiration from contemporary French engraving masters, and he had plans for travel for some months in Italy to further develop his talent as a medal engraver. He answered af Wirsén that he was flattered to take on the commission, but that it would not be possible to perform this task in time to make medallions available for the first prize ceremony in December 1901. This caused a conflict with af Wirsén. However, Lindberg got his way and at the first prize award ceremony temporary medals were handed out, later to be exchanged for the proper ones. In order to resolve the diverging opinions of motifs to be used Lindberg met with representatives of the prize-awarding institutions in November of 1901. An agreement on the motifs to use was finally reached. There were also discussions of the place for the name of the prize recipient. Lindberg proposed to have it on the reverse whereas the representatives of the prize-awarding institutions preferred the edge of the medal. Lindberg finally had his way.

Lindberg also came to influence the design of the medal given to the Nobel Peace Prize recipients. The famous Norwegian sculptor Gustav Vigeland had been selected to propose a suitable motif, but he needed advice since he lacked experience in the production of medals and engravings. The Peace Prize medal naturally has its own inscription. It reads "*Pro pace et fraternitate gentium*" meaning "For the peace and brotherhood of men." Even though this carries a masculine tone it presumably has its origin in "man" representing the sex-neutral "human being." However, the motif is a group of three men forming a circle holding each others' arms. The name of the Peace Prize recipient is not given on the reverse of the medal but on its edge. A final comment should be made about the gold content of the medals handed out to the prize recipients. Until 1980 23-carat gold was used for a medal with a diameter of 66 mm weighing approximately 200 grams. The following year the weight was reduced to 175 g and the gold content reduced to 18 carat.

It may be added that the Swedish Academy embossed medals of Alfred Nobel in 1926 and in 1979. The inscriptions used were respectively *mala corporis animiqve scientia vincet* — Science will defeat the evils of body and soul — and *hominibus prodesse gentes reconciliare* — fraternization of people to the benefit of mankind. 100 years after Nobel's death the Swedish Academy of Sciences also memorialized him by embossing a medal. The inscription on the front reads *creavit et promovit* — he created and rewarded.

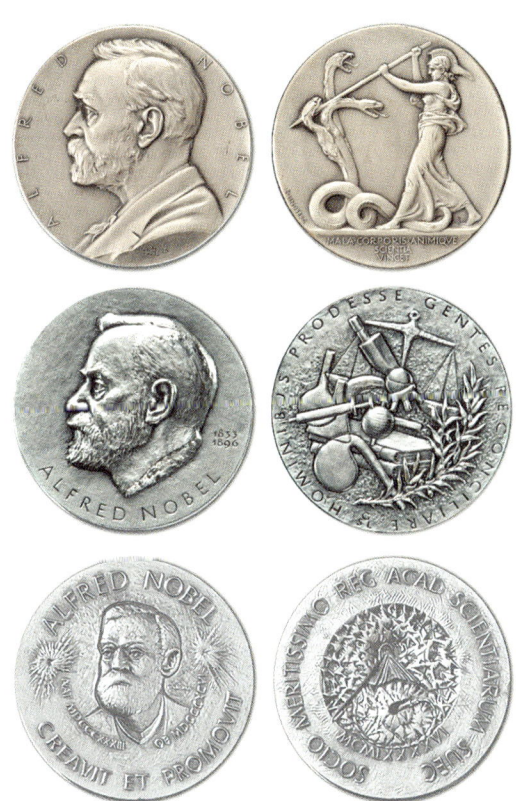

Medals embossed by the Swedish Academy in 1926 and in 1979 and by the Royal Swedish Academy of Sciences a hundred years after Nobel's death.

Coda

After this excursion into the world of Nobel Prize medals it should be added that Rous lived for four more years after he received his prize. He died of cancer, the disease to which he had devoted his exceptionally long scientific life. At the time he was writing a memoir of his colleague Shope, who was seminal in reactivating his involvement in the virus and cancer problem as discussed above. What Rous never got to know was that in the near future major breakthroughs were to be made in the understanding of the role of cellular genes in the development of cancer.

Dulbecco[10] has cited a conversation he had with Rous in the 1950s during which he suggested that a comparison might be made between the lysogenic state of viruses in bacteria studied extensively by Lwoff[4] and the interaction of a tumor virus with its cell. However, such a bold suggestion did not fit well with Rous's own ideas. Although he showed a certain curiosity about the rapidly advancing insights into genetics in particular due to the developments in molecular biology, he failed to assimilate them. He failed to understand that viruses could introduce new genetic material into cells and that mutations in somatic (non-germline) cells increasingly were being demonstrated to play an important role in the development of cancers. To quote Dulbecco it might be said about Rous that "he outlived his era of discoveries about cancer but what he did for science endures." It was for the next generation of scientists to develop the field of molecular oncogenesis using progressively introduced new techniques to study genetic phenomena in more detail. The virus that was used in the first breakthrough experiments was Rous virus! This story will be told in the next chapter.

Chapter 3

Rous Virus and the Elucidation of the Genetic Nature of Cancer

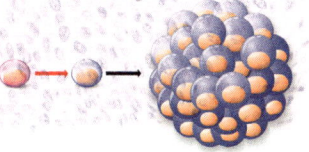

ORIGIN OF CANCER
MISDIRECTED EVOLUTION
TREATMENT IMPROVING

In the middle of his Nobel lecture[1] Rous stated

> "What can be the nature of the generality of neoplastic changes, the reason for their persistence, their irreversibility, and for the discontinuous, step-like alterations that they frequently undergo? A favorite explanation has been that oncogens (a term preferred by Rous for carcinogenic substances, my remark) cause alterations in the genes of the cells of the body, somatic mutations as these are termed. But numerous facts, when taken together, decisively exclude this supposition."

As emphasized by J. Michael Bishop[2] this formulation implies that Rous, in spite of his life-long involvement in studies of cancer, had completely missed the essential fact that it is a disease of genes controlling the growth characteristics of cells. Further developments came to reveal the existence of a completely unknown world of growth-controlling factors and the presence of defense mechanisms to prevent the emergence of abnormally dividing cells. It was the later studies using, in fact, Rous virus that pried the door open to the discovery of genes controlling cell division. When modified or over-expressed such genes could lead to the development of cancer. They appropriately came to be named *oncogenes*, a word very similar to the term oncogens used by Rous in his lecture for substances causing cancer. But, before analyzing the breakthrough discovery

85

by Bishop and Harold E. Varmus, there is a need to reflect on the developments of the intertwined fields of studies of oncogenic viruses and of developments of tools for use in molecular biological studies. In the 1960s two high-profile scientists came to the conviction that cancer viruses provided an important, possibly even all-encompassing explanation for the emergence of tumors and that hence there were opportunities for a generalized successful management of this family of conjectured infectious diseases, to bring about their conquest.

A Premature General Attack on Cancer

Originally the term "oncogene" was speculative, introduced in 1969 by Robert Huebner and George Todaro, both at National Institutes of Health (NIH), Bethesda, Maryland. The term was collectively used as a name for virus genes potentially involved in the development of cancer. During the late 1950s and the 1960s there had been a hunt for tumor viruses in experimental animals, as we have seen. Viruses similar to Rous virus had been identified in several species and they seemed to have a capacity to persist. It should be remembered that viruses contain either DNA or RNA and in this case they had been found to contain the latter kind of nucleic acid. Hence it remained an open question by which mechanism they might persist in cells. If they had contained DNA instead of RNA it could readily have been speculated that they were capable of associating with the host cell DNA genome. Electron microscopy of thin sections of infected cells revealed that enveloped particles of different morphology could be identified in different kinds of virus-associated tumor cells. These particles based on their morphology were originally referred to as types A, B and C. In particular the type C RNA tumor viruses were found to be prevalent and seen in many experimental animals like mice, rats and chickens. These findings were used to develop the *oncogene hypothesis* by the two colorful scientists Huebner and Todaro[3]. First a few words on their respective personality and career. Both of them, not least Huebner, had a forceful personality and they left a strong impression among people who had met them. I only met them peripherally on a few occasions.

Huebner was the firstborn in a family of five sons and four daughters. He tried out studies in different disciplines but finally settled for medicine at Saint Louis School of Medicine. He was almost expelled from school when it was discovered that, in violation of the rules of the school, he had taken work, including a job as a bouncer at a brothel, outside school. However, he finally graduated in 1942 and two years later he became affiliated with the NIH. He contributed to the discovery of a number of previously unidentified infectious agents, both rickettsia and viruses. During the golden era in the beginning of the 1950s when practically useful tissue culture techniques had become available, he pioneered the isolation from human samples of an important herpes virus, cytomegalovirus, and of several adenoviruses from adenoid and tonsil tissues. These kinds of viruses were found to cause tumors and during the 1960s various spurious additional findings were used to construct the hypothesis of

Robert J. Huebner (1914–1998). [From Ref. 38.]

viruses as a general cause of cancer. Like his father, Bob, as he was generally called, had nine children, and when the family moved from the NIH campus to a farm near Frederick, Maryland the whole family became involved in breeding Black Angus cattle. Bob was as intense in his professional as in his private life, but sadly he experienced the tragic loss of two of his sons. One son Edward was accidentally killed by another hunter at the age of 38 and the oldest son James fought his battle with gastric cancer unsuccessfully and died soon after his father's death. Bob's last ten years of life were darkened by the fact that he became afflicted by Alzheimer's disease.

George Todaro. [From Ref. 38.]

Todaro had a separate and distinct odyssey. He is a first-generation Italian-American who grew up in the Bronx, New York City and received a medical degree at New York School of Medicine. Together with Howard Green at the Department of Pathology of the same University, he developed an immortal mouse embryoblast cell line, NIH 3T3 cells, which became widely used in studies of tumor viruses and also for many other purposes. These kinds of cells which show an unlimited capacity to divide under laboratory conditions, so-called *established* cell lines, provide

Rous Virus and the Elucidation of the Genetic Nature of Cancer 87

a very useful alternative to *primary* cell cultures which are derived directly from animals, for the study of viruses and their properties. The most famous cell line, HeLa cells, described earlier[4], has been extensively used in studies of viruses. These kinds of cells are highly contagious and often when scientists in the past had believed that they had established a new cell line it has often turned out that the cells unconsciously represent contaminating HeLa cells. As was mentioned in Chapter 1 there have been concerns about the use of cells possibly containing oncogenic viruses for production of vaccines against virus diseases. The most attractive choice of cells for this purpose is so-called diploid cell lines[5]. They retain their normal set-up of chromosomes for a considerable number of divisions in the laboratory before they senescence and die. Thus these kinds of cells can be recovered in numbers substantial enough to allow production of large volumes of vaccines. Todaro's second major involvement in science concerned the development of the Huebner-Todaro oncogene hypothesis. Once the initial enthusiasm for that had been lost because of developments about to be described, he moved on to involve himself professionally in the applications of the new tools of biotechnology to manage, for example, the serious genetic disease cystic fibrosis and on the personal side the very different field of race horse ownership.

The viral oncogene hypothesis[3] postulated that normal cells would carry the genome of type C tumor viruses. During the 1960s a number of different rodents, primarily mice and rats, were found to host different tumor viruses, many of which had similar morphology as previously mentioned. It was proposed that the genome of such viruses included both genes allowing the production of virus and separately genes, of unidentified origin — oncogenes — responsible for the transformation of cells. It was further postulated that under normal conditions there should exist some enigmatic suppressor that kept these oncogenes under control. Altogether the underpinning of the hypothesis was very weak. It did, however, spur some scientists on to major experimental pursuits to test the validity of the hypothesis. One of them was Sol Spiegelman, another colorful narcissistic scientist introduced in one of my previous books[4]. His intense involvement in addition was fired by recent suggestions from Temin's laboratory that an RNA tumor virus could rewrite — reverse transcribe — its RNA into DNA and hereby integrate a copy of its genetic material into the cellular genome. We will soon return to this dramatic discovery. Spiegelman's important contribution was to develop techniques to compare if two different separate single-stranded nucleic acid molecules — either two forms of DNA or a DNA and an RNA molecule — contained the related kind of information, viz. whether they displayed nucleotide sequence homology. This kind of analysis was referred to as *hybridization* experiments and will be further discussed in Chapter 8. It did not take long before Spiegelman found traces of viral nucleic acid sequences in many human tumors. However, later follow-up studies demonstrated that the results presented were flawed. Spiegelman had pushed the limits of his hybridization experiment so far that the experimental results did not reflect reality. The kind of viruses that he assumed he had been examining were later found to be associated only with certain forms of T cell leukemia in people living in special habitats, which will be discussed towards the end of this chapter.

Mary Lasker (1900–1994).

The mood of the scientific community had been boosted in the 1950s and 60s by successes in treating acute lymphoblastic leukemia in children by use of so-called folate antagonist introduced by Farber, mentioned in the previous chapter. These encouraging developments together with the Huebner-Todaro oncogene hypothesis combined with Spiegelman's enthusiastic presentation of his preliminary data had major political consequences. Could it be that cancers had a uniform mechanism of origin and if that were the case, potentially a shared method for cure? Inspired by the success of the "man on the moon" project initiatives were taken to initiate a "war on cancer." A major figure in this initiative was the well-known philanthropist and socialite Mary Lasker. Together with her husband she had started the influential Lasker Foundation, to be further discussed below. In 1969 she proposed the establishment of a Commission on the Conquest of Cancer and managed to have what eventually was called a Panel of Consultants staffed with people predicted to be positive to the project. Farber was selected co-chairman. In the winter of 1970 a final report was produced, not surprisingly advocating the establishment of a new independent cancer agency.

In March 1971 Ted Kennedy and Jacob Javits presented a bill to the Senate based on the recommendations by the panel. At the beginning of July a modified version of the bill was passed by an overwhelming majority of senators.

When the cancer bill reached the House problems started to accumulate. There were many hesitant voices both in and outside the House, not least from reputable scientists, that begun to be heard. Nixon started to be restless as the election in 1972 was getting closer. The war in Vietnam dragged on. He needed some positive news to attract voters. An all-encompassing cancer project, copying the successful Apollo — moon landing — project, might be helpful. The processing of the proposed initiative in the House dragged on but in December it voted almost unanimously for the bill, the content of which, however, had been considerably watered down. Still the annual budget for the National Cancer Institute at the National Institutes of Health was increased from $230 million to $940 million for the coming three years. However, what promoted the developments forward was not massive clinical trials with different kinds of early candidate anti-cancer drugs performed by use of the expanded resources but groundbreaking basic science in a few laboratories. A completely new perspective on the genetic fundamentals of cancer development was to emerge.

The American "Nobel Prize"

The Lasker Foundation was established in 1942 by two philanthropists Albert and Mary Lasker as a means of promoting medical research. It has developed into the most respected foundation when it comes to prizes in this field in the U.S. A considerable fraction of the financial resources came from Albert Lasker's success in the advertising business. My wife who spent five years as a young girl in the U.S. in the mid 1940s still remembers the successful radio advertisement "L.S.M.F.T. — Lucky Strike Means Fine Tobacco!" which successfully inspired many, not least women, to start smoking. It is ironic that wealth generated by this success in advertising later was later used for initiatives within the framework of the American Cancer Society employing the same medium to mitigate behaviors that could lead to the development of cancer.

The different Lasker Awards in medical sciences, distributed since 1945, have become famous because of the high quality of the process to select prize recipients. Today they are awarded in three separate categories; Basic, Clinical and Public Service/Special Achievement (alternating every second year). Up to 2015, 87 recipients of the Lasker award have gone on to receive a Nobel Prize in physiology or medicine, an impressive agreement on quality assessment between this Foundation and the Nobel Assembly of the Karolinska Institute. There is a difference between the prizes from the two kinds of institutions in that the number of annual recipients of Lasker awards, not least in the category of Basic Medical Research, can be larger than the maximally three recipients of a Nobel Prize in physiology or medicine in a single year. The Lasker Prize most similar to the Nobel Prize in physiology or medicine is the latter kind of prize which recognizes "a fundamental discovery that opens up a new area of biomedical science." The additional two Lasker awards mean a wider involvement in recognizing developments in the medical sciences. The Clinical Medical Research Award recognizes "a major advance that improves the lives of many thousands of people." On the condition that this advance is based on a distinct discovery, a Nobel Prize could be considered for candidates to this prize. This Lasker Award has the name DeBakey attached to it.

Michael DeBakey was a medical statesman at the Baylor College of Medicine in Houston, Texas. He pioneered heart surgery and made many critical contributions to development of the heart-lung machine, to coronary artery bypass surgery and to carotid endarterectomy and also to the use of Dacron grafts to repair blood vessels. He remained remarkably professionally involved throughout his long life dying at the age of 99 years. I met this impressively active surgeon in his mid-eighties when he visited the Karolinska Institute and I was his host in my function as the Dean. He was on his way to Moscow to supervise the quintuple bypass surgery to be performed on the Russian President at the time, Boris Yeltsin. The third kind of Lasker Prize alternates between the Special Achievement Award in Medical science recognizing "research accomplishments and scientific statesmanship that engender the deepest feelings of awe and respect" and the Public Service Award, for which there are five different possible motivations. The Special Achievement Award has since 2008 had the name Koshland added to it to honor the demise of Daniel E. Koshland, Jr. He had received the prize ten years earlier and had been a statesman of basic sciences, pioneering insights into the protein metabolism of bacteria. He also served the important function of editor of the journal *Science* for ten years. Koshland entered science out of curiosity having read the two books mentioned in Chapter 1, *Microbe Hunters* and *Arrowsmith*. In reality there was no need for him to work since he might have relied on his private fortune derived from his preceding generations of Jewish relatives who established and managed Levi Strauss & Co. Prize recipients in this third category of Lasker Prizes would be difficult to recognize by a Nobel Prize in physiology or medicine.

There are certain similarities as well as differences in the way recipients of Nobel Prizes and Lasker awardees are selected. An important component in the process to select Nobel Prize recipients is that only invited individuals, sometimes as a member of an academic institution, can make a nomination before January 31, in the year of the prize to be awarded. This is not the case for nominations to the Lasker awards. Anyone can nominate or send a letter supporting a nomination at the latest on February 1 for the prize in a certain year. The number of nominations for the Lasker awards is much smaller, roughly by a factor of four, than those for Nobel Prizes in the natural sciences. It applies to both kinds of awards that nominations, supporting letters and protocols of meetings of the committees are kept secret for 50 years. There is a single committee handling all the different Lasker awards and it is composed of 22 members. The committee is led by a chairman, at the time of writing Joe Goldstein[4], who kindly provided me with the information about the work of the committee. Each year about two members are rotated out and replaced. About one-third of the members are Nobel Laureates.

A short list of finalists is prepared for each category and each finalist is judged by a jury of five committee members. One of the five members provides a formal presentation of the candidate at the single full day meeting of the committee at which final decisions are taken. Strong candidates, who are not selected for the prize at a meeting in a certain year, are carried over from one year to the next. A new jury member is selected as presenter at this second time of discussion. In contrast to the Nobel Prize selection process there is no overriding body controlling the work of the committee. The Lasker committee has the final say. It can be noted that the lack of written reviews is a disadvantage for future historians. For the sake of clarity it should be emphasized that there are no formal contacts between the respective committees responsible for the Nobel Prize in physiology or medicine and for the Lasker awards. However, representatives for the Lasker Awards have participated in the Nobel festivities on a number of occasions. Once in 1989, when I addressed the Nobel Prize recipients in physiology or medicine, as we shall see, I sat at the main table. My table companion was a charming woman Mrs. Alice Fordyce. She was the sister of Mary Lasker and was involved in the development of the Lasker awards, also by contributing her own donations. For decades until December 1990 she served as the Executive Vice President of the Foundation and was the Director of the awards program. She was also highly supportive of the cultural life in New York. She encouraged me to go and listen to music on a barge anchored at the southern end of Brooklyn Bridge.

I did of course do this the next time I visited New York. It was a high quality experience of chamber music under unpretentious conditions together with the cultural elite of New York, to a large extent of Jewish origin.

A Critical Amendment of the Central Dogma

Temin, a scientist of unique qualities, was briefly introduced in the previous chapter and has been referred to in all my three previous books on Nobel Prizes[4,6,7], in relation to his 1975 prize in physiology or medicine together with David Baltimore and Renato Dulbecco. Temin was an independent scientist and a highly original thinker. When he arrived at the California Institute of Technology (Caltech) in the early 1950s he rapidly switched from genetic studies of fruit flies to studies of Rous virus. Together with Rubin he developed a technique of studying cell transformation by Rous virus in the test tube in chicken embryo cultures as mentioned in the previous chapter. This was an imaginative approach to studies of critical changes of cell behavior in cell cultures. But Temin's thinking took him further. He speculated that the RNA-containing agent he was studying shunned the cycle of full replication of virus leading to a synthesis of new particles and instead stayed behind in some form and caused the abnormal growth of the cells. He knew that the virus nucleic acid was RNA and he boldly conjectured that this RNA could be copied into DNA and hence potentially remain associated with the genome of the cell and be the cause of its abnormal behavior. The unsteady way of experimenting that finally gave solid support to this concept was described in Temin's Nobel lecture[8]. A drug called actinomycin played an important role in Temin's studies. This compound had been observed to block the synthesis of RNA from DNA. RNA-containing viruses in general, like the measles virus that I studied in my Ph.D. work, replicated happily in the presence of this drug, but DNA-containing viruses could not replicate under the same conditions of treatment of the cultures. However, the drug blocked the replication of the RNA-containing tumor viruses

Howard M. Temin (1934–1994). [From Ref. 38.]

and hence it could be speculated that there must be some flow of information from the virus RNA to a copy in the form of DNA. This was a truly heretical proposition.

It took Temin another decade until he, possibly to his own surprise or at least to that of the scientific community, was able to demonstrate that the central dogma of DNA to RNA to protein, had one hitherto unrecognized important characteristic. It turned out that information could flow not only from DNA to RNA by transcription but also in the opposite direction. This logically came to be called *reverse transcription*. Temin pursued his work throughout the 1960s at the McArdle Laboratory at the University of Wisconsin, Madison, but had great difficulties convincing colleagues that his hypothesis was even worth testing. I remember in the later 1960s when I was chairing a session at a conference in Cambridge, England, to which Temin contributed, that I had great difficulty in making head or tail of his lecture. The fact that he was a very shy and not very articulate speaker added to the vague impression of his presentation. It was in experiments together with a Japanese postdoctoral student Satoshi Mizutani, a highly talented biochemist as it turned out, that the first evidence of the existence of an enzyme which could transcribe RNA into DNA was found. The enzyme was referred to as *reverse transcriptase*, and a number of years later virologists named the group of viruses containing this kind of enzyme *retroviruses*. They came to play a major role in developments in the rapidly expanding field of virology. But Temin was not alone.

As often happens in science, unknowingly, similar important results had been obtained by another scientist, David Baltimore at the Massachusetts Institute of Technology (MIT). Baltimore has one of these brilliant minds and applies strict logic in his approach to daunting challenges. I have had the pleasure of meeting him in many contexts, we were born only one year apart, and I have come to admire his lucid way of dissecting and reducing the complexity of a problem and approaching it experimentally. In his Nobel lecture[9] Baltimore described how he came to search for a reverse transcriptase enzyme in an RNA tumor virus. It all started with studies of small viruses, Mengovirus and poliovirus. In June 1946 a

David Baltimore. [From Ref. 38.]

hind-leg paralysis was observed in a captive rhesus monkey at the Yellow Fever Research institute in Entebbe, Uganda. The monkey came from the Mengo district in the neighborhood and the virus isolated from the animal therefore was referred to as Mengovirus. Later studies demonstrated that it was a small virus containing RNA. It was then shown that poliovirus shared properties with Mengovirus and they were both allocated to a family named *picorna*viruses, from *picos* meaning small (Latin *beccus*, beak modified in Spanish to *pico*, meaning both beak and small) and *rna*, the nucleic acid in the virus. In his critical early work Baltimore first used Mengovirus and then poliovirus.

These viruses at the time had just been shown to contain a single strand of RNA. In order to make more copies of this molecule Baltimore reasoned that it has to be copied back and forth. In order to keep track of the image molecules the terminology "plus" for the RNA in the virus particle and "minus" for its mirror image was introduced. Baltimore could demonstrate how replication could take place and also identify which of the two kinds of strands were used as the messenger-RNA for synthesis of virus proteins. It turned out to be the form in which RNA is present in virus particles, the plus strand. This RNA could associate directly with the cellular ribosomes and initiate the synthesis of critical virus proteins including more of the critical polymerase enzyme driving the replication by copying minus RNA on plus strand RNA. It should be added that in the normal cell there is no enzyme that can copy RNA from RNA. The early Mengovirus work served as basis for Baltimore's Ph.D. thesis, delivered in 18 months, under the tutelage of Richard Franklin at the Rockefeller Institute. At about the same time in the mid 1960s the already mentioned technique of nucleic acid hybridization had been introduced. Comparisons could be made both between sequences in two single strands of DNA or of RNA and also between a DNA and an RNA strand. Together with Franklin Baltimore discovered the existence of an RNA-dependent RNA polymerase in the virus-infected cells. This was a unique finding.

Baltimore then returned to MIT and switched to studies of another RNA virus, vesicular stomatitis virus (VSV), an agent infecting different animals and potentially also man. The choice of virus was fortuitous. It was brought to the laboratory by Alice Huang, a student of Robert R. Wagner at Johns Hopkins University. Studies of this virus demonstrated that different from poliovirus the RNA in this enveloped virus did not serve directly as the messenger-RNA. It was the mirror image, the negative strand that carried this function. Hence this kind of infectious agent was later referred to as a "negative-strand" virus. However, since the form of virus RNA present in virus particles, the minus strand, could not associate in a functional way with cellular ribosomes it was required that, in order to initiate protein synthesis, the virus carried along an enzyme that could bring about a conversion of its negative RNA strand to its positive version. The latter molecule could then serve as the required messenger-RNA. The predicted enzyme, an RNA-dependent RNA polymerase, was promptly found when searched for. It may be intercalated at this stage that Alice Huang later became Baltimore's wife and life-partner.

The logical next step was to look for an enzyme in an RNA tumor virus. In his Nobel lecture[9] Baltimore paid tribute to Temin in the following way: "Although his logic was persuasive, and it seems in retrospect flawless, in 1970 there were few advocates and many skeptics. Luckily, I had no experience in the field and so no axe to grind — I also had an enormous respect for Howard dating back to my high school days when he had been the guru of the Summer School I attended at the Jackson Laboratory in Maine." In order to start the work a suitable virus and some resources were needed. Peter Vogt, whom we will soon meet again, provided some Rous virus, but for unexplained reasons the initial experiments did not work. Help was then given by Todaro who provided resources out of the Special Virus Cancer program of the National Cancer Institute described above and sent samples of Rauscher mouse leukemia virus, a member of the type C RNA tumor viruses, like the previously mentioned Friend virus. Using preparations of this virus the tests worked like a charm. Baltimore thus independently discovered the reverse transcriptase.

The results from Temin's and Baltimore's laboratories were published back to back in *Nature*. At this stage the critical difference between the enzyme in negative strand viruses and that in RNA tumor viruses needs to be emphasized. In the former case the enzyme produces more RNA, which can be used as messenger RNA to produce proteins and drive virus replication. The situation in RNA tumor viruses is completely different. Using the reverse transcriptase the virus RNA is copied into DNA. This DNA can then be transported to the nucleus of the cell, where it may become integrated into the host cell chromosomal DNA. Hereafter the virus DNA

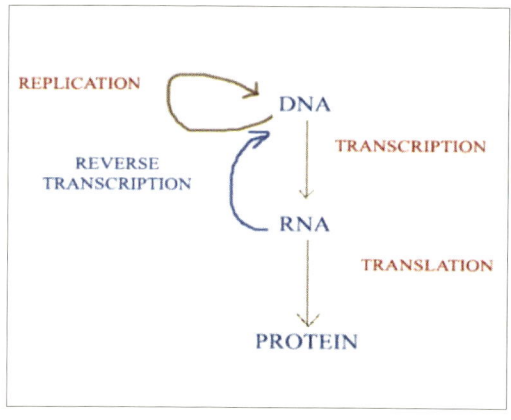

The complete version of the central dogma.

can be transcribed into RNA, using the regular molecular machinery of the cell, to drive the continued replication of the virus. Thus in practice the RNA tumor viruses operate like a DNA virus. Even more important was perhaps the fact that the discovery of the reverse transcriptase gave an important new perspective on the central dogma formulated by Francis Crick and others. Temin's and Baltimore's discovery meant that information could flow not only from DNA to RNA, but also in the reverse direction.

Temin's and Baltimore's data were presented at the Tenth International Cancer Congress held in Houston, Texas in May 1970. I was present at this meeting and shared the same feeling when the new data were made public and it dawned on the audience that a major discovery had been presented. The appreciation that RNA could be a mold to produce new DNA was momentous and it was at this meeting that Temin and Baltimore found out that they had simultaneously identified the critical enzyme. The Central Dogma of information flow from DNA to RNA to protein in cells from then on needed to be presented in a radically new perspective. This was bound to lead to a Nobel Prize in physiology or medicine and I was a young member of the committee when it was awarded. I had the pleasure of accompanying Temin in the procession entering the stage, but it was the more experienced member Peter Reichard, professor of biochemistry, who gave the laudation at the 1975 prize ceremony[10]. He emphasized the importance that Baltimore's and Temin's observations might have for the persistence of certain kinds of RNA viruses in animals and in man but also stated "…we still lack conclusive evidence for viral involvement in any form of human cancer." However there was accumulating evidence for the involvement of a herpesvirus in Burkitt's lymphoma, which will be discussed later. It should be added that Renato Dulbecco was included in the prize because he had previously shown that the genetic material of DNA tumor viruses, like papilloma viruses, became integrated in the genome of the cells in the process of their transformation. The full insight into this historically important cancer prize as evidenced by available archives will not be available until 2026. But the story does not end here.

Renato Dulbecco (1914–2012). [From Ref. 38.]

The Secrets of Rous Virus Finally Unraveled

In 1989 a prize equally important to the 1975 prize in the field of cancer research was awarded. The critical scientists recognized in this year were J. Michael Bishop and Harold E. Varmus and I had the pleasure to be the presenter at the prize ceremony[11]. A good insight into their work can be found not only in their Nobel lectures[12,13] but also in Bishop's already mentioned book, which had the enticing title *How to Win a Nobel Prize (An Unexpected Life in Science)*[2]. Bishop had his original training as a scientist at NIH and his mentor was Leon Levintow. His initial research concerned poliovirus, but progress was slow. However, he received a permanent position at NIH and was sent by Karl Habel to Hamburg in Germany for a year. After this non-productive year Bishop decided to leave NIH and to try San Francisco, where Levintow had now settled. He continued to work on poliovirus, but stimulated by the work of a colleague in the next-door laboratory, Warren Levinson, he had become curious about Rous sarcoma virus and the way in which it replicated. Bishop together with Levinson and Levintow had hardly started their work on this virus when the news broke of the finding of reverse transcriptase by Baltimore and Temin. Bishop felt he had missed an opportunity to make a contribution. However, he soon overcame his misanthropy and realized that knowledge about reverse transcriptase offered new opportunities for studies of tumor retroviruses.

He started to build a research group and a preeminent member of that became Harold Varmus, who joined it in late 1970 as a postdoctoral fellow. It allowed the formation of a very productive partnership. They decided to widen their studies focusing on Rous virus. Their approach to the work was simplified when it was demonstrated by others in 1970 that the virus contained a special single gene that appeared to be responsible for initiation and maintenance of cellular cancer growth. The gene was named src (the spelling of this three letter acronym varies from all capital letters, an initial capital letter to all lower case letters — I have chosen the latter as used in many textbooks) — because it caused *sarc*omas. It was then found, using the new tools of molecular biology, that the virus genome only contains four genes, three of which were needed to allow virus replication, and then src as a fourth gene. Bishop and Varmus decided to search for the origin of this extra gene. Before discussing their approach to this daunting problem Varmus deserves to be introduced.

Like many other American Nobel laureates Varmus has an Eastern European Jewish ancestry. He grew up in Freeport, New York on the southern

J. Michael Bishop and Harold E. Varmus. [From Ref. 38.]

coast of Long Island. His academic qualifications led him onto a literary scholarship at Harvard University, but then he was lured into medicine and began studies at Columbia College of Physicians and Surgeons, New York. In 1968 he joined Ira Pastan's laboratory at the NIH. Pastan was a central figure in molecular research on cancer. In 1970 he founded the Laboratory of Molecular Biology as a new separate branch under the largest Institute at NIH, the National Cancer Institute (NCI). Varmus was uncertain why he was accepted in the laboratory, but speculated that it was because his wife was a poet, a profession also successfully upheld by Pastan's wife. At Pastan's laboratory Varmus learned to use molecular hybridization techniques, originally developed by Spiegelman as mentioned above and to be further presented in Chapter 8, searching for nucleotide sequence homologies between pieces of heterologous forms of DNA and between DNA and RNA. This came in handy when he joined Bishop in the hunt for the origin of the src gene. Their encounter was accidental.

Varmus was on a back-packing tour in California and Rubin, mentioned in the previous chapter, advised him to seek out Peter Duesberg at Berkeley. However, Duesberg was not available at the time and hence Varmus fortunately ended up at Bishop's door. A unique temporary twinship of scientists was formed. Of crucial importance to the joint initiatives to be taken was of course the already mentioned demonstration by Temin and Baltimore of the existence of reverse transcriptase. The work progressed slowly but then a very useful new reagent became available. The already mentioned Vogt, originally of Czech origin but trained in Germany, had done some very interesting analysis of Rous virus at the University of Southern California. Vogt later moved to the Scripps Research institute in La Jolla where I have had the pleasure to meet him many times and to learn that besides being an excellent scientist, he is also a good artist in his own right. Together with an associate Vogt had managed to isolate a mutant of Rous virus, which could replicate efficiently but was incapable of transforming cells. This variant of the virus had a genome that was 15% shorter than that of the original virus. Bishop and Varmus deduced that it would be possible by use of this defective virus to prepare a probe, radioactive DNA, specific for the sequences deleted from the mutant virus[14]. By use of such a probe it might be possible to trace the origin of the transforming genetic material in Rous virus. Vogt generously shared his virus variant with Bishop and Varmus, an illustration of the importance of open collaboration in many scientific endeavors.

One obvious puzzle was why src was represented in the virus genome when it apparently had no influence on the replication of the infectious agent. Could it be that it had a cellular origin and might it possibly represent one of the oncogenes postulated on very vague grounds to exist by Huebner and Todaro? Finally after many preliminary tests Bishop and Varmus, using a specific probe, were able to make the critical analysis and look for src genetic material in normal cells. The first positive results were recorded on a Saturday night by a visiting French colleague, Dominique Stehelin. In an Open Letter to the Nobel Committee for physiology or medicine (November 10, 1989) he wrote:

> "The intensity of the emotion I experienced and the intellectual clarity induced by the situation at that moment were very special… The fantastic results came out in the night of Saturday October 26th, 1974: Normal DNA contained sequences related to the src gene of the transforming virus … I suspect that few have the privilege of enjoying such a moment when one is intensely and profoundly aware that a major step forward in Science has been made, and that one has contributed to it."

This letter describes well the joy of being the first person ever to identify a new kind of critical knowledge. It was of course written because Stehelin thought that we should have included him together with Bishop and Varmus in the prize. However, one needs to put this discovery in its proper context. The initiative to look for a possible cellular origin of the src gene came from Bishop and Varmus but it was Stehelin who had the pleasure of being a part of the implementation of experiments to test this hypothesis and he also was privileged to read the first results supporting the hypothesis of a possible cellular origin of src on that particular Saturday night. It may be superfluous to mention in this context that scientists often do not have regular working hours. The design of the experiments frequently does not harmonize with preselected times for ordinary daytime work.

These early results did not make it decisively clear if the gene that had been found in chicken cells also was present in genomes of other species. In order to examine this, an evolutionary biologist at the University of California at Berkeley became involved in the studies. He recommended examination of cells from ratites, a group of large flight-less birds, which include ostrich, emu and certain other related species, representing the most primitive species among birds and hence most divergent from chickens. It could then be shown that cells also from these distantly related birds contained src genetic material[14]. The final sentence of the critical publication was visionary: "We are testing the possibility that src (in normal cells) is involved in the normal regulation of cell growth and development or in the transformation of cell behavior (to a cancerous state) by physical, chemical or viral agents."

The next question thus concerned if src might also occur in mammalian cells. Another young colleague carried the main responsibility for testing this issue. It did not take long before src had been detected in the genetic material from humans, mice, cows and fish. At the time of these experiments the techniques to examine selected pieces of DNA were expanded by introduction of the recombinant DNA technology[4]. It now became possible to demonstrate that the src gene of Rous virus was a cellular gene that sometimes during the evolution of this particular virus accidentally had been picked up and become integrated in the virus genome. It did not seem to provide any advantage to the survival of the virus. The, at the time, bold conclusion was that src was a normal cellular gene carrying an unknown but obviously important function, since it was conserved throughout evolution. The concluding evidence for this was obtained when the technique of molecular cloning could be applied. Under certain conditions this normal cell gene — a *proto-oncogene (c-src)* — could express increased and uncontrollable functions leading to cell transformation. It had become an *oncogene (v-src)* associated with the virus genome.

Hereafter the following obvious question to answer was by what mechanism src in the form in which it occurred in Rous virus could lead to cell transformation. Work in the San Francisco laboratory and also in other laboratories, as presented in Varmus Nobel lecture[13], led to the identification of the protein as a so-called *kinase*, which means that it could enzymatically add phosphor groups to other selected proteins. The discovery of this central phenomenon in regulation of cell metabolism was recognized by another Nobel Prize in physiology or medicine only three years after the prize to Bishop and Varmus. The 1992 prize honored Edmund H. Fischer and Edwin G. Krebs "for their discoveries concerning reversible protein phosphorylation as a biological regulatory mechanism." Fischer's Nobel lecture[15] gave a good account of the development of the field. It was in 1978 that Bishop, Varmus and collaborators and simultaneously also another group led by Raymond L. Erickson at Harvard University published that the src gene was a protein kinase. Soon thereafter Tony Hunter and Bart Sefton at the Salk Institute, La Jolla, CA, found that the gene product was different from that generated by other proteinases known at the time in that it exclusively modified residues on the amino acid tyrosine.

It still remained to explain the difference between the proto-oncogene, c-src and the oncogene, v-src. Saburo Hanafusa's group at Rockefeller University were able to demonstrate that the kinase activity of the v-src was markedly enhanced because of certain critical amino acid changes, which they could identify. The importance of this discovery was emphasized by the fact that when Bishop and Varmus received the Lasker Award seven years prior to their Nobel Prize, Hanafusa as well as Erickson was included in the prize. There was even a fifth recipient recognized by the 1982 Lasker Basic Medical Research Award. It was Robert C. Gallo, who had discovered growth factors that allowed the first identification of two human retroviruses, which we will discuss further towards the end of this chapter. The individual Lasker awards unlike the Nobel Prizes can recognize more than three recipients on a single occasion, as mentioned above. Another example of this was that when Rous in 1958 became recognized by the same award, there were five other prize recipients. As science has progressed it has been learnt that there exist about 2,000 cellular kinases, which have been identified in the human genome and more than 90 of these are different protein tyrosine kinases.

The importance of these kinds of kinases was also highlighted by another Nobel Prize in physiology or medicine. This was the prize in 1986 to Stanley Cohen and Rita Levi-Montalcini "for their discoveries of growth factors"[16]. Such growth factors play an important role in embryological differentiation. However, if they are expressed in an uncontrolled fashion they may also drive the development of tumors. Cohen had discovered a growth-stimulating factor named *epidermal growth factor* (EGF). The receptor for this factor on the surface of cells was also shown to have a potential tyrosine kinase activity, which was activated when a substance had bound specifically to the receptor. Later studies have shown that also in this case many similar receptors can be demonstrated in different forms of cancer. Activation of signaling pathways by kinases thus can occur at different levels in cells, via receptors at their surface or inside cells by modification of different molecules in the chain of proteins involved in the signaling.

Bishop and Varmus extended their original finding of the existence of oncogenes. It remained to analyze whether the occurrence of src was a unique accident of nature, a curiosity, or if other viruses, similar to Rous virus, might also contain hijacked cellular DNA. Another of the younger collaborators in the San Francisco laboratory approached this problem by using a chicken virus named MC29, which could cause tumors in the animals. Another oncogene, named myc because it could cause *my*elocytomatosis — a tumor in bone-marrow derived cells containing granule with certain staining characteristics — was identified. Again it could be shown that it had its counterpart in normal cells. The same results were obtained at the same time by Stehelin, who at this time had moved back to France and established his own laboratory. In fact he discovered three different new retroviral oncogenes. myc was also found in another laboratory by the already mentioned scientists Vogt and Duesberg.

The introduction of the term proto-oncogenes literally transformed the conceptual perspective on mechanisms of development of cancer cells. Irrespective of where the cancer cell originated it started to lack obedience to growth controlling signals by mutations in its own genes. Cancer indeed was caused by damage to critical genes in a cell and such damage could be expanded by accumulated mutations involving other genes resulting in altered functions favoring an abnormal growth behavior. Such critical genes potentially causing cancer might be transferred from one individual to another under experimental conditions as originally demonstrated by Rous in his studies of the avian virus. This important new perspective emphasized the possibility that in the development of a cancer, aggregated damage to the cellular genome leading to progression, a term introduced by Rous, into an increasingly more malignant cell, eventually with a capacity to spread and potentially kill its host. Completely new ways of understanding the behavior of cancer cells had become available. New insights into the control of the replication of cells were needed to understand the inappropriate behavior of cancer cells and to develop rational means suppressing their development and spread. As a truism it should be emphasized that proto-oncogenes indeed do not exist in cells to be picked up by retroviruses and endow them with a cell-transforming capacity. They existed in their original healthy form to serve many critical *normal* functions in cells.

Oncogenes Take the Stage

The list of cellular proto-oncogenes increased rapidly. In his Nobel lecture[12] Bishop mentioned some 20 different genes potentially involved in cellular transformation. This *cornucopia* of biology as Bishop called it has continued to grow and the number of identified proto-oncogenes now exceeds 30[17]. The immense complexity of regulation of cell growth in highly developed vertebrate organisms has become progressively unraveled and continues to be researched. In addition to src, only two selected oncogenes will be discussed, illustrating principles of regulation different from the one represented by this oncogene. These are an oncogene called ras and the already mentioned oncogene myc. The three-letter designations of oncogenes have often been derived from their original discovery in different laboratories using various experimental animal tumor systems. Among the different tumor viruses discovered during the 1960s there were two forms of sarcomas in rats originally described by Jennifer Harvey and Werner Kirsten. They were found by other researchers to contain an oncogene of another kind than src. The fact that the tumors from which it originated were *rat s*arcomas resulted in the designation ras. This oncogene has a completely different function than src. It is a so-called G protein, a group of proteins which have a dominating importance for signaling in cells. The G proteins, their structure and function have been recognized by several Nobel Prizes.

The original recognition was by the 1994 prize in physiology or medicine to Alfred G. Gilman and Martin Rodbell "for their discovery of G-proteins and the role of these proteins in signal transduction in cells." The name G protein derives from the capacity of this large family of proteins to bind *g*uanosine-nucleotides, one of the building blocks of nucleic acids. Gilman's and Rodbell's Nobel lectures mostly focused on the central role of this group of proteins in regulation of various hormonal functions. However, they play a role also in many other contexts and as many as eight additional Nobel Prizes have been awarded which directly or indirectly include examination of this kind of signaling system. Not surprisingly they can play a role also in tumor emergence, growth and spread. The ras oncogene was the first one to be identified in human cancers. Many research groups were involved in this work including Robert Weinberg's at MIT, the author of the comprehensive book on the biology of cancer already referred to[17]. Ras is the oncogene most frequently encountered, in about 20 to 30% of all cases, to be activated in tumors in man.

Robert A. Weinberg.

In contrast to src and ras the oncogene myc is not involved in a signaling system of intermediary metabolism in cells. It is a *transcription factor*. A transcription factor is a protein with a capacity to bind to a stretch of DNA with a certain base sequence and thereby control the formation of RNA by copying from this DNA, the transcription. This kind of factor has a very central role in regulating the progression of the cell cycle, in apoptosis and in other corresponding contexts. When inappropriately expressed they may lead to transformation of cells. Myc controls the synthesis of gene products that contribute to the regulation of the expression of other genes in many different contexts, it is pleiotropic. Hence it is not surprising that genetic alteration of this factor may lead to over- or under-expression of a particular gene or sets of genes and consequently contribute to transformation of cells. In particular tumors in man it has been found, to give just one example, that the "drivers" involved are genes responsible for the expression of immunoglobulin heavy- or light-chain genes. The myc gene was discovered to be of importance for tumors in man by studies of Burkitt lymphomas as will be further discussed below. Factors controlling signaling systems and transcription of genes also play a central role in another remarkable, but in this case a physiological condition. This is the formation of an embryo from a fertilized egg. Hence the discoveries by Edward B. Lewis, Christine Nusslein-Volhart and Eric F. Wieschaus "for their discoveries concerning the genetic structure of early embryonic development" recognized by the 1995 Nobel Prize in physiology or medicine, may be surprisingly, also are of relevance in developing an understanding of the mechanisms of cancer formation.

In summary there are several mechanisms by which oncogenes can lead to certain genes being inappropriately over-expressed in time and in place leading to a transformation of cells. A particularly important field is the one involving the highly complex signaling networks connecting the outside of a cell to its interior or furthering communication between the different compartments within the cell. If the functions of these networks of signaling malfunctions and signals are inappropriately kept on, the cell, or rather its host, may be in trouble. Similarly the expression of individual genes or sets of genes have their own complex and carefully balanced control system. If a gene or a set of genes unphysiologically are retained in an activated state the result also may be the development of cancer.

The identification of oncogenes has allowed the development of new chemotherapeutic treatments of different forms of cancers. One oncogene named HER2/ERBB2 has been shown to be closely related to the EGF receptor mentioned earlier. Amplification of this oncogene has been found to be critical in many cases of breast cancer. This observation has led to the development of a humanized antibody against HER2. This remedy with the name Trastuzumab (HerceptinR) has markedly changed possibilities for survival of many patients with breast cancer. The disclosure that oncogenes can display tyrosine kinase activity has also led to the development of important new drugs. One of them is a drug with the name imatinib (Gleevec R in the U.S. and Glivec R in Europe), which effectively blocks BRC-ABL, an oncogene playing a major role in chronic myeloblastic leukemia. The use of this drug has led to a major improvement in survival rates for this form of cancer.

Tumor Suppressor Genes

But there is much more to the complex story of cancer emergence and development. If a cancer can be due to a dysfunction (hyper-expression) of a normal gene one could also consider the opposite mechanism. This would apply to hypothetical genes which as their normal function would carry a specific activity to *prevent* the emergence of defective or abnormally dividing cells. Such genes can potentially be of many different kinds as we shall see. They may have a normal function in the management of the large number of cells that are continuously phased out because of aging or indicated dysfunctions. The existence of such genes of particular significance in the control of cancer cells was identified by very special experiments.

In my early Ph.D. studies of measles virus, two properties associated with the envelope surrounding the virus were studied. One was the capacity of the virus to attach to cells and the other the potential of the lipid-containing envelope surrounding the virus particle to fuse with the cell membrane. This fusion allowed a deposit of its centrally-located genome-containing structures inside the cell facilitating the initiation of replication and expression of the genetic material of the infectious agent. Both these early steps in virus multiplication could be mimicked by use of red blood cells from African green monkeys. These cells were agglutinated and also lysed (broken open) as indicated by a release of hemoglobin. In cell cultures the capacity of the virus to melt together its envelope membrane with that of a cell may lead to a fusion of neighboring cells with each other. Multinucleated cells can be formed, so-called *syncytia*. This capacity of certain viruses to fuse cells was exploited by a scientist who wanted to evaluate the potential dominance of different properties of normal cells and of cancer cells. His name was Henry Harris.

Henry Harris (1925–2014).

Harris was Jewish, born in Russia, but raised in Australia. He developed most of his scientific career at Sir William Dunn School of Pathology at Oxford University. At this school he became professor of pathology, succeeding Howard Florey, also originally from Australia, who was recognized by a Nobel Prize in 1945 for his purification of penicillin. On one occasion at the end of the 1960s Harris and I were speakers at the same symposium in Helsinki, Finland. I described the importance of having fully immunogenic representation of both the major surface components of certain envelope viruses in the development of candidate inactivated vaccines. The examples given were measles virus and respiratory syncytial (RS) virus, the latter agent capable of causing serious capillary bronchitis in young children. It had been found that the use of incompletely immunogenic inactivated vaccines might lead to unusual complications in connection with virus exposure, like an atypical pulmonary disease in measles and an aggravated disease after infection with the distantly related RS virus. Measles has now been brought under control in many countries by use of an effective live instead of an inactivated vaccine. It is a disease that will be the target for global eradication once polio has finally been eliminated from our world. RS virus infections in young children, predominantly during the early part of their first year of life, remain a major health hazard. However, recent developments allowing the stabilization of the fusion protein in a critical immunogenic form by use of molecular biological techniques may eventually lead to the availability of a much-needed, effective vaccine. This kind of vaccine will be of great importance to immunize pregnant mothers aiming at providing the newborn child with protective levels of maternal antibodies. But let us now, after this digression, focus instead on the presentation by Harris.

Under the theme of virus-induced cell fusion he described some very important new findings. He had tested what would happen if, by use of a non-infectious virus preparation, a normal healthy cell was fused with a cancer cell. The outcome was unequivocal. In the fused cell the characteristic traits of the cancer cell fusion partner were suppressed. The important conclusion of the outcome of this kind of experiment was that normal cells must have genes that can suppress the expression of genes potentially causing cancer. Later such genes, referred to as *cancer (tumor) suppressor genes* and also *antioncogenes*, have been discovered to be of a number of different kinds. A *defective* expression of such genes is an alternative to the over-expression of oncogenes as a cause of cancer. The potential role of tumor suppressor genes in cancer development represents a highly complex and diversified field of growing knowledge. Only a selected number of examples of their importance will be discussed in the present context. In particular such genes will be discussed as they may have significance in relation to Nobel Prizes in physiology or medicine awarded for discoveries regarding certain critical cell physiological phenomena.

Alfred G. Knudson (1922–2016).

In our genomes, genes are represented in two copies, there are two alleles. Thus, in order to knock out the effect of a tumor suppressor gene, in principle both these alleles need to be inactivated. This was illustrated by the important findings made by Alfred. G. Knudson. He examined a rare genetic form of eye cancer called retinoblastoma. In studies of the occurrence of this tumor, which can develop at different ages and in different forms, he concluded that both alleles carrying the specific gene must be affected for a disease to develop. If one allele was damaged from birth, tumors could already develop early in life and there was an increased risk for development of additional tumors. If, however, the two alleles were normal at birth, development of disease could only occur after both of them had been affected by mutations. Hence in this case the disease developed much later in life. The publication of these data in 1971 made it clear that in addition to activation of oncogenes an alternative to development of cancer was inactivation of tumor suppressor genes, but that generally this should be a two-hit phenomenon.

Many different kinds of such suppressor genes have later been identified. They all relate to fundamental cell biological properties, many of which have been recognized under the heading physiology with Nobel Prizes in physiology or medicine. One fundamental aspect of the life of cells is the cell cycle, the different carefully-controlled steps in the division of a cell to form two new identical copies of itself. The Nobel Prize in physiology or medicine in 2001 recognized Leland H. Hartwell, Tim Hunt and Paul M. Nurse "for their discoveries of key regulators of the cell cycle." The fundamentals of the cell cycle, which has its evolutionary basis many billions of years ago was unraveled in studies of yeast cells and of sea urchin cells. It would take too long to describe the fundamentals of the cell cycle, but to illustrate the significance to cancer research we can turn to the Nobel lecture by Hartwell[18]. In his early studies he worked with tumor virologists Dulbecco and Marguerite Vogt, whom we have met before. The object of his studies was examination of cancer cells with the intention of getting an insight into DNA metabolism in mammalian cells. However, it turned out that the fundamentals of the cell cycle were more readily investigated in yeast cells.

Key regulatory genes were identified in such cells and these were referred to as cell division cycle (CDC) genes. Examples of important gene products were cyclin dependent kinase and cyclin. Insights into the basic mechanisms of functions of their products allowed an understanding leading to research into means of controlling abnormal cell growth. There is a section in Hartwell's lecture entitled "Cancer therapeutics." In the work described a particular focus was given to the cell cycle checkpoints and also to the mechanisms of DNA repair. The techniques to study yeast genetics were used in a drug discovery program. Genetic damage frequently encountered in cancer cells was introduced into the yeast cells. A broad spectrum of drugs was then tested to evaluate their capacity to compensate for the introduced defect in cell division. Several interesting compounds were found including one called Cis-platin. This compound has been found to be highly effective against solid tumors, in particular in many cases of testicular and ovarian cancers. It was originally approved for use in these forms of cancers in 1978 but has since then been tried also for treatment of many forms of solid cancers. This field continues to develop and involve alternative candidate drugs referred to as "checkpoint damage inhibitors." We will return to a brief summary of cancer chemotherapy later.

To finish off the discussion of the importance of tumor suppressor genes in cancer formation, the role of the so-called p53 gene needs to be described. This gene represents an exception to the "two-hit" rule formulated by Knudson. It appears that p53 acts as a "dominant negative" meaning that the incorrect expression of one defective allele by some mechanism suppresses the expression also of the second "healthy" allele of the gene. In the late 1960s a new technique for separation of unfolded proteins from virus particles was introduced by Jacob V. Maizel. It was referred to as the Maizel technique or as SDS-polyacrylamide gel technique. In the original electrophoresis technique introduced by Tiselius[4] folded proteins were separated by the electrical current applied. However, when the protein mixture was treated with the anionic detergent sodium dodecyl sulphate (SDS) the protein chains unfolded and because of the attached detergent they migrated roughly in proportion to their length. The introduction of this technique overnight changed possibilities to examine the different proteins composing for example a certain type of virus particle. The size of the different protein chains that could be identified was determined by use of reference proteins chains of known size.

Thus for example in our studies we could identify a 79k — k stands for kilo, viz. 1000 — Dalton (the unit for specification of a molecular weight) protein representing the hemagglutinin of measles virus. The already referred to tumor suppressor p53 gene hence was a gene coding for a protein with a rough molecular weight of 53,000 Dalton. Eventually, the true molecular weight of the protein turned out in fact to be only 43,700 Dalton. This specific protein is rich in the amino acid proline, which makes it behave abnormally upon SDS-polyacrylamide electrophoresis. It has a very central role in maintaining the health of cells and has sometimes been referred to as "the guardian of the genome." It was originally interpreted to be an oncogene, but work by Bert Vogelstein and Arnold Levine showed that it is in fact a tumor suppressor gene. It is mutated in more than 50% of human cancers, hence playing a major role. The p53 protein binds to DNA and serves many functions. Failure of these functions potentially has a great importance in the process of cancer development. Among these can be mentioned repair of damaged DNA; causing arrest in the cell cycle, allowing critical repair to be made; initiate apoptosis and finally respond to the presence of short complementary pieces at the end of newly synthesized DNA, the telomeres, soon to be presented. UV light stimulates p53 synthesis, events which can lead to tanning. The role of excessive exposure to UV light for the possible development of skin cancer might be mentioned in this context.

The function of the p53 gene may be negatively influenced by many factors acting like mutagens, such as carcinogenic chemicals, radiation and potentially transforming viruses. The papilloma viruses represent a certain case in point. These viruses code for a certain protein called E6, which has the special property of binding to the p53 protein and inactivating it. In the case of an infection in the skin the consequential activation of cell division may lead to the development of warts, which in contrast to normal skin expose live cells at their surface. Hereby virus particles may be released from the tip of the papilloma for transfer to other individuals. However the wart is a benign form of accentuated cell division. By way of contrast certain serotypes of the virus, for example types 16 and 18, cause infections preferentially in the genital tract potentially leading to changed growth behavior in the cervical mucosa. This may lead to the development of cancer, which we will return to below in the discussion of tumors caused by viruses.

Aging Cells and Cancer

In 1933 the first independent female recipient of a Nobel Prize in physiology or medicine Barbara McClintock (Laureate in 1983) postulated in the very early phase of her career that there were some special structures at the end of chromosomes. A similar observation was made by the 1946 Nobel Prize recipient, Hermann Muller[4] and he named this part of the chromosomes *telomere* from Greek *telos*, meaning end and *meros*, meaning part. When techniques had been developed to allow molecular studies of the replication of DNA in chromosomes it became obvious that unless there existed additional mechanisms, the chromosomes would become shorter and shorter with each cell division. Since this is not the case there must exist some compensatory mechanism. This turned out to be the existence of a particular enzyme *telomerase*. The discovery and characterization of this enzyme was identified by the Nobel Prize in physiology or medicine in 2009, which recognized Elizabeth H. Blackburn, Carol W. Greider and Jack W. Szostak, with women for once dominating. The prize motivation was "the discovery of how chromosomes are protected by telomeres and the enzyme telomerase." The enzyme turned out to be a reverse

Elizabeth Blackburn, Carol Greider and Jack Szostak.

transcriptase, in principle like the one found in retroviruses as introduced earlier in this chapter. Instead of virus RNA the telomerase uses its own RNA molecule of cellular origin as a template. The level of telomerase activity is relatively low in resting cells in our body, but it is high in stem cells providing the source of renewed tissues. Certain tissues that show a continuously high rate of renewal like lymphatic tissue, intestinal tissue, skin and sperm cells all have a marked telomerase activity. A particularly high activity is found in the intensively dividing cancer cells. Blocking the enzyme has therefore been discussed as a possible means of restricting the growth of cancer cells. It can finally be added that identification of the telomerase enzyme has relevance to discussions of the origin of life, to be briefly discussed in the last chapter. Szostak is one of the leading scientists in this exciting field of science.

Aging of cells can be studied by recording the highly variable half-lives of cells of different tissue origin in our body. It can also be examined by growing cells in tissue cultures. As already mentioned in Chapter 1 seeding of normal cells dispersed by treatment with, for example, the enzyme trypsin and suspended in a nourishing medium will attach to a glass or plastic surface and start dividing. They stop dividing when they come in contact with other cells. This phenomenon of contact inhibition (of growth) leads to their forming a mosaic of cells, a single cell monolayer. Cells in such a culture can be detached by binding of calcium and the solution of suspended cells can be transferred to a number of new vessels again to grow and divide to form subsequent generations of monolayers. This technique is used extensively by virologists in their studies of the infectious agents they are concerned with. Cells with a normal chromosomal set-up cannot be split an unlimited number of times.

For example epithelial or fibroblastic cells taken from a legally aborted human fetus can form new cultures for many generations but they stop dividing after some 30–50 reseedings. They have reached what has been called the Hayflick limit, from the researcher Leonard Hayflick, who discovered this phenomenon. The telomerase activity of these cells is low. Cells of this kind retain a normal setup of chromosomes and hence have been used extensively for the production of virus vaccines[5] as mentioned in the first chapter. The situation is quite different if cells of cancer origin are propagated in cell cultures. Such cells can display a perpetual capacity for cell division forming an unlimited number of consecutive cultures. Tons of such cultures can be produced. They display a high level of telomerase activity and their chromosomes are irregular and in a varying degree of disarray. These have general hallmarks of cancer cells. A typical example is the already mentioned HeLa cells originating from a severe form of cervical cancer in Henrietta Lacks.

This discovery of the existence of the telomerase enzyme is of major importance in many different fields of medicine. There are rare genetic diseases leading to dysfunction of this system. The process of aging is related to the progressive shortenings of the end of chromosomes. In patients with a genetically deficient telomerase, the aging process is accelerated. In contrast, cancer cells in our body, like such cells established in cell cultures, have a capacity to divide endlessly and hence they are also characterized by *increased* telomerase activity. Logically there is therefore a search for drugs that can reduce this accentuated enzyme activity, hopefully leading to an amelioration of the cancer disease but without interference with the normal process of aging.

Aging cells and cells showing signs of abnormal function need to be continuously eliminated from our body. This is managed by use of different mechanisms. They include controlled cell death by *apoptosis*, a poetic term derived from Greek, meaning "falling off of autumn leaves," and *pyroptosis*, a special inflammatory form of programmed cell death with the first part of the word derived from a Greek word for fire. To these control functions can be added the phenomenon of *autophagia*, literally self-eating. These kinds of fundamental cell biology phenomena have also been recognized by Nobel Prizes in physiology or medicine. In 2002 the prize recognized Sydney Brenner, H. Robert Horvitz and John E. Sulston "for their discoveries concerning genetic regulation of organ development and programmed cell death" and in 2016 Yoshinori Ohsumo "for his discoveries of the mechanisms for autophagy." The phenomenon of apoptosis carries a considerable importance in the development of different organs in the body, not least the brain. Thus the formation of many organs during embryonic development represents a controlled balance of creation of a precursor mass of cells and a trimming to the final organ, like when there is a development from a flipper-like structure to the formation of our hand with its five fingers. Similarly controlled cell death is critical also for example in the development of our complex brain. Horvitz's Nobel lecture[19] contains the following paragraph:

> "Conversely, some human disorders involve too little cell death. The number of cells in our bodies is defined by equilibrium of opposing forces: mitosis adds cells, while programmed cell death removes them. Just as too much cell division can lead to a pathological increase in cell number, so can too little cell death. Certain cancers, including follicular lymphoma, which can be caused by misexpression in B cells of the CED-9-like proto-oncogene Bcl-2, are clearly a consequence of too little programmed cell death."

Similarly it was stated in the award ceremony speech to highlight Ohsumo's discoveries[20] "Impairments in the regulation of autophagy area linked to a variety of human diseases, such as diabetes, *cancer* (my italics), infections and severe disorders of the nervous system." However, the importance of autophagy in cancer is complex. In the early phase of a carcinogenic development it seems to provide preventive effects. However, in another context in the case of a more advanced form of cancer the phenomenon may even be to promote the cancer. Thus there is a need for more knowledge about conditions of autophagy before it can be considered for a possible use in cancer treatment.

Chromosomes and Genes in Cancer

The first speculation that development of cancer might have its basis in genetic changes dates back as long as over 100 years. Already in the late 1900s a German pathologist David von Hansemann noted the presence of abnormal cell division — mitosis — in biopsies of cancers. Then in 1914 his fellow-countryman the biologist Theodor Boveri wrote a very visionary book *Zur Frage der Entstehung maligner Tumoren* (On the question of the origin of tumors). He predicted that genetic changes were at the root of cancer, but it would take an extensively expanded insight into the nature of the genetic material and the interplay of its many different components before his hypothesis could be validated. In current terminology it is referred to as the *somatic mutation theory of cancer*, the all-encompassing theory that Rous during his long life as a cancer experimentalist never accepted, as mentioned.

The Swedish chromosome researcher Albert Levan was introduced already in my first book on Nobel Prizes[6]. In the tradition at Lund University in the early 1950s he was involved in examining chromosomes in plants. Using material of this origin a technique of "squash and smear" was introduced to facilitate the visualization of chromosomes appearing in dividing cells. In the early 1950s it was realized that this technique might also be applied to animal cells. During a visit to the mouse geneticist Theodore Hauschka at the Institute for Cancer Research in Philadelphia Levan became interested in studying mammalian tumor cells. After his return to Lund, he continued comparative studies of normal and transformed mammalian cells. In passing it can be mentioned that, together with his co-worker J. H. Tjio, Levan made the remarkable fundamental discovery that human cells did not, as I had learnt in school, contain 48 chromosomes but only 46. It is salutatory to learn that the higher number, authoritatively, had been accepted for some decades by chromosome researchers around the world. Under the yoke of this dogma it was concluded that the identification of only 46 chromosomes meant that two must have been lost in the processing of the sample!

Albert Levan (1905–1998).
[Copyright Creative Commons BY-SA]

Studies of mammalian chromosomes were facilitated during the 1950s by the introduction of convenient techniques to cultivate human peripheral blood cells and also to grow mammalian cells in the laboratory. It started to be understood that there were ubiquitous chromosomal variability both during the time of development of a tumor and also between tumors of similar organic origin. As a consequence Darwinian terms began to be applied to describe the evolution of tumors. Critical mutations in cells were accumulated to endow them with a competitive advantage in relation to surrounding cells and also to facilitate their spread and survival even when a larger size had been reached, requiring independent provision of blood vessels to transport the substances necessary for the high-level metabolism of the tumor.

In the early 1960s, as already mentioned, I got to know Levan, a very charming scientist, who instructed me about how to examine the effect of virus infections on cellular chromosomes. At this time he had become deeply involved in studies of possible specific chromosomal changes in the development of cancer. I remember that at the time I was somewhat skeptical about this approach. How could gross changes in chromosome structures teach us something about changes in a few critical genes among the many thousand genes each one of them might contain? But I was wrong. It turned out that there were a number of preferred chromosomal rearrangements that had such a prominent influence on development of certain forms of cancer that they could be morphologically identified by the appearance of particular new structures.

In 1959 David A. Hungerford at Fox Chase Cancer Center and Peter Nowell at the University of Pennsylvania School of Medicine discovered specific chromosomal changes in cases of human leukemia. The changes were highly prevalent, occurring in 95% of cases of chronic myeloid leukemia, but also in certain cases of acute leukemia. Because of the home state of the two scientists, one of the critically changed chromosomes structures observed was referred to as the Philadelphia chromosome, later abbreviated to Ph. For a long time the meaning of this finding remained a conundrum. Specific reproducible abnormalities in the chromosomal patterns were encountered also in various kinds of experimental tumors, although this generally occurred against a background of a plethora of fortuitous random chromosomal changes. This duality of evolving chromosomal abnormalities was analyzed in more depth by studies of some 200 primary sarcomas induced by Rous virus in various experimental animals, mice, rats and Chinese hamsters[21]. Two different kinds of abnormalities were identified; nonrandom changes mainly involving certain chromosomes and then general massive random changes afflicting all chromosomes as a background noise. For a long time it remained a challenge to identify the origin of different chromosome fragments forming new and characteristic abnormal structures. However, in 1970 a new technique to stain chromosomes was introduced. The central person in this discovery was Caspersson, repeatedly encountered earlier and also figuring in all my three previous books on Nobel Prizes as both a reviewer and personally a serious candidate for a prize[4,6,7].

Towards the end of the 1960s he had established collaboration with Farber at Harvard Medical School. The aim was to identify possible specific targets at chromosomes for cytotoxic drugs. Caspersson liked advanced instruments and he had constructed one that could measure fluorescence emitted from isolated chromosomes. One of the collaborators in Farber's laboratory was the chemist Edward Modest. He recommended that a chemical compound combining mustard and quinachrine (QM) should be tried. The latter compound is also known as mepacrine, a drug used for treatment of malaria since the early 1930s. Later this compound was considered for other uses, for example in treatment of autoimmune diseases. The proposed construct was prepared and Caspersson carried it back home to examine, together with his collaborator Lore Zech, its effect of staining chromosomes examined by his instrument. They were in for

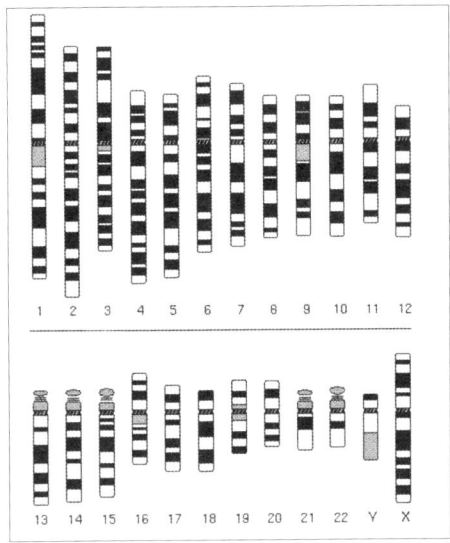

The 23 different human chromosomes identified by the banding technique.

a surprise. The QM did not stain possible targets for cytotoxic drugs. Instead it gave a pattern of bands on all chromosomes, unique and reproducible for each one of them. This is good science, an unexpected result opening a field for advances into new important knowledge. For the first time it became possible to map out the specificity of each chromosome. They could now be individually recognized and numbered as pairs 1 through 22. In addition the X and the Y chromosomes could be identified. This finding completely changed the possibility for cytogenetic examinations. The new staining method was referred to as the *chromosome banding technique*. It provided unique possibilities to separately identify different chromosomes and parts of them participating in the possible formation of new structures. For example the origin of the Ph chromosome could be determined as we shall see.

The discovery of this technique was of a quality motivating it to be discussed in connection with a Nobel Prize in physiology or medicine. However, neither Caspersson nor Levan, who had discovered the correct number of human chromosomes, were ever to be recognized by a prize. Still the banding technique truly revolutionized possibilities for studies of chromosomes in congenital human diseases and also in cancer cells. In the latter category the first object to be examined was the Philadelphia chromosome. Its origin was demonstrated to be in part chromosome 22, from which a critical piece had been deleted and transferred to chromosome 9. This so-called *translocation* occurred at the ends of the chromosomes leading to a longer version of chromosome 9 and a shorter version of chromosome 22. Later it was found that this shuffling of genetic material led to juxtaposition in the latter chromosome of two genes which were the cause of transformation of cells. This association was remarkably precise with connection of DNA at the exact nucleotide position. One gene was named abl because it had been found in one of the many mouse leukemia viruses, the Abelson leukemia, and the other bcr (*b*reakpoint *c*luster *r*egion). The abl gene like the ras gene codes for a tyrosine kinase. Under normal conditions the bcr variant of enzymes are regulated in some kind of auto-inhibitory way, but when this gene became located adjacent to the abl gene, it was perpetually turned on. It had become an oncogene.

After this original discovery many different kinds of specific chromosomal rearrangements were identified in a wide range of different tumors in man. One of the kinds of tumors showing specific chromosomal changes was the Burkitt lymphoma. The name of this lymphoma derives from the surgeon Dennis P. Burkitt, who discovered its existence while working in equatorial Africa. This form of lymphoma was discussed earlier[4]. It is associated with an infection with a human herpes virus named Epstein–Barr virus (EBV) from its discoverers, later renamed human herpes virus 4. In industrialized countries this virus was serendipitously discovered to cause mononucleosis, a disease generally occurring in the upper teens leading to development of large tonsil tissues and also other symptoms from internal organs like the liver. The mechanism of the disease has been interpreted to be a conflict between the two dominating kinds of lymphatic cells, the T and the B cells. Children infected with EBV and simultaneously with malaria in Africa may develop very different kinds of diseases; various forms of B cell lymphomas and nasopharyngeal carcinoma. In this case the pathogenic mechanism derives from a juxtaposition of a constitutively active gene region and an oncogene. This physical association leads to a deregulation of the controlled expression of the oncogene myc, already described above. The drivers of myc are genes responsible for the expression of immunoglobulin heavy- or light-chain genes. The original discovery of the myc oncogene was in fact made in studies of Burkitt lymphomas. It was demonstrated to be associated with specific chromosomal translocations, frequently involving chromosome 8. A persistently (constitutively) expressed myc has later been found also in many other kinds of tumors. Tumors in African children caused by EBV are treated by cytotoxic drugs of the kind used to manage leukemia in general.

In later studies in excess of 2000 solid tumors in man were examined, which allowed the recognition of some 200 reproducible structural chromosome changes. The recognition of these repeatedly encountered specific changes raised the question of what kind of pathogenic interaction of genes they might allow. This was resolved when the cytogenetic studies were supplemented by molecular genetic investigations. Thus in the 1980s the advent of molecular genetics once more revolutionized cancer research. By the introduction of DNA-sequencing techniques with rapidly advancing efficacy, so-called deep-sequencing of massive parallel reading technologies, it has now become possible to compare the detailed genetic characteristics of normal cells and transformed cells. Furthermore the changes in the genetic characteristics of an individual cancer can be followed during its development. By this new kind of approach, the search for deviating characteristics is not made by reference to chromosomal morphological structures but instead represents a global analysis of molecular genetic changes. These studies have led to identification of one more mechanism of gene modification of potential disease-causing importance. This is a shortening, truncation, of genes. Such a loss of material may lead to inactivation of tumor suppressor genes.

It is now possible to screen a large number of different tumors and the more extensive the materials, the larger the number of gene fusions or gene truncations that can be identified. In a 2015 review[22] the total number of gene fusions in a large material of cancers was 9,928 and 330 of these were confirmed to occur repeatedly. The fact that a large majority of the changes seen were not recurrent highlights that a considerable number of them may be incidental and of lesser or no relevance to the understanding of the disease process. In the future it will be important to filter out this background noise. Techniques for genome analyses are progressively refined and in the future it should be possible to retrieve the most relevant information for the unique cancer in a particular patient. Using the best practice available a pathology atlas for the human protein expressions — the transcriptome — in cancers has been published[23]. Already today identification of certain kinds of gene fusions has developed to be of use in clinical practice and the number of applications to different kinds of tumors continues to expand.

Genetic studies have also been found to be of great value in following the development of tumors. Bert Vogelstein at the Johns Hopkins School of Medicine, a frequently cited pioneer in cancer genomics, has shown in longitudinal studies of colorectal cancers the importance of a step-wise accumulation of genetic changes accompanying the progression of the disease. Sometimes premature proposals with a visionary strength have been published. Carl O. Nordling was a Finnish-born architect and urban planner. However

Bert Vogelstein.

he had many strings to his bow. Being a skilled statistician he daringly published an article "A new theory on the cancer-inducing mechanisms" in 1952. Its summary stated "The theory is put forward that cancerous cells contain not one but a number of mutated genes." This was a speculation ahead of its time, not least since the chemical background to mutations as yet had not been identified. In a way Rous's introduction of the term progression could be seen to apply to accumulated changes, although as mentioned he did not interpret the alterations of properties of cancer cells as having a genetic background.

The growing insight into the role of different specific genetic changes in the appearance and development of cancer immediately raises the question of the possible role of inherited genetic dispositions to cancer development[24]. As a general figure it has been estimated that there are preexisting genetic mutations which confer an increased susceptibility to development of certain forms of the disease in 5–10 percent of cancers in man. The prototype disease in this context is the already mentioned hereditary retinoblastoma. Knudson's two-hit hypothesis had already been developed at the beginning of the 1970s and some 15 years later it became possible to isolate the critical gene, named RB1. The product of this gene also plays a contributary role in certain other forms of cancers. The possibility of identifying a specific gene predisposing to the development of cancer allows the use of prenatal or even better preimplantation diagnosis to secure the development of a healthy child. Among breast cancer patients, the dominant form of cancer in women, it is estimated that again about 5–10 percent of the patients have a hereditary predisposition to development of the disease. Studies in the U.S. have led to identification of three predominant mutations, two in a gene named BRCA1 and one in another gene named BRCA2. These mutations are of Eastern European Jewish origin. They also may predispose to a tumor development in other organs, like the ovary. Repeated screenings are recommended for early diagnosis. Occasionally individuals with the critical genetic predisposition may choose to have a prophylactic removal of the critical organ.

The Li-Fraumeni Syndrome is another example of a hereditary form of cancers, predominantly osteosarcomas. It is a rare form of disease, but afflicted families need to be watched carefully. Other examples are familial adenomatous polyposis in the large intestine. Again awareness of this genetic predisposition are inductive to a vigilant follow-up and a potential preventive removal of parts of the critical organ. There is also a hereditary nonpolyposis colorectal cancer, which has been demonstrated to be associated with certain inborn mutations. Further additional examples could be given and since genetic information will be handily and inexpensively identified in the future the significance of identifying them will increase with time.

Immune Defense and the Development of Cancer

The many different arms of the immune system are central in the management of invading pathogens, but they also serve many other functions of importance for the balance between the different populations of cells in our body, the homeostasis. The immune system is one critical guardian of the integrity of multicellular organisms. The relationships are complex with many actors, besides antibodies, cell-mediated responses, interferons and a host of lymphokines and the many kinds of cells in action, in addition to B and T lymphocytes, NK cells, macrophages, dendritic cells, etc. Once an infection has been brought under control the inflammation associated with the process subsides, but if this is not successful a chronic infection may be the outcome. Such a persistence of inflammation may, as we shall see, be of importance and lead to cancer formation. But the immune defense has a much more general importance for the way in which the body tries to counteract the emergence and development of tumors. Besides the importance of activated oncogenes and inactivated tumor suppressor genes in the emergence and development of cancers, the way the immune system manages potentially emerging abnormally dividing and spreading cells is central for the outcome of an incipient cancer disease. This also implies that it should be possible to use the accumulated deep insights into the various immunological defense mechanisms involved to control and potentially even eliminate a cancer.

If Nobel Prizes specifically mentioning cancer are rare, they are all the more common in the field of immunology. Altogether some 13 prizes in the field of physiology or medicine have been awarded, highlighting most of the major developments in immunology except perhaps the original identification of cellular immunology, the T cells, as one of the two dominating subdisciplines, together with humoral immunology executed by the B cells[7]. The only field that to date has been recognized at an even higher frequency is neurobiology with 15 prizes[4]. In 1960 a Nobel Prize in physiology or medicine recognized the already mentioned Burnet and Peter B. Medawar "for the discovery of acquired immunological tolerance." They provided an explanation for the fact that we do not respond immunologically to our own tissues, although the immune system at the time of late embryonic differentiation acquires a capacity to recognize in principle any kind of structure serving as an antigen. It is of course fundamental that we have this *tolerance* and only under abnormal pathological conditions may react to ourselves, a phenomenon referred to as auto-immune diseases. Cancer cells are different from our normal cells and thus immune surveillance mechanisms in our body potentially should be capable of identifying and eliminating such cells. Hence it is important to evaluate to what extent the emergence of a cancer is due to some paralyzing effect on one or more of the different arms of our immune defense and conversely if it would be possible by changing the conditions again to mobilize one or more of these arms as a means of counter-combating the pirating cancer cells. The role of the immune defense in the development of cancer needs to be evaluated in relation to different kinds of tumors, in particular if a specific cancer involves any of the different kinds of cells representing the immune system itself, as in different kinds of leukemia.

One insight into the relative role of the immune response is to study the relative frequency of occurrence of different kinds of tumors under the special conditions of immune suppression established in patients receiving organ transplants. This was discussed in zur Hausen's Nobel lecture in 2008[25]. One particular form of tumor called Kaposi's sarcoma showed a 200-fold increase in appearance. This was the kind of tumor which was encountered at a high frequency in the 1980s in patients with an advanced form of HIV/AIDS, a stage of extensive immunosuppression. This tumor was later found to be associated with an infection with a certain type of herpes virus, to be further mentioned below. There was also about a 20-fold increase of cancers in the vulva and the penis. Some of these most likely were due to papilloma virus infections, which we will also return to below, but the remaining cases can be speculated to be due possibly to infections with some other as yet unidentified virus. Paradoxically tumors in the rectum and in the brain occurred in a reduced frequency in immune-suppressed patients. This may highlight some special complexity of the role of the immune response in different kinds of tumors.

There are several different ongoing attempts to harness the immune response to interfere with the growth of different kinds of cancer cells. It appears likely that as the gene expression is fundamentally changed when a cell stepwise develops from a normal cell to a cancer cell, the transformed cells may expose new antigen targets for the immune system to attack. However, as repeatedly stressed a tumor cell is actively evolving and when it encounters an immune attack against some of the new antigens it may present, it will try to selectively escape this attack by the evolutionary means it continuously applies in its drive towards replicative advantage. This does not only imply the appearance of modified tumor antigens, the target, but possibly also an interference with one of the many potential inhibitory pathways which are available in the complex immune network system. This needs to be kept in mind when attempts to use immunotherapy in the management of different forms of cancers are initiated. The field of cancer immunotherapy is large and it will only be briefly alluded to in the present context.

Encouraged by the success of developing vaccines against infectious agents, discussions about the possibility to vaccinate against cancer were initiated at an early stage. However, the initial results, were poor due to ignorance of the elaborate mechanisms of possible immune reactions against cancer cells at the time. In the late 1950s a Swedish scientist by the name of Bertil Björklund initiated some very premature attempts to immunize against cancer. He reasoned that if there was some tumor-specific antigen, the presence of this antigen might be enriched if tumors of different origins were pooled and homogenized. It was speculated that using such a homogenate might possibly induce a general tumor-specific response. Very soon fellow scientists reviewing the vague data presented concluded that there were obvious major flaws in the hypothesis and in the experiments performed to seek data to support it. My predecessor as professor of virology at the Karolinska Institute, Sven Gard, who at first glance had been intrigued by the hypothesis very soon revealed the quackery of experiments performed to examine it. That should have been the end of the story but it turned out that Björklund through his mother was closely associated with the Bonnier publishing house, responsible for circulation of some of the main daily newspapers in Sweden. This led to the press becaming involved in the heated discussions, which eventually also mobilized some politicians in the Swedish Parliament. A colorful and committed member of the ruling Social Democrats at the time, Nancy Eriksson, persuaded colleagues in the parliament that it should take control of the heated debate. Björklund was appointed associate professor at the National Bacteriological Laboratory and given means to continue his research. In addition, to please the academic community, Astrid Fagraeus a leading female immunologist introduced earlier[7] was appointed professor at the same institution, but with an extra affiliation at the Karolinska Institute, a promotion that was highly overdue. The moral of the way this "affair" was resolved is that parliamentarians should never mingle in scientific debates, like when Nixon, based on tainted information launched the already mentioned "conquest of cancer" program. Of course nothing came out of Björklund's cancer research, but Gard became very depressed for some years. So what is the rational attack on cancer using immunological means?

Cancer cells by definition will evolve genetically to provide a number of tumor-associated antigens that different arms of the immune system potentially can identify. However, as in a chess game, the cancer cell registers the mobilizations of certain critical immune checkpoints as a host defense and in its turn it tries to escape by exploiting additional spontaneously-occurring mutations These are all very complex relationships and way beyond what can be discussed in the present book. However, what should be mentioned is that scientists taking advantage of the wide range of sophisticated molecular genetics techniques can get involved in an analysis of the chess game between the cancer cell and the immune response of a certain patient or a group of patients afflicted by the same kind of tumor. Progressively new tools for immunotherapy are being developed and important advances have been made[26-29]. The tools used are diversified. They include monoclonal antibodies against selected critical antigens; collection of immune cells from the patient and "educating" them in the laboratory to react towards selected antigens specific for the tumor; blocking of immune checkpoints by specific antibodies or by interfering with certain ligand-receptor interactions, etc. For example it has been possible to demonstrate improved survival in patients with advanced melanoma using antibodies against a certain antigen in T lymphocytes, called CTLA-4. In 2017 the Food and Drug Administration in the U.S. approved a cancer therapy, called CAR-T, based on the use of the patient's own immune cells genetically modified in the laboratory. These important developments have also been noted by Nobel Committees.

At the time of writing the final parts of this book the 2018 Nobel Prizes were announced. Interestingly both the prizes in chemistry and in physiology or medicine had bearings on possibilities for immunotherapy of cancer. In fact the motivation for the prize in physiology or medicine read "for their discovery of cancer therapy by inhibition of negative immune regulation." This is the fourth time that the word cancer has been included in a prize motivation. The prize recipients were James P. Allison, (Houston, Texas) and Tasuku Honjo, (Kyoto, Japan). Both of them took advantage of discoveries of surface structures on T immune cells which serve as the source of brake signals in the cells. In Allison's case it was the CTL-4 structure already mentioned above and in Honjo's case another structure with similar properties, PD-1. Intense studies now concern possibilities to further advance "immune checkpoint therapy" targeting either one or possibly even both these surface structures. It is a matter of balance so that not too overactive immune reactions are elicited. In such cases severe autoimmune reactions may be elicited. The results of use of the checkpoint inhibitors have been especially encouraging, in particular in cases of melanomas and lung cancer. Survival rates have increased markedly and in many cases the effects appear to be long-lasting and extended even into years after the treatment has been finished.

The prize in chemistry in 2018 concerned new techniques to bring about directed evolution of enzymes in the laboratory with half the award to Frances H. Arnold, (Cal Tech, Pasadena, CA) and the other half for a corresponding evolution of antibodies to Gregory P. Winter, (MRC Laboratory of Molecular Biology, Cambridge, UK) using the phage display technique developed by his co-recipient of the half award, George P. Smith, (University of Missouri, Columbia). The antibodies developed by Winter have found a major use in the treatment of autoimmune diseases, but also in the treatment of cancer, occasionally with spectacular effects, with regress also of metastatic forms of the disease.

The Human Microbiome and Cancer

In recent decades it has been discovered that the human body contains about ten times more microorganisms than the nucleated cells that compose it. We are a walking community. A detailed analysis of the microbial flora on different mucosal surfaces and possibly other compartments has become important and data are accumulating rapidly. Instead of cultivation, characterization of the microbes by their signatures, identified by nucleic acid sequencing, has become possible to perform at an ever-increasing speed. The change in occurrence of representative species of microorganisms and also of associated viruses infecting them — the *microbiome* and the *virome* — can now be readily identified. Interesting data of considerable medical relevance are now accumulating at a rapid pace. The avalanche of new information is continuously

Karen Nelson. [Private photo.]

subjected to analyses and a completely new perspective is emerging. Microorganisms and viruses are ubiquitous. A considerable fraction of these data have relevance to our understanding of the concept of health in general. An important part of these new insights are of significance also to the formation and development of cancers. Besides the indigenous capacity of certain viruses carrying an oncogenic capacity, to be discussed below, there are in principle three different mechanisms by which the microbial flora may influence cancer development. Firstly, genotoxins may be released possibly causing damage to host cell DNA. Secondly, bacterial toxins and possibly other metabolites may result in inflammatory reactions which in their turn may contribute to cancer development. Thirdly, the resident microbiome may result in immunological dysregulation of consequences also for the development of cancer cells[30]. I have a long time affiliation with the J. Craig Venter Institute, at the time of writing as the Vice-Chairman of the Board, and one of the leading groups in the field of microbioms and viromes is headed by Karen Nelson, who is also the President of this organization.

A total of ten bacteria and viruses have been registered by the International Agency for Cancer Research in Lyon, France, in its report of 2014, to be causative agents of cancer. A particular case is *Helicobacter pylori*, an agent colonizing the stomach mucosa in about half of all individuals in the world. The identification of this agent is a saga illustrating the importance of heterodox thinking in the pursuit of science. In 2005 the Nobel Prize in physiology or medicine was awarded to Barry J. Marshall and J. Robin Warren "for their discovery of the bacterium *H. pylori* and its role in gastritis and peptic ulcer disease." In their studies Warren even fed himself material from a culture of the bacterium studied and was able to induce a transient gastric infection. Their discovery overnight changed the management of gastritis and not least gastric ulcers. These conditions were resolved by treatment with antibiotics. The major surgical interventions used in severe cases of stress-related gastric ulcers all of a sudden were no longer necessary. These operations were referred to as Billroth I and II, from the famous Prussian-born Austrian surgeon Theodor Billroth, who introduced them at the end of the 19th century. In passing it can

Barry J. Marshall and J. Robin Warren.

be mentioned that Billroth was also a competent amateur musician and had a major influence on Johannes Brahms' development as a composer.

The chronic inflammation established by the bacterium has also been found to be of importance for the development of stomach cancer. However, experiments with germ-free experimental animals have shown that the bacterium alone is a rather weak cause of cancer, but this can be markedly amplified by the presence of additional bacteria, highlighting the complexity of interactions of microorganisms and cells in our body. In this context it can be added that the first-year secretary of the committee at the time, Bengt Gustafsson, was one of the pioneers in the development and use of germ-free animals. These animals are delivered by caesarian section and brought up in a germ-free environment[7].

Chronic inflammation has also been found to be of importance in another form of cancer in the gastro-intestinal tract, namely colon cancer. In this situation it is another bacterium that plays a major role, namely *Fusobacterium nucleatum*. This bacterium is found to be associated with about half of all cases of this form of cancer. Surprisingly it has been found that it can travel with the tumor cells as they spread for example to form metastases in the liver. This unexpected finding has been confirmed by experiments in animals demonstrating that antibiotic treatment reduces the replication of tumor cells not only at the original site of its residence but also in other organs to which its cells have spread. Further data in this young field of research needs to be provided and access to germ-free animals will be an important asset in these studies. Additional carcinogenic agents in the microbial world are viruses.

Viruses and Cancer

As we have seen the identification of viruses causing tumors in experimental animals led to the discovery of more than thirty cellular oncogenes with a range of different functions. It was these observations that led to insights into fundamental mechanisms of the development of cancer cells by accumulated critical damage to the cell genome. In the 1980s it became possible, by the already mentioned range of molecular biological tools, epitomized by the recent introduction of deep genomic sequencing, to examine the role and the mechanism of virus-driven tumorigenesis in man. This in-depth examination was made against the background insight that it was difficult to reconcile a general hypothesis of cancer as an infectious disease with the accumulated knowledge of the biology and epidemiology of human cancer. The final conclusion from these studies, perhaps somewhat surprisingly, was that retroviruses of the kind causing the development of most experimental tumors examined in laboratory systems *did not* have significance for human cancers. Thus viruses that were later found to be of importance for cancer formation in man did not belong to the experimental RNA tumor viruses discovered to be capable of hijacking cellular protooncogenes and turning them into oncogenes. However, it should be emphasized that this does not mean that the extensive understanding of the critical role of different kinds of oncogenes did not have any relevance to the human condition. On the contrary it provided a very rich critical broad insight into fundamental mechanisms of carcinogenesis as a process in which there is an accumulation of damage to the genome of a healthy cell. These detailed insights have provided important opportunities to develop drugs that specifically block the process of carcinogenesis. Interestingly the absence of importance of the kind of retroviruses, starting with Rous virus, that have played such an important role in understanding the process of carcinogenesis also does not imply that viruses of this kind do not occur in man. They do. There is in fact a host of *human endogenous retroviruses* (HERV) revealed to be present in our genomes by use of nucleic acid sequencing. In addition four types of infectious human retroviruses, human T cell lymphoma viruses (HTLV-1 to 4), have been demonstrated. These viruses need to be discussed separately.

The original oncogene theory proposed by Todaro and Huebner was wrong. But more recent genome studies have revealed that the human genome does in fact contain a large number of endogenous retroviral sequences as mentioned. Only a few percent of the three billion nucleotide bases in the human genome code for the roughly 22,000 proteins that manage the complex intermediary metabolism and serve the required structural functions in cells in our body. In addition the genome codes for a plethora of different forms of RNA[31], the function of which has been clarified only in part. Research into the euphemistically-named non(protein)-coding RNA remains a hot target for experimentalists attempting to understand the total function of the human genome or whatever genome for that matter. We will return to this question in Chapter 8.

About 45% of the mammalian genome is represented by monotonous transposable elements, referred to as LINE and ALU, but as much as 8% clearly derive from retroviruses, remaining as fragments of the original virus genomes introduced into the cellular genome at the dawn of evolution[32]. To a variable degree the critical viral genes env (virus *envelope* genes), pol (viral enzymes, preeminently the reverse transcriptase, the *polymerase*) and gag (internal structural components, *group-antigen*) in variable vestigial forms can be recognized. The possible physiological or pathological role of these endogenous retroviruses (ERVs) has been and continues to be discussed. It has been proposed that their capacity to bring about cell fusion may possibly serve a physiological function in the formation of the multinucleated syncytial mammalian placenta.

The remnants of vestigial infectious retroviruses naturally have also been considered possibly to be of importance in the development of cancers. This remains an open question, but at present there is no hard evidence linking these relics of viruses to emergence of tumors in man[33]. Instead, as discussed at length it was found that it was pathological alterations of the *normal* cellular gene products that were the potential culprits driving the oncogenic process. These products were inappropriately over-expressed or under-expressed by many different mechanisms resulting from chromosomal rearrangements and in a number of cases virus infections were documented to play a role in this process of transformation. However, there is a need to further evaluate the quantitative and qualitative role of different viruses, proven or indicated to have an involvement in cancer formation.

Originally three hypothetical situations were considered. These were that viruses were the cause of all cancers, that certain viruses were the cause of certain forms of cancer and finally of course that it was the development of a tumor that led to activation of a dormant, latent virus that then appeared in connection with the tumor. Alternatives one and three can now be set aside and as to the second proposition the current state of knowledge indicates that there are different kinds of infectious processes by which specific viruses cause cancer. However, the mechanism of oncogenic action may vary. As in the case of certain bacteria mentioned above the virus infection may become persistent and cause a chronic inflammation, which may lead to cancer. Further it can be that the virus infection by itself may cause severe immunosuppression, which in turn may allow the expansion of tumor cells emerging in the individual. An example of this has already been given above in the description of the situation in which an HIV infection caused the emergence of Kaposi's sarcoma. The effects of the virus can also be more direct with virus genetic material interacting with host cell DNA perturbing growth regulation genes. It should be noticed that this may occur both with viruses that have DNA as their genetic material, but also by the effects of RNA in retroviruses, which in operative terms function as a DNA-containing virus. Finally the virus that is the focal point of this and the preceding two chapters, Rous virus, as we have seen acts by carrying an oncogene of host cell origin. However, as already emphasized, this seemingly is *not* a mechanism of importance for the emergence of tumors in man.

The human virus first identified to be potentially oncogenic was HTLV-1. This retrovirus was isolated in 1980 by Robert Gallo and colleagues[7]. A parallel discovery of this kind of virus was made in Japan. The latter kind of agent was found to infect about one percent of the population in Japan's southernmost island, Kyushu. Similar infections were later observed in the Caribbean islands and in the aboriginal people of Australia. The uniqueness of the latter population should be emphasized. They reached Australia and associated islands as early as 55,000 years ago. Hence this infection must have been transferred from mother to child through more than one thousand generations. Infections with HTLV-1 lead to the development of adult T-cell leukemia in some 3-4% of the infected individuals. Data on the global occurrence of HTLV-1 is incomplete, but speculatively the figure may be 0.1%. The infection also may lead to a neurological disease in the spinal cord, spastic paresis caused by myelopathy. Molecular analysis demonstrates that the oncogenic effect of HTLV-1 is probably not due to insertion mutagenesis, as we have seen with experimental RNA tumor viruses. However, integration of virus genetic material in tumor cells has been demonstrated. Still the preferred theory is that it is virus products which activate the synthesis of cellular growth factors that initiated the oncogenic process. These growth factors in their turn provide a hyperstimulation of bone marrow cells leading to the appearance of hematogenous tumors. Even though HTLV-1 infects a sizable number of individuals, it does not appear justified to attempt to prevent this infection by immunization against the virus. There are other means of treatment of more general use in the management of leukemia that should be tried. Regarding the other types of HTLV it can be added that type 2 may also be oncogenic although with a smaller impact than type 1 and that there is no evidence for oncogenic effects of types 3 and 4.

Besides HTLV agents, viruses representing six different families have been documented to be involved in cancers in man. These are several different types of papilloma viruses, hepatitis B virus, hepatitis C virus, adeno-associated virus, certain forms of herpes virus — the already mentioned human herpes virus type 4 (Epstein–Barr virus) and in addition a type 8 — and a special kind of polyoma virus. Many other viruses have been candidate tumor viruses, but solid documentation has not been forthcoming. There is an agreement that 15–20% of all human cancers are due to virus infections as already mentioned. The identification of an etiological relationship between an infection with a certain kind of virus and the development of cancers is very important, since it implies that if the infection can be prevented by vaccination, and this is motivated by its epidemiological occurrence, there will also be a prevention of the emergence of cancer cells. This is well illustrated by a half Nobel Prize in physiology or medicine awarded in 2008.

Harald zur Hausen. [From Ref. 38.]

The Nobel Prize in this year recognized Harald zur Hausen "for his discovery of human papilloma viruses causing cervical cancer." As already mentioned this was the third time among the hitherto 110 prizes awarded in this field that the word "cancer" was used in the prize motivation. The many aspects of possible relationships between infection by particular viruses and development of cancer cells was extensively discussed by zur Hausen in his Nobel lecture[25] and also in earlier reviews[34]. Major progress has been made in studies of papilloma viruses since it was first suggested that they were correlated to cancer in man by zur Hausen in the early 1970s. A large number of different serotypes of papilloma viruses have now been identified, but only certain of them dominate among infections in the cervical tract. These are types 16 and 18. Since these infections were interpreted to be the cause of cervical cancer later in life, development of vaccines was initiated. It was found that the components of the virus shell could form separate particles that did not contain the nucleic acid. Hence they lacked capacity to cause infections but they retained an ability to induce an effective immune response. This new principle of vaccine production was recognized by the Lasker DeBakey Clinical Medical Research Award in 2017 to John T. Schiller and Douglas R. Lowy of the National Cancer Institute at NIH. These vaccines have now come into broad use in industrialized countries primarily to immunize girls in the early teenage years. Three different versions of the vaccine are currently used. They contain two, four or nine types of the virus, however always including types 16 and 18. It has later been found that also certain tumors in males, albeit at a lower frequency than in females, originate from papilloma virus infections. Whereas on a global scale roughly 50% of all infection-linked cancers in women are estimated to relate to papilloma virus infections, the corresponding figure for men is given as 5%. Still vaccination programs have progressively started to also include teenage boys. When fully implemented this general immunization, as in the step-wise introduction of rubella vaccines in the 1960s first to girls and later also to boys, will lead to a full epidemiological quenching of the circulation of virus. The irrevocable complete elimination of a cancer-causing virus, like the case of the 1968 eradication of smallpox, would represent a major achievement.

The second kind of virus infections clearly associated with development of cancers are agents that may cause chronic hepatitis. Three kinds of agents have been discussed in this context. The first one is hepatitis B virus. Like HTLV-1 it occurs in an increased frequency among Australian aborigines. In fact before Baruch Blumberg, the 1976 Nobel laureate in physiology or medicine, could demonstrate that what he had discovered was hepatitis B virus[6] he called it "Australia antigen." Effective vaccines against this virus, which cannot be made to replicate under laboratory conditions, have been developed. At first Maurice Hilleman and collaborators at Merck, Sharp and Dohme painstakingly retrieved viral antigen from the blood of persistently infected individuals, but later after the advent of the recombinant DNA technology it became possible to produce the critical virus surface antigen in yeast cells. The second agent of importance in this context is hepatitis C virus. This virus has a genome composed of RNA and is thus distinctly different from the larger DNA-containing virus of type

B. It has not as yet been possible to produce a vaccine to prevent the blood-borne infection by type C virus, but an effective antiviral compound has been developed. Because of the current high costs, the drug is at this time used only in selected cases. Possibly there is also a third situation when a virus may lead to the development of liver cancer. This is infections with adeno-associated virus type 2 (AAV-2). This agent needs to parasite on a concomitant infection by an adenovirus. In particular cases when the chronic infection in the liver does not lead to connective tissue scar formation, cirrhosis, the emergence of liver cancer may occur. Genetic material of AAV-2 origin has been detected in the tumor cells, but its role in cancer formation needs to be further studied.

Hepatocellular carcinomas occur worldwide but with a marked dominance in Asian countries. The viruses associated with this kind of tumor may spread iatrogenically with contaminated blood either in the form of blood transfusions or the use of unclean syringes by intravenous drug abusers. However, the way the virus originally survived in nature was by spreading from a mother with a chronic infection to her newborn child at birth. There is also a certain risk of becoming infected when practising unprotected sex. Recommendations for immunization take into consideration these particular types of spread. It is projected that a proper use of the vaccine in particular in Asian countries will lead to the eventual disappearance of the disease. This will also lead to protection against the development of liver cancer that the chronic irritation by the virus infection in this organ may lead to.

There are several other situations in which a certain virus infection may indirectly result in the development of a cancer. In these cases the cancer is a result of immunosuppression, already discussed above. The most extreme cases of such a situation were seen at the time of the development of patients with advanced infections with HIV leading to a complete wipeout of all immune functions. In such a situation Kaposi's sarcoma can develop, as also described earlier[7]. This kind of tumor is correlated to an infection with a certain type of herpes virus, referred to as type 8. The oncogenic potential of what was originally named EBV, now herpes virus type 4, was already discussed at length above. Hence it will not be returned to in this context.

Polyoma virus was briefly introduced in Chapter 1 as a member of the papovavirus family. Another member in the family, SV40, was for a long time a possible candidate for transformation of cells in humans, but eventually no such correlation was found as already mentioned in Chapter 1. It has however now been proposed that another related virus named Merkel Cell Polyoma Virus (MCV) might be the cause of so-called Merkel cell carcinoma which involves cells of neuroendocrine (hormonal) origin, a highly aggressive form of skin cancer. The role of the virus in this cancer is at the time of writing the object of ongoing studies.

Before leaving the field of viruses and cancer it deserves to consider the fact that attempts are made to develop viruses into means of destroying cancer cells. This is referred to as *virotherapy*. Once the existence of different viruses as the cause of specific infection was identified it was noted that infections by some of them coinciding with the development of certain tumors, in particular lymphomas and leukemias, might lead to regression of the cancer. This capacity to selectively destroy tumor cells by the *oncolytic* effect of a virus infection was confirmed in experimental studies and attempts are now being made to exploit selected kinds of viruses for oncolytic purposes in man. Sometimes the viruses used are genetically modified to amplify their preference for growth in the kind of cancer cell targeted. At the time of writing a herpes simplex type 1-based oncolytic vector has been approved as a cancer therapeutic in the U.S. Even my favorite virus, measles virus, which I studied in my Ph.D. thesis, is presently tested as a potential specific oncolytic agent[35]. One particular problem in this context is that the cancer patient may already have been sensitized to develop an immune response against the selected oncolytic virus. Various attempts are made to circumvent this problem.

There are a number of circumstances that may favor replication of a potentially oncolytic virus in the tumor cells. One is that the virus may exploit the preferential presence of receptors on tumor cells to destroy them. Another possibility is that the tumor cells have preferentially activated genetic control elements, which the virus may also take advantage of for its replication. Further it could be that the virus has a general preference for replication in the tumor cells because of their much higher metabolic rate compared to the surrounding normal cells. Finally it is possible that the virus infection may contribute to the expression of particular immunogenic structures or alternatively of certain radiosensitive sites providing opportunities for removal of the cancer cells by immunological reactions or by radiation.

The Hallmarks of Cancer

The field of cancer research has developed to become very complex. As repeatedly emphasized genetic studies have revealed the importance of a step-wise accumulation of genetic changes in the progression of the disease. Attempts have been made to summarize the state of the art of fundamental cancer research using the title of this section both in the year 2000[36] and as a follow-up also in 2011[37]. Originally the occurrence of six step-wise developments of cancer cell independence was proposed. The initial critical steps towards self-sufficiency were autonomy in response to growth signals, insensitivity to antigrowth signals and a capability to evade apoptosis. As the independently replicating cell managed to break out of its organ of origin three additional phenomena were defined to be instrumental to allow disease progression by metastasis. They were the formation of new blood vessels (angiogenesis), capability to invade unrelated tissues and finally unrestricted capability of cells to duplicate. In the later follow-up review two additional etiologically critical phenomena were added. These were reprogramming of the energy metabolism of cells and development of capacity to evade destruction by the immune system. As was emphasized above discovering malfunctioning genes in an ever increasing number has spurred a search for specific inhibitors and examples of an increasing number of successes, sometimes of a major nature, have been given. But let us once more by way of a summary revisit the highly complex field of cancer. This will be done using as a general background the fact that cancer represents the dark side of evolutionary developments.

The development of the field of cancer research after recognition in 1966 of Rous's and Huggins's discoveries by a Nobel Prize in physiology or medicine has been dramatic. Starting with an examination of the oncogenic potential of Rous virus and then an extended analysis of a host of experimental tumor virus systems allowed the identification of a plethora of oncogenes and antioncogenes. They turned out to be of cellular origin and their presence was revealed because they could be hijacked by the RNA tumor viruses studied. The focus shifted from examination of viral genes to normal cellular genes. A lot was learnt about the physiological control of cellular development in embryogenesis and in maintaining homeostasis in different kinds of tissues in the body of the adult individual. Furthermore, major insights were gained into how cells communicate with their surroundings and how different kinds of signals were transmitted into and inside a cell. Many of these fundamental advances in an understanding of the life of cells had a general importance for the interpretation of the abnormal behavior of cells in the process of cancer formation. Understanding the normal physiological events is a prerequisite to interpret dysfunctions that may lead to disease — pathology is the distorted mirror image of physiology. This was wisely understood by Alfred Nobel when he decided that there should be a prize in *physiology or medicine*. Hence, as illustrated above, to date there are only four Nobel Prizes, which have been awarded with a motivation that includes the word cancer, but there is a host of such prizes that widen our understanding of how cells control their life of division and growth that have been recognized. Not infrequently when journalists inquire about the medical importance of a prize for discoveries of fundamental functions of cells, the reply has included a reference to a possible importance for understanding the problem of cancer.

The disease cancer needs to be viewed in an evolutionary perspective as repeatedly emphasized. It has now become clear that it has its origin in changes in the genetic material of cells. Hazardously a certain cell starts to compete with neighboring cells in an abnormal way and replicates at their cost. This probably happens frequently in our body, but because of the presence of a host of different safe-guarding mechanisms the misbehaving cells are taken care of and eliminated. On rare occasions the rebellious cell may survive and continue to divide forming a tumor. Still at this stage the story may end well, the tumor may remain benign as in the case of the common wart. However, in other cases further stepwise genetic changes give the cell incrementally increased opportunities to replicate on behalf of surrounding cells. These steps in the development of a certain cancer can be monitored by use of molecular genetic analysis. Modern technology allows an amazing collection of data on the genomic changes in tumor cells. The technique of so-called deep (genomic) sequencing serves as a basis for what is currently referred to as *precision medicine.*

A cancer may develop taking many different courses depending upon the tissue origin of the parental cell. A critical step in managing cancer is an early diagnosis by use of as sensitive means as possible. Starting with conventional X-ray, a phenomenon identified by the first Nobel Prize in physics, there has been an impressive development of our abiity to see through the intact body. Ultrasound is another convenient technique that can be remarkably informative. It was never recognized by a Nobel Prize in physiology or medicine, but the two other techniques that have completely revolutionized medicine were rewarded. These were the prizes in physiology and medicine in 1979 "for their discoveries of computer assisted tomography" recognizing Allan M. Cormack and Godfrey N. Hounsfield, and in 2003 "for their discoveries concerning magnetic resonance imaging" to Paul C. Lauterbur and Peter Mansfield. Attempts have been made to find biochemical markers of emerging tumors, but this has been less successful.

Once a tumor has spread, metastasized, the situation becomes much more complex and urgent. One of Rous's important contributions was to emphasize the possibilities for appearance of different stages in cancer development, the progression. This has become even more important to evaluate when we, by the already mentioned use of modern tools for characterizing genomic changes, have come to appreciate the enormous dynamics of the progressive evolution of cancer cells, with their unique characteristics reflecting the organ in which it originated and its evolution since then. The complexity encountered in the studies by deep sequencing of the DNA of tumor cells is almost overwhelming and still we have to learn to exploit this information in the best possible way for the patient. The enormous advance in our understanding of the role of oncogenes, tumor suppressor genes and different immunological mechanisms involved in the normal attempts of our bodies to fence off the disease is critical. Knowing about the different signaling systems in cells, the way our DNA is safeguarded to remain undamaged, the way the body mobilizes the many arms of the immune system to curtail the spread of cancer cells has given us a completely new armament to treat cancer and it can be predicted that much more will come.

The techniques of physical removal of tumor cells by surgery or irradiation will always have a place in the treatment of cancer. However, as illustrated by a few examples in this chapter, there are an increasing number of new modules of chemotherapy and immunotherapy being introduced at present. The new information that is forthcoming could readily fill a fourth chapter in this book. It is a truism to note that we use these drugs to treat a disease which has its origin in cells that unintentionally use the entire means put at their disposal to challenge their surrounding of cells and body fluids exploiting the powerful tools of evolution. Random mutations will always put sharp weapons in the hands of the cancer cells. It seems that a cancer often has an extra arrow left in the quiver. However if we reverse the perspective we can note that to develop resistance to one drug may not be that difficult, but if we add a second drug all of a sudden it becomes much more difficult for cancer cells to evolve a resistant variant. Chemotherapeutic drugs preferably therefore may be used as a part of a battery. The possibility for escape of a resistant variant of the tumor cell should be less of a problem when immunotherapeutics are used since they generally rely on multiple interactions. However, monoclonal antibodies have a more narrow specificity but there are still a number of examples of their successful application in certain forms of cancer, like melanomas. It can be added that a vaccine induces such a broad immune response that it will never be possible for the virus involved to mutate to become resistant to this response. The spontaneous occurrence of different serotypes of a certain virus is a matter in itself which requires a polyvalent vaccine for full protection as in the case of papilloma vaccines. Fortunately it is rare that a certain virus has a capacity to change its surface characteristics to not be reactive to immunity produced against an earlier strain of the agent, either derived from an infection by the virus or by an earlier vaccination. The full and long-lasting protective efficacy of a virus vaccine depends on the antigen stability of a particular infectious agent.

Modern medicine to a major extent is dependent on the success which the medical community has in transferring information about the choices of lifestyles to the public at large. It turns out that they have a major impact, not only to quit smoking, but also in many cases of other behavioral matters, the food we eat and the exercise we take, etc. Applying these options to the way we live brings major benefits, but still a shadow of the risk for the appearance of cancer will remain. Certain cancers like, for example, those emerging from the pancreas are very difficult to identify at an early stage and the long term prognosis is bad. More and more there will be improved opportunities, not only to identify a cancer at as early stage as possible, but also once it has been established to find the right balance between different treatment modalities — surgical, radiological, chemical, immunological and possibly also the use of oncolytic viruses. In this chapter only a few selected examples which have been highlighted by Nobel Prizes in physiology or medicine recognizing discoveries of central cell physiological mechanisms have been described. It needs to be noted that cancer, the foe from within, remains a major health challenge to a considerable proportion of the population, increasingly with longer life spans. And still knowledge gives power. The more we can get an insight into the immense complexity of the metabolism of cells and in particular the intricate mechanisms of controlled cell division, in part unique to each kind of cells, the better the opportunities are for a cure. The originally blunt instruments of physical removal by surgery and by irradiation, although still of great value in their proper context, are being progressively replaced by chemical and immunological scissors and knives. All this will help to improve our chances of a positive outcome in the lottery of life. There are reasons to be optimistic that in the future the diagnosis of cancer will more and more have a reduced dramatic impact.

It is now time to move from the more speculative forward-looking ideas on how to manage the very complex disease of cancer to the physiological and chemical processes that allow us to see. We will start this intellectual excursion on firm ground.

Chapter 4

The Rock Foundation of Nobel Prize Developments

SEEING IS BELIEVING
IN DAYLIGHT OR PENUMBRA
COLORS ADD QUALITY

The world we live in regionally may give the impression of providing a stable ground, terra firma. However, this impression sometimes needs to be modified depending on where we are on the globe and also which reference of time we apply. One of the first problems to be approached by the newly created Royal Swedish Academy of Sciences in 1739 was the question of the progressive alterations of the coastal profile of the country. Significant changes in the water level could be observed even within a lifespan of a human. Valuable natural harbors in the archipelagoes could turn unusable in a few generations, affecting conditions of fishing. The question was if the land was rising or if the water level was being reduced? When it was discovered that the changes noted clearly varied along the extensive Swedish coast the answer was obvious. It was the land that was rising. This was soon deduced to be due to the pressure that the heavy inland ice existing some 10,000 years ago had applied. Not only did the ice suppress the rocks, in some places it also polished them to a silk-like texture. In some areas the height that the land has risen in recent times is quite impressive. At the island of Blidö in the northern part of the Stockholm archipelago, where my family has its summer houses, it is about 60 centimeters in 100 years. Thus what were two rocks that barely reached above the water level when I was a child has now developed into an island with grass and bushes. It can be added that the expected rise of the water level because of global warming will have essentially no effects in some parts of the east coast of Sweden because it is compensated for by the land elevation.

Besides this vertical change in the level of the land we have learnt that there are also horizontal changes in the relative position of all land masses. They are dramatic but play out over such very long periods of time that they remain invisible to the human eye. Thus just to give only one example it can be mentioned that the South American continent and the African continent, which have a complementary coastline profile because they were originally one land mass that broke apart, progressively become more distant from each other at about the rate at which our finger nails grow. The introduction of new techniques for geomagnetic measurements in the mid twentieth century — demonstration of magnetizations in rocks and the pattern of magnetic field variations over the ocean ridges — revealed that over extended periods of time there had been dramatic shifts in the shape and location of continents. It was the German Polar researcher Alfred Wegener who introduced the hypothesis of plate tectonics, the existence of continental drifts, in the early part of the 20th century. The supporting evidence he brought forward was the remarkable complementary fitness in shape between certain continents, suggesting that they, at some earlier stage, had formed one continuous land mass. He also brought additional evidence in the comparison of fossils at possible break points of continents and also the nature of mountain ridges proposed to be broken up during the separation of continents. His scientific colleagues at the time considered the idea of continental drift as wild fantasy and Wegener was never to experience being proven right eventually. He died at the age of 50 during an expedition to explore the Greenland ice in 1930.

When two continental plates collide, mountain ridges can be formed. One such mountain ridge that interested Wegener was the one that was later named the Caledonian ridge. It was formed by the collision of two plates some 400 million years ago and then broken up into pieces. In the present world these pieces include the Scandinavian mountain ridge, the highlands of Scotland — Caledonia was an old Roman name for Scotland — Greenland and the Appalachian mountains in North America. The Caledonian ridge originally was a part of a supercontinent Laurasia which later fused with Gondwana to form the all-encompassing continent named Pangea which existed some 300 million years ago. Remarkably it has been possible to trace the repeated formation and separation of land masses way back into the prehistory of Earth. Thus prior to Gondwana at the time of about 1.8 to 1.25 billion years ago there was another supercontinent called Columbia, which arose by a gigantic fusion of several large land masses.

One of these masses, named Baltica, involved what is now north-eastern Europe and included a very old mountain ridge, formed already some 1.8–1.9 billion years ago. This is the Svekofennidian mountain ridge stretching from southern Finland westwards towards Sweden. At about 1.65 and 1.5 billion years ago granites of different compositions were introduced into the Svecofennidian rocks. Granite is a collective name for granular rocks of volcanic origin. The name derives from Latin *granula*, a grain. The rocks have different compositions and colors. The oldest of the intrusions have been found in southern Finland, with a decreasing age distribution over the islands of Nagu, Korpo and Åland to central Sweden in the west. For those of us who have had the privilege to sail waters surrounding these islands there remain vivid memories of the extraordinary colors at sunset of the red rocks represented in these archipelagoes, the rapakivi granite. The Finnish name rapakivi is derived from *rapa* to grease and *kivi* rock. This special form of granite is coarse grained and often has a porphyritic appearance. In April 2018 I had the privilege of spending three weeks at Villa San Michele at Anacapri to work on this book, including the present chapter. One of the unique hallmarks of this villa is

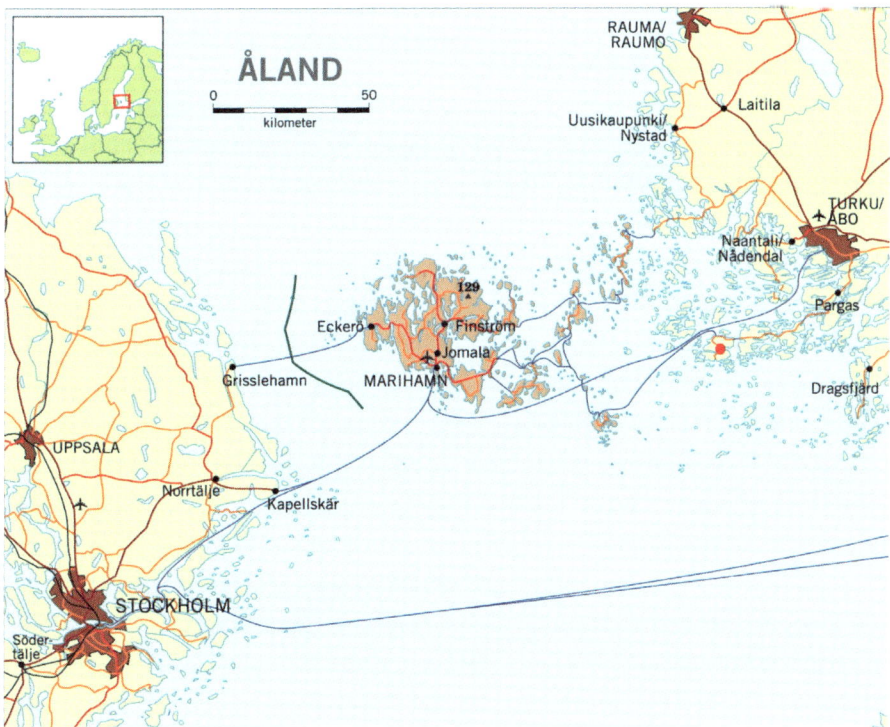

Map of the band of arcipelagoes between Sweden and Finland highlightening the location of Vikminne (the red spot) on the Nagu Island.

The Rock Foundation of Nobel Prize Developments 141

an Egyptian sphinx, dating back more than 3,000 years, which has a bird's eye view over the magnificent mountainous island of Capri. It is made of granite.

A Harmonious and Challenging Upbringing

On the above-mentioned Finnish islands there were some thrifty individuals, traditionally surviving on farming supplemented by fishing, who in the nineteenth century realized that involvement in the expanding international shipping industry at the time was an alternative way of sustaining their living. One of them was Jeremias Michelsson on the

The author at the sphinx at Villa San Michele. The inset shows the fine structure of the granite from which the sculpture is cut.

island of Korpo. He developed a successful business using sailing boats that he and his fellow islanders built. In these boats he circumnavigated the world as captain. The inhabitants of these islands belonged among the minority of Finnish people that were Swedish-speaking and the tradition, as in Sweden, which in 1809 had lost Finland in the war with Russia, was to use "son"-names. Thus Michelsson meant the son of Michel. This tradition was not liked by the Russian Empire rulers of the Grand Duchy of Finland. When making a choice for his new name Jeremias went to the local priest. He said that he wanted a hard name and the priest then said "Granit is hard." Thus the family name was changed to Granit. In Swedish there is an equal stress on both syllables and not as in the English granite on the first one. It would seem that carrying a name like this, as in the case of Jeremias' grandson Ragnar, the main actor in this chapter, might put a demand on personality traits. In fact in the dictionary the English granite has the alternative meaning "great hardness or rigidity." We will see how this name might have suited the Nobel Prize candidate to be presented. Could it even be that Ragnar was "engrained" to become a successful scientist and a Nobel Prize recipient? Be that as it may, but he did not engrave his Nobel Laureate status on his tombstone, which of course is made of granite.

However, in 2017, fifty years after his prize, a bronze plaque was fitted on the tombstone to note his status as a Nobel laureate.

By way of introduction it might be emphasized that this much endowed person, born in 1900, in the parish of Helsinge, Finland, as the oldest son of the Crown forester Arthur W. Granit, developed a range of different talents. It could be interjected that geologically speaking the Nobel Prize activities in general are on firm ground, and that this also came to apply to one of the 1967 laureates in physiology or medicine, Granit. He was a qualified humanist and writer. Already in the early 1940s, after having been immobilized by damage to a knee caused by a bicycle accident, he summarized the charm and challenge of being a young and enthusiastic scientist in a book called *Ung Mans Väg till Minerva* (The Road of a Young Man to Minerva)[1]. Minerva was the Roman Goddess of wisdom. Some 40 years later, at the age of 83, Granit wrote his autobiography, entitled *Hur det kom sig. Forskarminnen och Motiveringar* (How it came to be. A researcher's memories and motivations)[2]. His rich life has also been summarized by one of his younger collaborators and published as memoirs by the American Philosophical Society[3], and also by his successor as professor at the Karolinska Institute (KI) Sten Grillner, published as a biographical memoir by the Royal Society[4].

Ragnar, or Raggen, which became his nickname, grew up in very harmonious conditions spending his summers at a place called Vikminne, literally the "entrance of the bay." The Swedish word "minne" normally is translated as memory, but in this special case it is synonymous with "mynning" which translates as "estuary." Vikminne is located at Korpo, a large island in the west of Åbo (Turku) land archipelago. This village had been the home of ten preceding generations of farmers, who because of their chosen habitat also profited by the fruits of the sea. This environment provided a rich summer holiday environment for Ragnar. Fishing and sailing with cousins and other friends filled the days. The contacts with his energetic and dynamic grandfather were also of particular importance, not least since his own father was often absent for longer periods of time prospecting in the northern parts of Sweden and other

Ragnar Granit, the skipper. [From Ref. 2.]

distant places. Ragnar remained a qualified sailor through most of his life as we shall see. He also matured intellectually at an early age and began to read and write at the age of four. Later in life he explained that the core values he carried along from his upbringing included identity, loyalty and affection. Ragnar's father was a forester, as mentioned, and because he developed well professionally establishing a firm dealing with silviculture and forest products, the family were able to settle in the capital of Finland, Helsinki. His parents ensured that he received a high quality classical schooling.

In the gymnasium, the high school equivalent, Latin, Greek and the Arts dominated the teaching. He did well and the school taught him, according to his own remarks, logical reasoning, but he learned relatively little about the sciences. Just as his studies were nearly finished, in 1918, they were interrupted. After the October revolution in Russia, Finland declared its independence on December 6, 1917. Ragnar became involved in the traumatic civil war that ensued. He made some major personal contributions and was decorated. One of his classmates died and two were wounded in the war. During these challenging events he carried along in his backpack the bestselling book by Ernst Haeckel called *The Riddle of the Universe*. In 1919 he was able to return to finish his studies and start at the University of Helsinki the same year. His chosen disciplines were theoretical and practical philosophy, seasoned with some biology and chemistry. He had a much respected teacher of philosophy, Eino Kaila, who later became professor of this discipline, and they retained friendly contacts for decades. In his autobiography[2] Granit referred to Kaila as belonging to the category of humans driven by ideas and not by power, a categorization introduced by a philosopher at Lund University, Hans Larsson. However, Granit qualified his attraction to people carried by ideas by stating that their urge needs to be combined with a critical approach to their chosen discipline. It was the exposure to Kaila's simplified experimental approach to studies of the sense of seeing that initiated Granit's interest in eye physiology. During this time his father provided resources that allowed him to study abroad. He made his first trip to another country and he chose England. He liked the early contact with this country and it came to mean a lot to him throughout his life. In school the first foreign language was German, but already at an early stage Ragnar also started to learn English.

In 1923 he received the correspondence of a bachelor's degree having studied the subjects of philosophy, psychology, aesthetics and chemistry. As regards his tentative interest in understanding vision and sensory processing he realized that this was not possible within the framework of the psychology discipline. He became skeptical about the school of "gestalt psychology," which he personally learnt to know about during a visit to Germany, and also about Freud's theories of psychoanalysis and dream interpretations. Natural sciences appeared to be the way to go and hence he enrolled as a medical student. Although a humanist he was less attracted to do clinical medicine. During the studies he had the privilege of interacting with the talented Finnish physiologist Carl Tigerstedt and he also had some incidental contact with Carl's very famous father Robert, a world authority in the same field. He learnt a lot about the culture of science in general, but he had to manage the problem he had chosen on his own. There were many years of learning and of hard-won experience, but Granit kept his focus. He was very impressed by the early studies of the structure of the retina, by the Spanish histologist Ramon Y. Cajal, the recipient of the 1906 Nobel Prize in physiology or medicine[5]. His description of the retina, the innermost light sensitive membrane of the eye, as a "true independent nervous center" appealed to Granit. Studies of the retina might be a prelude to gaining insights into general brain physiology. In 1927 he presented his PhD thesis. It was written in German and the title was "Farbentransformation und Farbenkontrast. Experimentelle Beiträge zur Theorie der Transformation (Transformation and contrast of colors. Experimental support for the theory of transformation)."

Training to Become a Neurobiologist

Early on Granit developed very important international contacts. He successfully befriended the foremost authority in the field of research on the central nervous system, Charles S. Sherrington in Oxford. He was to stay for six months in 1928 in his laboratory and early on was caught by his magic wand. This turned out to be decisive for his development and he returned for a more extended stay four years later. During this time Sherrington was to receive his shared 1932 Nobel Prize in physiology or medicine at the mature age of 74, as presented at length earlier[5]. Although Sherrington focused on problems very different from those Granit wanted to study, being a part of his laboratory was a great learning experience. Granit aimed at measuring excitatory and inhibitory processes in the retina, something he was successful in doing in later studies in the mid-1930s. The stimulus he received from interaction with many other scientists working in Sherrington's laboratory was of considerable importance for his development. Among those he befriended should be mentioned in particular John Eccles, the forthcoming recipient of a shared Nobel Prize in physiology or medicine in 1963, also presented in detail earlier[5]. Their friendship was initiated already in 1928 and was to remain throughout their long lives. They exchanged many letters starting in July 1928 and throughout May 1985 discussing diverse aspects not only of science but also of life in general. Half way through their correspondence "My Dear Granit" was changed to "My Dear Ragnar." There were many personal encounters between their families in different corners of the world, including also Granit's summer home at Vikminne. Granit later came to have a decisive influence on events preceding the prize to Eccles as already described[5].

John C. Eccles (1914–1997), the recipient of one-third of the Nobel Prize in physiology or medicine in 1963. [From Les Prix Nobel en 1963.]

There is nothing in Eccles's correspondence with Granit indicating an undue influence in any direction on the materialization of their respective Nobel Prizes in 1963 and in 1967. In one letter Eccles shared the embarrassing situation when the press in Australia had got wind of the rumor that he together with Horace Magoun was expected to receive the 1960 Nobel Prize in physiology or medicine. As previously described[6] they had a strong minority vote but the prize finally went to Peter Medawar and MacFarlane Burnet, another Australian. Also the leak in 1961 that the prize would recognize Eccles, Anthony Hodgkin and Andrew Huxley, which was the majority recommendation by the committee, eventually overruled by the College of Teachers selecting Georg de Békésy, was noticed. When Eccles finally did receive the prize in 1963 his comments to Granit were truly spontaneous. He wrote "You could not possibly have arranged a more wonderful climax to my scientific life, than to be linked in a Nobel award with Hodgkin and Huxley." Later on he wrote "…because of the wisdom and skill expected by you scientists in Stockholm, so that you have become the acknowledged arbiters of scientific standards for the whole world." And of course the year after his prize, Eccles, as in the case of the 1960 prize to Burnet, was elected the Australian of the Year.

Returning to Granit, it may be noted that 1929 was to be an important year for him. He became an associate professor — "docent" in Scandinavian countries — and he married. His future wife in what became a 60-year-long marriage was Marguerite (Daisy) Bruun, with whom he had been acquainted for a number of years. The young couple sailed for the U.S. on their honeymoon since, upon Sherrington's recommendation, Granit was to take up a position as visiting scientist at the Johnson Foundation in Philadelphia. There he was to interact with two scientists of major importance for the development of his future work. One was H. Keffer Hartline, his future co-laureate almost 40 years later, and the other William Rushton, a British scientist whom he came to collaborate with much later in a number of studies. It should be noted that at this time the U.S. had not as yet developed into the dominating nation in biomedical sciences. This happened to a great degree after the Second World War, following which Nobel Prize recipients in the natural sciences from this country came to dominate. What Granit enjoyed in particular in the U.S. was the atmosphere of high ambitions, belief in progress and generosity.

Granit was inspired by work already performed at the time by Edgar D. Adrian, the one generation younger co-laureate of Sherrington, who had managed to make recordings of single cell electric activities from the isolated ophthalmic nerve in the eel. However, it took time before he himself was able to start using the technique developed for measuring nerve impulses. In these studies he had shown a sort of spatial summation from measurements of different parts of the retina in experimental animals. In later studies he managed to make similar studies with the human retina, establishing the important interactions of excitatory and inhibitory mechanisms in this extension of the brain. This work was developed when Granit returned to Sherrington's laboratory in 1932–33 for 18 months on a Rockefeller scholarship. It required the development of an electronic amplifier that could be used in this organ, measurements referred to as electroretinogram (ERG). This kind of study had originally been introduced as early as 1865 by the professor of physiology at Uppsala University, Frithiof Holmgren.

Granit's work progressed well and he grew to be an authority in this field. A seminal finding was made in 1934 when Granit could demonstrate that exposure to light elicited electrical impulses in certain conditions but in others instead could inhibit them. This so called *lateral inhibition* later turned out to be a very important phenomenon. As we shall see, this together with the physiological minute fluctuations of the eyeballs is critical for the ways we form a picture by use of signals from light-activated receptor cells. It was time for him to become professor and chairman, which at that time, with a limited number of positions available, could sometimes be cumbersome and loaded with potential inter-collegial conflicts.

Fundamentals of the Process of Vision

Before moving on it would seem appropriate to introduce some fundamental concepts of relevance to the way we see. Charles Darwin already noted in his 1859 book *Origin of Species* "To suppose that the eye, with all its inimitable contrivances... could have been formed by natural selection, seems, I freely confess, absurd in the highest possible degree." Those who do not believe in evolution, the creationists, usually end their quote here, but Darwin continued his text in the following way "Yet, reasons tell me, that if numerous graduations from a perfect and complex eye to one very imperfect and simple, each grade being useful to its possessor, can be shown to exist... then the difficulty of believing that a perfect and complex eye could be formed by natural selection, thought insuperable by our imagination, can hardly be considered real."

Later, accumulated insights into the structure of visual sensory organs containing photo-receptive proteins (opsins) occurring in different species have revealed that eye structures may have several independent evolutionary origins and that structures of this kind may have appeared even before brain structures. Primitive seeing organs have been found to occur already in animals that developed some 540 million years ago. Development of different kinds of opsin molecules, sensitive to light of different wavelengths, has allowed the emergence of color vision. This capacity developed relatively early during evolution, some 300–400 million years ago, and in a more primitive form probably even earlier. As we shall see we humans can normally distinguish three kinds of color spectra, unless we are affected by the color-blindness carried by genes in the X chromosome that commonly may affect males. In fact most species of mammals only have such a dichromatic kind of vision. In summary there is a wide range of eye structures in different forms of life. This variation is influenced by whether life exists on land or in water and whether the animal displays the major part of its active existence during daytime or nighttime. Many animals display nocturnal activities simply because their opportunities to survive are improved under these conditions.

Our eyes are highly complex organs and in their rear part there is a membrane structure referred to as the *retina*, from Latin *rete* meaning "net." This structure is richly supplied by blood vessels and contains a number of different specialized cells including the light-sensitive cells. They are of two kinds, the *rods* and the *cones*. In the 1990s a third kind of photoreceptor cells were identified, named photosensitive ganglion cells. They support pupillary reflexes and circadian rhythms. In the human eye there are about 120 million

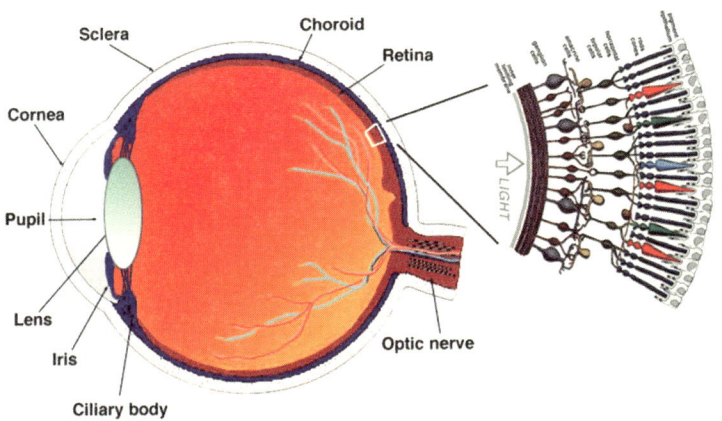

The anatomy of the eye and the structure of the retina.

rod cells and 6 million cone cells. In other species this relative representation varies extensively, depending upon if the animal is nocturnal or diurnal, as mentioned. The rods and the cones are in contact with bipolar cells which in their turn connect to ganglion cells. It is the latter kind of cells which collect signals to be transported via the optic nerve to the brain. The retina serves like an extension of the brain as a center for low-level integrated processing of visual signals. During transport of signals from the cones or rods to the ganglion cells there are additional cells such as the horizontal and amacrine cells which can restructure and concentrate the signals transmitted by the axons leading to the brain. The number of axons is only about one percent of the number of signal-collecting cells, highlighting the extensive integrative processing that takes place locally. Once the signals have been transmitted to the brain they spread over about one-half of this organ, allowing intermediate and high level processing. Paradoxically, generally the effect of light absorption is not to increase the production of a neurotransmitter substances by cells in the cerebral cortex, but to reduce it.

What is often overlooked is that the different successive evolutionary steps have led to the retina having, what appears in a technological perspective, to be a backwards arrangement. Thus after the light has passed through the cornea and the lens in our eyes, it hits a layer of nerve cells, the ganglion cells, then a layer of different coordinating cells, the amacrine, horizontal and bipolar cells until it finally hits the light-sensitive rods and cones. This arrangement is the reason for the fact that the nerve extensions from the ganglion cells which are accumulated to pass through all these layers at a special place in this inverted structure to reach the brain lead to the existence of a blind spot in our retina. Just to exemplify that evolution may lead to different solutions in the development of organs for seeing it can be mentioned that cephalopods, like squids and octopuses, have the reverse arrangement leading to the nerves extending from the back of the retina directly to the brain, obviously a more rational arrangement. These differences only serve to remind us that evolution meanders mindlessly.

The light absorbing parts of the rods and the cones contain the opsins, which are pigment molecules collectively named *retinal*, a derivative of vitamin A, as we shall see. This complex is referred to as *rhodopsin* when it occurs in rods and *photopsin* in cones. There is, as has already been mentioned one kind of rods but three kinds of cones, which respond differently to wavelength

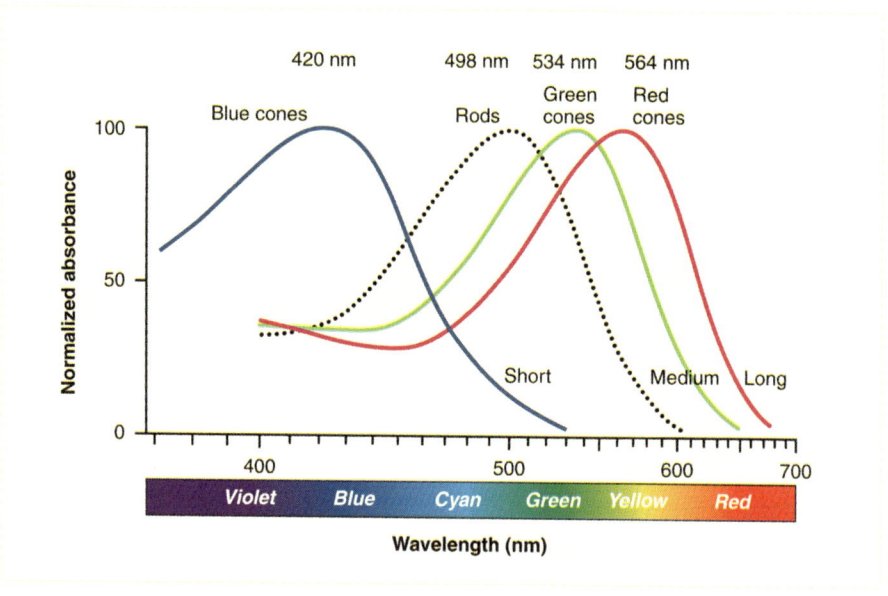

Wavelength sensitivity range of the three kinds of cones and of the rods.

and intensity. The rods are activated both in bright light and also at very low, so-called *scoptic* light. The latter term comes from Greek *skotos*, meaning "dark" and *opia*, meaning "a condition of sight." They retain activity after dark adaptation, a process which takes about thirty minutes, and lack inclusion of colors. The major role of the rods is to identify contrasts by recording the shades of gray. By use of slight regular movements of the eyes, so-called *saccades*, the image retrieved is sharpened. The framework of the complete picture is built up by different qualities of the signals collected. The eye does not operate like a camera and it is the processing in the centers in the brain that builds up the impression of an integrated image. In contrast to rods, cones are primarily activated in brighter — *photoptic* — light. Since they are of three different kinds they can distinguish lights of different wavelengths, predominantly in the blue, green and red parts of the spectrum. Integration of this information allows an almost limitless identification of different colors. The existence of the range of colors was systematically examined by Isaac Newton using prisms. Already in the early 1900s Thomas Young and Hermann von Helmholtz postulated a trichromatic theory of color perception.

It was a long evolutionary journey towards the color vision of modern man. Originally mammals, presumably living a burrowing existence and being active during the night, only had a bi-chromatic impression of their surroundings. Their retinas were sensitive to ultraviolet (UV) light and to the red color. Insects, fish and birds still use the UV spectrum of light. However, in mammals this had already changed dramatically some 30 million years ago due to a number of genetic changes. Many vertebrates had now developed four classes of opsin genes, which allow them to see the full spectrum of colors in visible light, but no longer identify UV light. Still there are variations. For example dogs and mammalian farm animals often have less diversified vision capacity. They use a two-color receptor system which limits their capacity to see orange and red. The brain of the group of mammals that includes us needs to use the latter colors to allow the identification of useful fruits. It was the origins of this sense of color vision that Granit's fundamental discoveries were to consolidate. Before leaving this presentation of the fundamentals of the mechanisms of seeing it might be added that we use our eyes — *des yeux inquisiteurs* — to understand how we can see, like we use our brain in general to think about how we think. It is like uroborus, the serpent that is eating its own tail, first introduced already in Egyptian iconography. Let us now return to Granit's development as a scientist.

The Rocky Road Towards a Stable Academic Position

Finally, in 1937 Granit was appointed professor of physiology at the Medical Faculty at Helsinki University. This was not uncontroversial. There was one professor in the subject who was Finnish-speaking and then Granit who spoke Swedish. The impeding challenge to Finnish independence, gained in 1917 from Russia, by the emerging threat of a war accentuated the development of pure Finnish movements. Swedish-speaking professors were less well seen and Granit did not have his own department to manage and he was not provided with any material resources from the university. His independence was secured by money from the Rockefeller Foundation that he managed to continue receiving.

During this time he made another major advance in his studies. Already since the early-mid 19th century it had been postulated, as already mentioned, that the eye should distinguish colors by use of three modules. This postulate was referred to as the Young–Helmholtz three-color theory. It had been consolidated in 1931 by Selig Hecht in mathematical studies of conditions of color blindness. There was a need to be able to measure the electrical activity of individual cells to further develop Granit's pursuits. A technically proficient younger colleague of his, Gunnar Svaetichin, who we will meet again later, managed to develop microelectrodes that could be used for this purpose. The electrodes were formed by pulling out glass tubes with silver cores in a flame to fine tips. Later the silver was exchanged for platinum. The signals recorded were registered as response of spikes heard in a loudspeaker. This was used to adjust the position of the electrode for optimal signal recording. Their first results of studies of the eyes of frogs gave encouraging results. There seemed to be three main regions for color recognition as would be expected. The basis for our capacity to distinguish colors appeared to have been experimentally verified! This original finding was expanded by use of many different kinds of animals, some of which have preferentially either cones or rods. Granit was invited to present the new data by Professor Gustaf Göthlin at Uppsala University and they were published in a local journal[7].

At about this time Göthlin was due to retire and Granit expressed an interest in succeeding him. However, again there were some complications. Swedish-speaking professors from Finland could not apply for positions in Sweden since that required Swedish citizenship. His personal situation was severely threatened when post-revolutionized Russia, the Soviet Union, once again tried to gain hegemony over the young nation of Finland by starting the "winter war" in 1939. For a second time Granit became involved in defending his nation, but this time as a civilian regional physician, with an associated military rank, providing help to island inhabitants in his home archipelago skiing over frozen waters. His personal and professional situation became ever worse. He was considering moving to Oxford, an academic environment he liked very much, but then he received an invitation to become Director for the

Laboratory of Ophthalmology in the Massachusetts Eye and Ear Infirmary at the Harvard Medical School. This of course was a very flattering invitation. Tickets for travel to the U.S. were arranged, but in the end this move never came about. An unexpected opportunity to remain in Scandinavia came up. It was due to an initiative taken by Carl Gustaf Bernhard together with some colleagues from the Karolinska Institute in Stockholm. In 1939 Bernhard had been a visiting graduate student with Granit on a grant from the Royal Swedish Academy of Sciences. Bernhard himself has described these events in his autobiography *Huset på höjden* (The House on the Hill)[8]. They have also been partly described by Granit himself in a book in 1960 celebrating the last 50 years of development of, at the time, the 150-year-old Karolinska Institute[9].

A Change of Homeland and Post-War Developments

The young and enthusiastic Bernhard contacted leading individuals at the Karolinska Institute and in the financial establishment of Sweden, including the head of the Knut and Alice Wallenberg Foundation, Marcus Wallenberg. Bernhard was successful and it became possible to secure the money needed. Hereafter developments moved swiftly. The College of Teachers, headed by the Vice-Chancellor Gunnar Holmgren and supported by the representatives of the Institute in a wide sense of the discipline of neurophysiology, Göran Liljestrand, Nils Antoni and Ulf S. von Euler, took the decision to support the initiative on June 4th; His Majesty the King gave his support on June 12th; the University Chancellor on June 19th and the following day the Vice-Chancellor of the Karolinska Institute invited Granit to head a newly-created Department of Neurophysiology established by financial support from the already-mentioned foundation. This was to be a rare example of rapid processing of a central academic matter. It took place at a turbulent time in the European war arena.

The young Carl R. Skoglund (1914–1990) and Carl-Gustaf Bernhard (1910–2001). [From Ref. 8.]

On June 14 German troops had occupied Paris. The first period of support was provided for five years and there was also important additional support by the Rockefeller Foundation. Granit's move to Stockholm had a major impact on the development of the discipline of neurophysiology at the Karolinska Institute as emphasized earlier[5,6]. The only qualified neurophysiologist at the institute at the time was Yngve Zotterman. He was stimulated to a new phase of creativity focusing on the neurophysiology of pain, taste and cold/heat. Also for Bernhard himself and his close colleague Carl Rudolf Skoglund (see p. 153), still in their early post-doctorate phase, it meant a lot to have an authority like Granit as their scientific colleague.

Yngve Zotterman (1898–1982). [Courtesy of the Karolinska Institute.]

The move to Stockholm allowed Granit to remain a part of Swedish culture, but now with Sweden as his home country, and also to retain links to the islands inhabited for a long time by his ancestors. In late August Bernhard and another professor from the Karolinska Institute, Adolf Lichtenstein, traveled to Åbo (Turku) and sailed from there with Granit in his boat via Vikminne in Korpoström to Stockholm. They had some information about how to avoid the existing underwater mine fields, but later learnt that they, fortunately without any incident, had crossed their periphery. Since the Stockholm archipelago was not available to non-Swedish citizens at the time, a military representative joined their boat when it reached Swedish waters. However, Granit soon became a Swedish citizen and could sail these waters on his own. He was in fact also called upon to perform Swedish military service. The pace of development of his science for natural reasons was somewhat slowed down by the move to Sweden and the shadow of the war. The fate of relatives and friends, in Finland during different phases of the war continually caused concern.

The end of the war in 1945 came as a huge relief. Granit and his family could now sail the Swedish-Finnish waters more freely. At the end of this year Granit traded his rather spartan, but very sail-worthy, 6 M R-yacht for a larger boat also made of wood. It was of the kind "stortumlare" (great porpoise) constructed by one of the most well-known boat designers in Sweden Knud H. Reimers. The boat was bought from Tage Hedqvist, who became the curator of the unique art museum, the Thielska Gallery. Several decades later, starting in 1979 onwards this museum and its curator at the time, Ulf Lindhe, hosted some fantastic dinners allowing the first encounters between the Nobel Committee and the prize recipients in physiology or medicine of the year as described earlier[10]. Granit came to love this boat, named Alone, and each late spring and early autumn he sailed it from Stockholm to his beloved homestead of Vikminne on Korpu and then back again. Sailing gives a unique opportunity to become a part of the elements and to come close to Nature, something I can testify to by my own experiences. As Granit himself expressed it in relation to passing a critical turning point on the journey, Björkö-huvud (the "head" of the island of birches), in springtime "The fragrance of lilies of the valley and freshly-opened-up birch leaves embraced the boat as it passed by the lovely island." [2]

In 1945 Granit traveled by air for the first time in his life. He went to England to visit the 88-year-old Sherrington in Oxford. It was a very emotional final encounter. The Nobel Prize in physiology or medicine in 1945 recognized Alexander Fleming, Ernst B. Chain and Florey[10], the latter being a collaborator in the early 1930s of Granit and Eccles in Sherrington's laboratory, as mentioned. The prize highlighted the discovery of penicillin and the development of methods to produce it in quantities for general use in patients, a process accelerated by the war situation. During his visit Granit was instrumental in retrieving information from his contacts in the U.K. which secured that Chain, eventually, and appropriately, was included among the prize recipients. It was Liljestrand who gave the introductory speech at the 1945 prize ceremony.

During his 44 years as secretary of the Nobel Committee at the Karolinska Institute Liljestrand carried out this task no fewer than seven times. The only professor at the Institute who has given more laudations is Karl A. H. Mörner. During the first 11 years that the prize was awarded he gave 10 of the

Göran Liljestrand, the long-time secretary of the Nobel Committee for physiology or medicine. [From Ref. 8.]

presentations. He did this in his role as Vice-Chancellor at the Institute and, in connection with this, as the chairman of the committee. In terms of numbers of presentations my predecessor as professor of virology, Gard, comes next, having given five presentations. In fact he should have given one more had it not been for the fact that in 1951 the out-going Vice-Chancellor Hilding Bergstrand grabbed the responsibility of being the speaker to introduce Max Theiler, the father of the yellow fever vaccine, as already told[10]. Considering Gard's authority in the committee there is no wonder that I became an adjunct member of the committee already in 1973 during my first full year as professor and chairman at the Institute. Two biochemists with major status in the committee have given four presentations; Einar Hammarsten[5] during the years 1931 and 1955 and Peter Reichard, who we will meet repeatedly in Chapters 6 and 7, during the years 1968 and 1978. Only one committee member has given three presentations. This was Johan E. Johansson and the presentations were given during 1920–24. In the early phase of his career he had been a collaborator of Nobel in Paris[10]. As a professor through the decade of the 1920s, he was an influential chairman of the Nobel Committee introducing the word *discovery* in the prize motivations[5]. Finally there are 13 professors who have given laudations on two occasions. Among these are Granit (1944 — only a short speech over the radio — and 1963[5]) and Bernhard (1961 and 1967, as we shall see).

The continuation of Granit's academic employment towards the end of the 1940s remained uncertain. Originally he had only been appointed for five years, which should be compared to a potential life-long appointment at Harvard University. The further developments have been presented earlier[5]. Securing a stable future for Granit became a part of a larger project to establish a Medical Nobel Institute, originally aiming at providing new laboratories for Hugo Theorell in biochemistry and Torbjörn Caspersson in cell biology. Establishing a permanent chair for Granit and a new department and a building for neurophysiology became integrated into this project. One of the main critical colleagues furthering these developments was Liljestrand. According to Granit[2] he should have the major credit for the establishment of neurobiology at the Karolinska Institute. The first professor and chair of neurophysiology was established in 1946 to secure Granit's future. Tage Erlander, originally minister of education and later prime minister for two full decades, came to play a major role in the post-war developments of Swedish science. Research councils were established, and Granit became one of the initiators of a research council for medicine.

A Shift of Focus in Science

The different activities were moved into two new buildings representing the Medical Nobel Institute in the autumn of 1947[5] and they were solemnly inaugurated at the time of Whitsuntide the next year by the Swedish Crown Prince. Granit now had his own home for scientific endeavors during the rest of his professional life. He built a strong team of collaborators recruiting colleagues from both Finland and Sweden. For some years it continued to focus on the functions of the retina, but then he was ready to try new hunting grounds. However, before leaving the field and also some time after this he summarized the accumulated results in two monographs and a textbook chapter of considerable influence[11-13]. During this time and partly into the 1950s he continued to guide qualified researchers in the field. The list of neuroscientists who were trained by him is long[5] and covered fields of both theoretical and applied clinical neurophysiological research. Some co-workers developed to become involved in clinical disciplines like neurology, neurosurgery and ophthalmology. Among those in the latter group that he trained was Christina Enroth-Cugell, who has written very personal memoirs already referred to[3]. She was in fact the daughter of one of the most popular clinical teachers during his medical studies. She wrote that he created "an atmosphere of dedication to research and willingness to share ideas and to help and cheer up others at times when everything seemed to go wrong." Further on she wrote "... at times he did appear to have a somewhat rough and occasionally 'grumpy' outer shell which I believe was erected to hide his shyness and a self-confidence that was not as unshakable as some may have thought."

Granit in his laboratory. [From Ref. 2.]

In 1948 Bernhard became professor of physiology at the Karolinska Institute and he moved into his own department building the following year. There were in fact some tensions developing between the Division of Neurophysiology of the Medical Nobel Institute and the newly established Department of Physiology at the Karolinska Institute also focusing on neurophysiology. Thus of Granit's early graduate students Bernhard Frankenhaeuser concluded his thesis work with him in 1949, whereas Svaetichin, who was a critical person in the early development of electrodes used by Granit, preferred to become a part of Bernhard's department. He became a respected collaborator and presented his Ph.D. thesis under the mentorship of Bernhard in 1951. He continued to make important contributions to the field, but according to Bernhard[8] he did not have pretensions to share in the Nobel Prize which Granit finally received. In 1955 Svaetichin moved to Caracas, Venezuela, where he continued his career at a newly established institute for the neurosciences.

In Bernhard's biography[8], there are no indications of the existence of tensions between the two departments. Bernhard once told me that one should not write a biography until one has turned 90. This was the rule that he himself applied, but he also applied another rule to his biography and that was not to discuss conflicts and their emotional dimensions. Hence it is not possible to read from his biography why, as we shall see, for more than a decade he developed such a negative attitude to Granit's science and to his candidacy for a Nobel Prize in physiology or medicine.

When Granit and his collaborators had left the field of retina research towards the end of the 1940s it was taken over by other qualified scientists. One of them was Granit's special friend Rushton who was to make continued important contributions. He has made a very complimentary comment on Granit's pursuit of science. This comment refers to his intuition, his subjective feelings for which direction to take in the work. Since the two of them had also become friends in sailing Rushton referred to this as Granit's capacity "to find the buoy for anchoring in complete darkness."

In his autobiography Granit has expressed his enthusiasm for science in reference to the "flow" he experienced in the mid-1930s. He wrote, translated from Swedish[2] "My experimental results gave me entry to the international brotherhood, where everyone without jealousy or secrecy presents his findings and

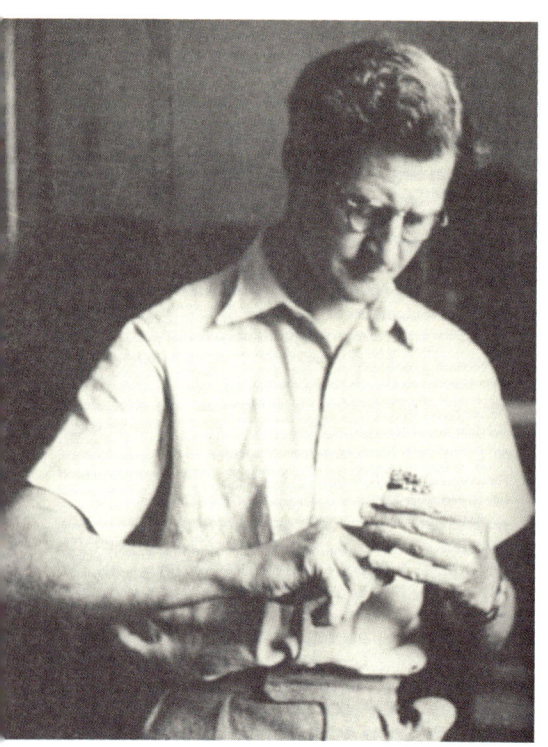

William Rushton (1901–1980).

thoughts and in return can take part in what his readers and correspondents think in their turn, probably the most decent brotherhood in the world." These are no small words! In another case he quoted Cajal's list of five qualities that sum up a good scientist. They are independent judgment; perseverance; striving to be independent; passion for research; and ambition (and vanity?). It should be added that Granit was also well aware of the challenges, in their most extreme form, the absence of creativity, the acedia. Already in the previously mentioned early book he raised this potentially paralyzing phenomenon and the impact it may have on academic researchers as well as representatives of other creative professions.

It took almost two decades after Granit had started phasing out his interest in studies of the retina before he received his prize for this early work. In this time he took up new important fields of neurobiological research and his institution continued to flourish. He persistently provided high quality training of the next generation of Swedish neurobiologists, neurophysiologists and clinicians in the discipline as already referred to. In addition the institute also became the temporary home of visiting scientists from many countries around the world. One main field of involvement was the muscle spindles, the stretch receptors that signal the tension of the muscle and allow control of its contraction and relaxation. Another field of observation connected to the more general functions in motor control, the balance of excitatory and/or inhibitory signals. All this work was in the rich tradition of Sherrington's original work and a tribute to this Nobel laureate as his "father" of science. As we shall see also this work on projects not related to seeing neurophysiology was referred to in nominations of Granit for a Nobel Prize.

In 1966 the Nobel Foundation decided to expand its domains of involvement. It needs to be emphasized that the room for expansion was limited, as already mentioned[10], since all the engagements of the Foundation defined by its statues should be relevant only to the furthering of the work by the prize-giving institutions. Within the framework of this restriction the Foundation decided to arrange selected international conferences and Granit was invited to initiate this new program. He decided to focus on the areas of neurophysiology which he pursued in the 1950s and 60s, not the work that came to be recognized by his Nobel Prize one year later. The 39 presentations at the symposium were collected in a book entitled *Muscular Afferents and Motor Control* and a foreword was written by (by now) Lord Adrian, who was also to honor the prize ceremony in 1967 with his presence as we shall see. Since Granit's conference there have been a number of Nobel symposia and conferences devoted to many

The Rock Foundation of Nobel Prize Developments 159

different topics arranged over the years. Only very recently the responsibility for these symposia arranged since their beginning by the Nobel Foundation has been taken over by the Royal Swedish Academy of Sciences.

Granit's already mentioned second autobiographical book[2] illustrates that he was a writer of class both in the sciences and also in broader cultural contexts. The second to last chapter of this book is of particular interest in relations to the work on Nobel Prizes. The title is Upptäckt och insikt — Discovery and insight. Discovery is a key word in the specification of the prizes in the natural sciences in Nobel's will. Granit noticed that research efforts may lead to new understandings. He qualified this stating that there is a scale of the magnitude of new findings. A new finding may come as a surprise even to the researcher and be truly new. It is such an unexpected finding that may be called a discovery, a key word in Nobel's will as repeatedly emphasized. A discovery should have wide-reaching consequences and an emerging involvement of many other scientists in the field leading to the appearance of a markedly increased number of publications in the newly opened area of research. He referred to a discovery metaphorically as a new tower of observation. Granit did not refrain from mentioning what he considered his own three major discoveries. One was the lateral inhibition which, however, was presented by Hartline earlier. The second was the finding that darkness adaptation was of electrical origin and not only due to a bleaching and regeneration of light-sensitive pigment in the cells sensitive to light. Finally he included the hypothesis that the generator potential in a sense organ mediated the transduction of impulses by being electronically conveyed to the sensory nerve.

The role of introduction of new techniques was also highlighted. An endless number of examples on this can be cited. As will be discussed at length in Chapters 6 and 7 the rapidly developing field of molecular biology provides one good example. Another is the improved tools for ultra-structural studies discussed in the final chapter. Granit naturally also emphasized the importance of defining priority to a major new discovery, a central target in the work of Nobel Prize committees. It remains a continuous responsibility to recognize the "eureka" moment of a particular discovery.

In addition to his two books in Swedish he wrote a number of books in English. Not surprisingly he wrote a biography of Sherrington[14], who he interestingly concluded had not made any single readily identifiable discovery. He also wrote widely available books summarizing the progress of the fields he knew the best, like *Basis of Motor Control*[15] and *The Purposive Brain*[16]. Granit did not support C. P. Snow's concept of the two cultures, the humanities and the natural sciences. He believed they belonged together and potentially could provide mutual support to societal advances. At the time of the 200th anniversary of Linnaeus's death the class for medical sciences at the Royal Swedish Academy of Sciences initiated a book project involving 12 authors[17]. Granit being the editor-in-chief wrote the introduction and also contributed one of the essays. Invited by the Editor-in Chief of the major Swedish liberal newspaper *Dagens Nyheter*, originally Herbert Tingsten and later Olof Lagercrantz, he made a number of contributions on diverse cultural topics.

Painting of Granit. [From S. Grillner.]

Because Granit received his Nobel Prize at the mature age of 68 years he had time to be recognized by a number of major prizes before that. He was also a member of many academies including the Royal Swedish Academy of Sciences, elected in 1944 and its President in 1963–65. Finally he was also recognized by honorary doctorates at a number of universities. He ended his long and rich life at the age of 91. His son Michael developed into a respected architect. He has three children all with families. The whole family continues to enjoy the rich qualities of summer life at Vikminne. Their father and grandfather was honored with an exhibition in 2017, celebrating that 50 years had passed since he was recognized by the Nobel Prize, an event that incidentally took place in the same year that Finland had existed as an independent country for 100 years. It is now time to see how Granit was judged by reviewers of his candidacy for a Nobel Prize.

The Early Enthusiasm of the Nobel Committee

The first nomination for a Nobel Prize in physiology or medicine to Granit was submitted in 1946. The proposal came from the world-renowned neuro-anatomist at the time, Stephen Polyak at the University of Chicago. It provided an

Ulf von Euler (1905–1983), recipient of a shared Nobel Prize in physiology or medicine in 1970. [From *Les Prix Nobel en 1970*.]

excellent description of Granit's contributions under the headings general remarks, evidence and comment. A rich list of references was provided. In summary the nomination argued that Granit was the first scientist providing a comprehensive presentation of the retina as a nervous center characterized by excitation, inhibition, spatial and temporal summation etc. Ulf von Euler was selected to make a full review. This was impressively detailed and covered 19 pages. Von Euler divided his analysis to present the developments in the fields of the electroretinogram, the physiology of photopic (well-lit conditions) and scotopic (degrees of darkness) seeing and the sense of distinguishing colors. A thorough historical perspective was provided to each field. In the description of the advances Euler often referred to "Granit and his school." Granit had a capacity to stimulate involvements of a number of enthusiastic collaborators.

It was emphasized that developments of the microelectrode techniques by Adrian and later Erlanger and Gasser was critical for the gaining of new knowledge. It was an important advance when Granit and collaborators were able to refine the technology by designing new kinds of microelectrodes. In his studies of the retinogram Granit managed to separate the phenomena observed into three different wave characteristics, named I, II and III. He could distinguish their respective role in so-called on-off phenomena. Phenomenon II seemed to play a central role in inhibitory effects. The relative role of rods and cones was analyzed by exposure to stimulus at wavelengths 0.500 and 0.579, which are the respective maxima for their separate and independent stimulation. Von Euler then discussed the first successful recordings from individual threads of the optic nerve, made by Hartline and Graham in 1932. They discovered that some nerve fibers reacted to light by an "on-effect" whereas others reacted with an impulse first when the light was turned off, an "off-effect." We will come back to this fundamental discovery in the next chapter. One limitation of the technique used by Hartline was that it could not be applied in warm-blooded animals. This was remedied when Granit in his collaboration with Svaetichin in 1939 managed to develop their already mentioned microelectrode technology. This technology allowed the opening of a new door to knowledge and a deeper understanding could be retrieved of the excitatory and suppressive phenomena and in particular major new insights into color vision could be gained, as we shall see.

Von Euler then considered light- and darkness vision. Interesting comparative work had been done on eyes from a large variety of different species of animals. Thus particular findings could be made in examination of the retina of the dove, which is dominated by rods, and of the owl, which by contrast is dominated by cones. Examination was made under conditions of daylight and after darkness-adaptation. The phenomena observed were correlated to the concentration of purplish-red pigment, rhodopsin, which will be discussed in depth in the next chapter. The major emphasis in the review was on the critical findings made by Granit and collaborators in developing insights into the sense of color vision. Comparisons were made by studies of cold-blooded animals, like frogs, and warm-blooded animals, like rats or cats. Sometimes exotic animals were used for the experiments like the grass snake (Tropidonotus). The eye of this animal only contains cones. Granit also examined fishes, which instead of visual purplish-red pigments have a pigment that allows them to see also in the ultraviolet spectrum of light. Some of the data obtained supported earlier work by Wald, which will be discussed in detail in the next chapter. The review finished by emphasizing the general impact of Granit's dominator-modulator theory. It was formulated in the following way (translated from Swedish) "A dominator, photoptic or skotoptic, is according to this theory a mechanism of sensibility of an end organ with a broad spectral sensibility, making a wide band of wavelengths accessible to seeing." It then went on to say, in an almost panegyric way (translated from Swedish):

> "Granit without doubt can be seen as being the one who at present is the foremost in knowledge about the physiological processes which take place in the retina. His domains of work encompass such fundamental problems as the nature of the electroretinogram and the importance of electrical discharges in the optic nerve, the physiology of photoptic and scotoptic seeing and color vision. In all these fields he has given particularly important contributions and by introduction of new techniques and new perspectives provided the basis for a view on the retinal processes which on essential points differs from the 'classical'."

Von Euler also pointed out that the new findings had been found to be of use in ophthalmological medicine.

Not surprisingly von Euler concluded that Granit's contributions were worthy of a prize in physiology or medicine. This conclusion was supported by the committee, who in this year decided to award the prize to Herman J. Müller for his discovery of the mutagenic effect of X-rays.

The following year Granit was nominated again, not surprisingly by his "father" in science, Sherrington. The letter finished by stating "The research bears witness to the boldness of conception, and to the great manual skill of Professor Granit." This time Zotterman (see p. 154) was selected to make a review. Like von Euler he was very impressed by Granit's contributions. In particular he emphasized the new perspectives given on color perception, which had been summarized in a comprehensive monograph *Sensory Mechanisms of the Retina* published by Oxford University Press in the summer of 1947[11]. Zotterman raised the interesting question about how an electrode with thread-diameter of 25 µ can measure electrical impulses of nerve fibers with an estimated size of 4–8 µ. Later on this has been explained by deducing that the recordings probably derived from contacts between the electrode and the ganglion cells. Zotterman emphasized the importance of Granit's conclusion that most of the functional elements in the light-adapted eye do not participate in the service of color discrimination, but that this is managed by a minority of cells with uniquely selective capacity to react. It appeared that Granit's findings implied a synthesis of the two separate theories by Young-Helmholtz (on trichromatic color vision) and by Hering (on how the two eyes interact). Since the latter integrated the phenomenon of color-blindness in his hypothesizing, the new interpretations also had an impact of clinical importance in the latter condition of a physiological defect.

Some of the results obtained by Granit and collaborators gave support to the possible significance of a special kind of cells that had been histologically identified first at about the same time. These were the *amacrine* cells, from Greek *a* — non — *makr* — long — *ine* — fiber —, which in spite of their name have short neurotic fibers connecting cells within the retina and serve inhibitory functions. This added layer of complexity accentuated the interpretation of the retina as an extension of the brain. Zotterman emphasized that the techniques introduced by Granit and collaborators have opened up possibilities for more in-depth studies of the complexity of the retina. However, he made the reservation that the studies so far had not included animals closer to man than the cat. It would seem urgent to examine primate representatives. The conclusion of the review was that Granit took a special position in his field of research because of his richness of ideas, his introduction of improved technology and his general jauntiness. He was considered without doubt worthy of a Nobel Prize. The committee agreed that Granit's (translated from Swedish) "discovery of the dominator-modulator mechanism for perception of colors" was worthy of a prize. It can be noted that in this year the 10-member committee for the first time included a woman, Nanna Svartz, mentioned already in Chapter 2. It was apparent from Zotterman's later review in 1951 that Granit in 1947 was very close to receiving the award. He wrote (translated from Swedish) "When the question (of Granit's candidacy) was discussed in the Nobel Committee in 1947 there was a complete agreement that he was worthy of the Nobel Prize and about half of the committee (5 members, my remark) voted for a recognition by an award this year, but since one preferred to have a more unanimous committee supporting the proposal to recognize, for the first time, a Swedish researcher, the committee finally agreed to propose the spouses Cori and Houssay" for the Nobel Prize this year.

In 1948 Sherrington repeated his nomination and there was also a nomination by one of his most important collaborators E. G. T. Liddell which finished "…Granit has composed a masterpiece of outstanding scientific artistry which well merits the addition of his name to the Roll of Honour of Nobel laureates."

This time the committee took the exceptional step of using an advisor from another university in Sweden. It was Professor Georg Kahlson. He was of Finnish origin and had become professor of physiology at Lund University 1938, a position he held until 1968. He had a much profiled personality and was one of the first in the academic establishment in Sweden to speak out against Nazism. He was also involved in the original creation of the Medical Research Council in Sweden. His review was as thorough as those of the two previous evaluators and also included a number of figures. The essence of the review remained the same as in the earlier evaluations, but there were differences in the emphasis. Towards the end Kahlson cited H. Hartridge the foremost sense physiologist at the time in the U.K. He had written recently "Of all the advances made in recent years in the physiology of vision possibly the least expected, and the most far-reaching, are the researches of Granit on mammalian retina." As cited earlier in the review Hartridge had been involved in following up Granit's theory of color vision and he had experimentally confirmed that it also applied to humans. Kahlson also concluded that Granit "…by his discovery of the selectively color-sensitive modulators and the broad dominators for the impression of clarity has achieved a break-throw on a central part of the frontier (of seeing physiology), where other researchers have got stuck." The final conclusion was that Granit well deserved a Nobel Prize in physiology or medicine. Again the committee confirmed the strength of Granit's candidacy. The composition of the committee in this year deserves to be commented on. It was composed of 11 members and included for the first time two new members selected from the College of Teachers. They were Sven Gard who had received a personal chair in virology research in the same year and Torsten Sjögren, professor of psychiatry since 1945. In addition there were two members who were *not* employed at the Karolinska Institute. There are no rules against such an arrangement, but it has been used only to a very limited extent over the years. The only example I can recollect during my twenty years on the committee was when one year we brought in an external expert on immunology, Peter Perlmann from Stockholm University. The two external members in 1948 were Georg de Hevesy, the recipient of the Nobel Prize in chemistry in 1943 (awarded in 1944) presented at length in my second book on Nobel Prizes[6] and Kahlson, who contributed the review on Granit this year.

Turmoil in the Evaluation Process

In 1949 Granit was nominated by Bernardo A. Houssay, the prize recipient two years earlier, in a proposal also including Haldan K. Hartline — his first nomination — and by a major authority of the field, the 1932 prize recipient Edgar D. Adrian. Von Euler was called upon to make another review of Granit and to also include Hartline. In addition the "permanent" secretary Liljestrand invited the recipient of the 1944 Nobel Prize in physiology or medicine Herbert S. Gasser[5] to give his opinions on both candidates and received a three-page letter response. Von Euler initiated his review by the remark that there were already three earlier reviews by different specialists in the field who had declared that Granit was worthy of a prize. Hence he concentrated on Hartline who had not been reviewed before and made comments to Granit's work only in reference to some recently added information, but still this latter section of the report covered the major part of the 16-page-long text. We will return to the evaluation of Hartline's work in the next chapter and here only make remarks regarding the added comments on Granit.

His recent work included extended analysis of the so-called on/off elements. The variability observed was taken as evidence for a possible importance of the above-mentioned amacrine cells. The on/off elements seemed to be in the majority in the kind of eyes containing predominantly cones. These elements were examined by use of polarized weak currents. Granit had earlier introduced the concept of modulators by observations in light-adapted eyes of compressed curves for responses to a certain frequency of light. He now took his work further by use of a technique introduced by other researchers, the polarization method in which two spectral colors were used. To cut a long story short it was demonstrated that the phototopic dominator was established as a physiological reality and in this work there was an overlap with Wald's studies of various chemical substances involved in collection of light impulses, as we shall see. We will come back to an analysis of the role of specific receptors in the absorption of light of different wavelengths in the discussion of Hartline's contributions. Suffice to note that the results added by him did not infringe on the importance of Granit's fundamental interpretation of his experiments of reactions to illumination with light of different wavelengths. His fundamental conclusions remained firm and the role of phototopic dominators and modulators combined with his analysis of color vision made him a strong candidate. Hartline's contributions were interpreted not to make him an obvious candidate at the time. We will return to this.

Gasser's letter of response to Liljestrand may represent the first external review in the archives of prizes in physiology or medicine in which the English language was used. The remarks given were formulated in general terms and Gasser did not dissect individual experiments, as Swedish committee members used to do in their reviews. He only gave an over-encompassing picture and drew general conclusions. Having freshened up his contact with this specific field he referred to it as "majestic in its unfolding." He noted that thanks to Granit's contributions, which carried both width and depth, it could be concluded that "…the retina is a synaptic system: a microcosm displaying properties of the central nervous system." He then praised in general terms Hartline's contributions and compared them with Granit's. We will return to these comments in the next chapter, but it can be noted already here that Hartline was given priority for certain fundamental observations. Gasser's respect for Granit's work was reflected by the following remark "each worker (Granit and Hartline) has handled his problems in his own way. Granit has provided an improved version of the retinal potential and combined the findings with those obtained with the micro-electrode. With the description of modulators he has placed color vision physiology on a new plane; and he has given more attention to the mechanism of inhibition." In summary Gasser supported Granit's candidacy

Erik Jorpes (1894–1973). [Courtesy of the Karolinska Institute.]

for a prize, and he concluded, as we shall see, that in his previous nomination of Granit he might have included Hartline, who was also considered worthy of a prize. But this was not the end of the story this year. An unexpected interference occurred.

Erik Jorpes was a competent and high-profile scientist who received a personal chair in medical chemistry in 1947 and then succeeded Einar Hammarsten in 1955 as presented earlier[5]. The surname of Jorpes' father was Pettersson and his given name was Johan. Upon the move of Erik, and later also some of his siblings, to Sweden a new family name was created by using the first two letters in their father's name and then combining them with the first two letter of the name of the archipelago farmstead from which they originated, *Per*sgården (named after an earlier ancestor). In order to make the name pronounceable an "r" and an "s" were added. Thus the new name Jorpes did not have a geological origin, but Jorpes' early upbringing was very similar to Granit's. He was a farmer-fisherman's son brought up at the barren rocks of Kökar, a group of islands in the southeastern part of the Åland archipelago (see p. 141). It is very likely that these two high-profile Swedish-speaking Finns did not interact smoothly. However it was not a matter of guarding the borders of the areas in which you were allowed to fish, a potential major source of violent skirmishes in the environment in which they were brought up. They came to pursue different "fishing-waters," at the time the non-overlapping fields of science, medical chemistry and neurobiology. Rather the difficult relationship between these two scientists may have its origin in contrasting political experiences reflecting their islands of origin.

The political affinity of Åland during the first two decades of the twentieth century was a matter of intense debate. Sweden was of the opinion that it should belong to that country and of course Russia and later the nation of Finland established in 1917 argued that it should remain a part of the latter country. The question was not resolved until 1921 by a decision of the League of Nations. To the chagrin of Sweden and many inhabitants of the island it was decided in 1921 that Åland should remain a part of the young Finnish nation. Thus there was a latent animosity between "Ålanders" and the inhabitants of the more eastern islands collectively referred to as the "Ny(new)landers." In the year when it had achieved its independence Finland endured a very painful civil war. The massive killing of the "reds," which lost the war to the "whites," remains a blemish on the escutcheon of the country. Granit represented the "whites" and Jorpes the "reds" — he was in fact one of the founders of the communist party in Finland. Disregarding what the speculative background might have been, it happened that on September 19 Jorpes on his own initiative sent a letter to the committee heavily criticizing Granit's work. At the time he had not himself been actively involved in the practical Nobel Prize work, but he had previously submitted a nomination of candidates for the prize.

Jorpes' letter was formulated to highlight that Granit's involvement in research appeared to represent a relatively one-sided point of view and also to have been praised excessively. Jorpes wanted to provide a necessary subjective guidance to an objective evaluation of the question whether Granit deserved the prize. The letter criticized the lack of originality of the techniques used by Granit and referred to the fact that Hartline had preceded him in collecting recordings from single nerve fibers. Jorpes questioned if Granit really had collected signals from single nerve cells. He then entered into polemics with Kahlson by citing his laudatory formulation "Granit's work concerning color perception (a contribution which Jorpes however to a certain extent seemed to respect) has most of the characteristics which are associated with classical publications." He believed that this was a presentation of a one-sided view citing one of Granit's previous collaborators W. D. Wright from Imperial College London. He also cited a recently published monograph which was referred to as stating that there was no support for the existence in primates of three different kinds of cones. Furthermore he referred to professor R. A. Morton, also cited by Kahlson, and underlined in a text cited from *Nature* "it was hard to believe that more than three, or at most four, light-sensitive substances could co-exist there at the same time." Finally Jorpes had also contacted an American scientist William R. Anderson. He cited the full letter from him including critical remarks about Granit's work. In the final paragraph Jorpes tried to diminish the impact of Granit's theory of color vision.

Of course the committee was somewhat perturbed by the letter from Jorpes. As previously mentioned[5] originally he had become less involved in the preparatory work for selection of Nobel Prize recipients, probably because of the dominating influence of another chemist, Einar Hammarsten, and of course also his favorite student Theorell. However, as we shall see Jorpes intermittently was a member of the committee during forthcoming years. In an attempt to stabilize the situation the committee called upon Bernhard, who as we have seen personally was highly instrumental in bringing Granit to Sweden, to make an additional late review. Bernhard had been appointed professor of physiology the previous year and served for the first time on the committee in 1949. However, the supplementary review over 8 pages which the committee received, surprisingly was not supportive. Without going into details of the unexpectedly critical review it may suffice to cite the conclusions. They read (translated from Swedish):

Carl-Gustaf Bernhard (1910–2001).

"In conclusion I thus would like to bring forward as my opinion that:

1. It is not possible at the present stage to draw conclusions concerning the specific spectral sensitivity of light receptors in such a way that Granit based on the modulator curve he has demonstrated, has led (us to) suppose.
2. The conclusions concerning the modifying influence that synapse stations in the retina has on the activities coming from the receptors are uncertain as long as one does not know the properties of the activity coming from the receptors, which have been modified in the ganglion layer."

Bernhard's conclusion was that Granit's contributions at the present stage should not be recognized by a Nobel Prize. No wonder there were contrasting opinions in the committee. It wrote in its final protocol (translated from Swedish):

"Concerning Ragnar Granit it was noted that an opinion had been submitted by Professor Jorpes (Attachment B) and professor Bernhard (Attachment C). It was the view of the committee that Ragnar Granit's work concerning the physiology of color vision and the electrophysiology of the eye should *not* (my italics) at present be considered for a prize. Professors von Euler, Liljestrand and Zotterman were of the opinion that Granit's and Hartline's discoveries concerning the electrophysiology of retinal elements and reactions to light of different wavelengths were worthy of a prize."

Granit's Candidacy Back on Track

There was no nomination of Granit in 1950 but he was proposed all the remaining years of the decade that followed. As we shall see Hartline was also nominated in 1951, but this was a proposal by the secretary Liljestrand to keep him available for an evaluation. Hereafter he was not nominated again until 1960. Granit was nominated by two German professors and also by his good friend John Eccles, the forthcoming recipient of a Nobel Prize in physiology or medicine in 1963. In 1951 another review of both Granit and Hartline was made by Zotterman. In addition the committee called upon an external major authority Lord Adrian. Zotterman commented that the critique from Jorpes and Bernhard in 1949 of Granit's work in parts missed the target and in other parts could be rejected. He tried to explain the difficulties of interpreting Granit's work with reference to three circumstances. These were (a) that the methods used in his research were intricate, (b) that the analysis of complicated synaptic organizations of the retina was not readily made understandable and (c) that "older" physiologists (Bernhard? My remark) studying the sense of seeing might have difficulties in following the new and revolutionary dimensions of Granit's electrophysiological contributions and hence correctly interpreting his results, which has led to a critique which has often been unjust.

Adrian wrote:

> "The committee may feel that it would be wiser to postpone judgment on Granit's claims until greater certainty has been reached. Since Granit and Hartline should not be separated this would mean holding up the proposal, possibly for several years. In my own opinion their claim at the present time is stronger than that of any other neurophysiologist, or physiologist with whose work I am familiar."

The ten-member committee came to the following conclusion (translated from Swedish):

> "Concerning Ragnar Granit and Haldan Keffer Hartline the committee found their discoveries of the electrophysiology of the retina and the reaction patterns of the retinal elements and sensitivity to light of different wavelengths worthy of a prize. The Professors Bernhard, Hammarsten and Jorpes (a member of the committee for the first time this year, my remark) were of the view that the contributions by Granit and Hartline should not at the present time be seen as eligible to a prize. Professor Kristensson (a professor of medicine, my remark) abstained from voting."

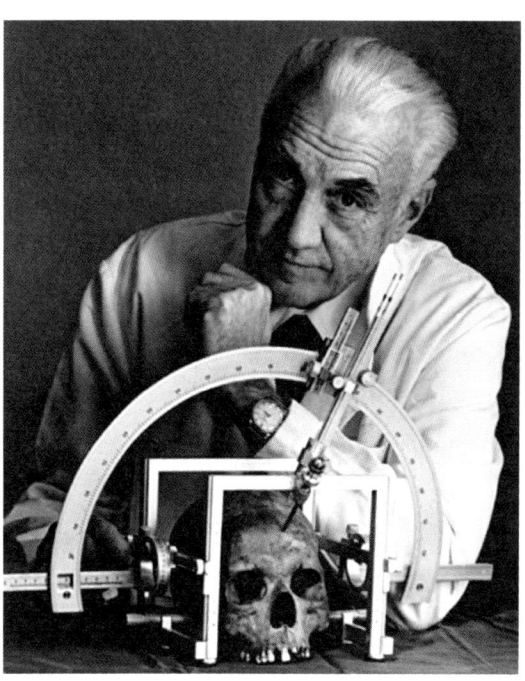

Lars G. F. Leksell (1907–1986), professor of neurosurgery at the Karolinska Institute and the inventor of the "radiation knife for surgery" of the brain.

During the coming years there were one or more nominations of Granit and these proposals often also mentioned his more recent work on muscle spindles and autogenic inhibition in the spinal cord. Proposals for this latter work also included additional candidates, Lars G. F. Leksell, professor of neurosurgery at the Institute from 1960 and Bryan Matthews, professor of physiology in Cambridge, U.K. It would take until 1956 before a new evaluation was made. This year Granit had been nominated by Matthews himself. This nomination unsurprisingly not only brought forward Granit's discoveries in studies of the electroretinogram (ERG), in particular his work on color vision, but also the more recent work on mechanisms of supraspinal control of the muscle spindles. This time it was von Euler's responsibility to defend Granit's colors, but since the more recent field of his research was also cited the professor of neurological diseases at the Institute from two years back,

Eric Kugelberg, was also used as an expert. Von Euler took notice that he had to review Granit alone since Hartline was not nominated. He preempted his follow-up analysis by citing the statement from Adrian's 1951 review mentioned above. In his analysis von Euler departed from a recent summary given by Granit, the Stillman Memorial Lectures 1954 published by Yale University Press[12]. He then referred to some interesting recent work by Ottoson and Svaetichin who in 1954 with further refined electrodes had managed to record the electrical activity of single ganglion cells. Their findings had raised the question of how much of the ERG reflected activity in the receptor layer and how much the overlying neuron structures contributed.

Erik Kugelberg (1913–1983), professor of neurology at the Karolinska Institute.

Von Euler then commented on comparative studies with different experimental animals with an eye structure dominated either by rods or by cones. It had been deduced for example that in mammals, which have eyes dominated by rods, it is natural that cone-ERG will be masked off by rod-ERG. Zotterman referred to studies that he himself had performed in the late 1940s together with the forthcoming 1961 recipient of the Nobel Prize in physiology or medicine, Georg von Békésy on the auditory nerve. The signals recorded in this case were concluded to derive from contacts between the electrode and large ganglion cells. Rushton working in Granit's laboratory soon thereafter also obtained similar results in studies of the retina. Von Euler also cited recent work supporting the theory that the rules defined by Granit from studies in experimental animals had been found to also apply to humans.

In addition von Euler returned to a discussion of the dominators and modulators, also collectively called broad band elements, in reference to another review by Granit, *Sensory Mechanisms of the Retina*[11]. These concepts were introduced based on studies of the reception and discrimination of light of different wavelengths on the retina. The dominator elements were also discovered earlier by Hartline in his work with the eye from a marine non-vertebrate. There is also an overlap between Granit's electro-physiologically defined sensitivity spectra and the presence of different naturally occurring or chemically synthesized compounds with unique capacity to absorb light. In this part Granit's work overlapped to some extent with Wald's, as will be seen. The skotopically and photopically dominating light sensitivity curves shown by Granit appeared to correlate characteristics of visual purple 497 and with the cone pigment (iodopsin) examined by Wald. At the time, work on other pigments, also referred to as "small band pigments" was ongoing. Finally Granit's work was also compared to recent work by Motokawa in studies of retina from carp fishes. He found four different absorption maxima interpreted to signify the presence of four kinds of receptors. It was summarized that forthcoming work with recordings directly from individual receptor cells would be an important supplement to Granit's more complex, assumed recordings from separate ganglion cells. Von Euler finished by noting that he already previously had considered Granit's work worthy of a Nobel Prize and that he saw no reason to change his opinion on this matter.

In Matthew's nomination an equal emphasis was given to Granit's "discoveries on the electroretinogram" and on "recent important discoveries of the mechanism of supraspinal control of the muscle spindles." Kugelberg made a careful review of this latter field. Starting with his summary it should be noted that he brought forward three major teams contributing to our understanding of the control of muscular tensions. This was work by Matthews, the nominator, by Stephen W. Kuffler[5] and his collaborator Carleton C. Hunt and Leksell. Matthews had demonstrated the functions of this spindle, and Granit's and his collaborators' contribution described the nature of the gamma system's control of supraspinal structures and the capacity of the system via the stretch reflex to be deeply involved in the control of motility and muscular functions. Kugelberg generously proposed that these combined contributions would be worthy of being recognized by a Nobel Prize in physiology or medicine, However, he was well aware of the fact that Granit was already considered worthy of the prize for his earlier work on the retina and he therefore simply noted that his work on the (translated from Swedish) "…gamma system further strengthens his position as a master within the field of sense physiology and as a serious aspirant to the prize." Since in the end Granit was recognized by a prize for his work on the retina, the interesting background material provided and the comparative analyses presented in Kugelberg's review will not be further refereed here. It remains only to note that to be considered worthy of a prize for contributions in two different fields of neurophysiology appears very impressive and says something about the unique general caliber of Granit as a scientist. The committee concluded in 1956 that (translated from Swedish)

"Concerning Ragnar Granit it was the view of the committee that his discoveries of the physiology of sensory organs was worthy of a prize but (it) recognized that with consideration taken to contributions made by other researchers, at the time not proposed for the prize, the forthcoming developments should be awaited. Professor Bernhard was of the view that Granit's work at present should not be considered for a prize."

We will have reasons to return to Bernhard's attitude of hesitancy later.
Nominations continued through the remaining part of the decade but only in 1959 did the committee made a comment to its concluding protocol. It wrote (translated from Swedish):

"Concerning Granit it was the view of the committee that a final decision on the (importance of) his work concerning the electrophysiology of the retina and the reaction patterns of the elements of the retina should be deferred until certain other, at the present time not proposed contributions, could be considered."

There is in fact a problem with this statement. It is in conflict with the rule that applies to the Nobel Prize work at the Karolinska Institute. This is that in cases when a prize is given to more than one recipient, each one of them should be eligible to carry the prize on his own. Thus in principle the candidacy of Granit should not be dependent on the availability of other prize-worthy candidates in the field. It is another matter if a committee wants to combine candidates, who each on his own could receive a prize, since their discoveries belong in a closely related field of research.

In 1960, as in the year of 1949, Granit and Hartline were again nominated together. No new review was made. This time the committee wrote that both candidates were worthy of a prize. There was no expression of a divergent opinion in the protocol, as in 1951 and in 1956. Hammarsten had retired at this time. Jorpes, who had been a temporary member of the committee in 1951 was out of it the following year. The reason probably was that he had been nominated himself for a prize that year for his work on the anticoagulants heparin and dicoumarol. Hammarsten, who evaluated this contribution, stated that if this kind of anticoagulant treatment developed to be of considerable clinical importance there might be reasons to return to Jorpes and also the contributions by Karl Link from the University of Wisconsin. The following year Jorpes was again back on the committee never again to return after that. He was later used as a reviewer, however. But the big question is why in 1960 there was no reservation by Bernhard. He had considered Granit not worthy of a prize in both 1951 and in 1956 and no essential information had been added in the intervening years. It could not be that Wald had been added as a candidate potentially providing a lever of Granit's candidacy, since after his first review in 1955 he was considered at the time not as yet worthy of a prize, as we shall see. There will be reasons to return to Bernhard attitude towards the three candidates, since he eventually become the person to introduce them at the prize ceremony.

In 1961 and 1962 Granit was nominated alone and in 1963 and 1964 he received no nominations. These nominations did not lead to any new reviewing and throughout these four years the committee did not mention him in the concluding protocol. It is likely that Granit himself declined being a candidate in these years because he prioritized his own involvement in the Nobel Prize selection work. Thus it may be appropriate to mention this work, which was already referred to my two previous books[5,6]. To repeat Granit was never a full member of the committee but he was an adjunct member in 1952, 1953, 1955 and 1961-63. He had a central role in two prizes, the 1961 Nobel Prize in physiology or medicine to Georg von Békésy and the 1963 prize to John C. Eccles, Alan L. Hodgkin and Andrew F. Huxley. Concerning Békésy he made the first review in 1950, followed up by another review in 1954. Hereafter the reviewing of Békésy was taken over by Bernhard, who was also the presenter at the prize ceremony in 1961, after Békésy somewhat unexpectedly had been selected to receive the prize this year. In the case of Eccles et al. Granit contributed important reviews in 1953, 1955, 1961 and 1963. He also gave the laudation at the prize ceremony in 1963. From 1964 and onwards he did not participate in the Nobel Prize work. Like Hammarsten[5] Granit wrote his reviews during late summer staying at Vikminne in the beautifully seductive environment of the Finnish archipelago.

After two years of silence Granit's case was reopened in 1965. He was nominated by R. J. Whitney from

The view from the room at Granit's summer house at Vikminne where he wrote his Nobel reviews.

London, not for his work on the retina, but for the studies of gamma-fibers and mechanisms of muscular control. The committee again made an uncommon move and requested a review by an external authority. It was Fritz Buchtal, of Jewish German origin, who in 1933 had moved to Copenhagen and became the founder of neurophysiology at the University of Copenhagen, a field he developed to a high international standard. He stayed in that city until 1982, when he moved to the U.S. During the end of the Second World War he fled to the University of Lund where he temporarily developed his research. He wrote his evaluation in a Scandinavian language, Danish. Since this concerned high quality experimental work within an area of neurophysiology, which was not recognized by the prize finally given in 1967, there are no reasons to present it in any detail. The review covered 11 pages and it was concluded by the following paragraph (translated from Danish):

> "Granit's and his collaborators work within this field (of research) includes a number of important and original contributions characterized by scientific stringency and strategy. The (total) contribution is characterized by (the fact) that by focused experiments it has stabilized earlier somewhat more vaguely formulized concepts and ideas. This work has formed schools for work in the field both in Sweden and in other parts of the world."

Buchthal did not make any judgment on whether Granit was worthy of a prize for the contributions he had examined. The committee's conclusion was that Granit was "not at the present time worthy of a prize for his work on mechanisms of muscle control."

Finally an Expanded Basis for a Prize

The following year all the three prize candidates, Granit, Hartline and Wald for the first time finally were nominated together. The nominator was Lord Adrian and he started his proposal by mentioning that he had earlier nominated Granit and Hartline, but that presumably it was considered that further supporting evidence for the correctness of Granit's color vision theory needed to be awaited. He then remarked that important data supporting Granit's theory had now been provided by Wald and further that Hartline's work had gained in importance. It is apparent from the nomination that Adrian had considerable respect for Granit's general importance in the field, seeing him as a statesman of science. There was also another nomination of the three researchers by the recipient of the Nobel Prize in physiology or medicine awarded in 1939 (the 1938 prize), Corneille J. F. Heymans, from Ghent, Belgium. He emphasized in particular how the recent work by Wald had consolidated Granit's theories. He recommended giving half the prize to Wald and to split the other half between Granit and Hartline. Heyman wrote the following:

> "Recently Brown and Wald have determined the light absorption curves for single cones from the human fovea. These curves show that a cone has a very restricted spectral sensitivity in red, green or blue in very much the same regions as the "modulators" found by Granit 25 years ago in the cat's retina. These nerve data prove that Granit's assumption (1943) that his modulator received information from a cluster of cones, each displaying a relatively narrow spectral sensitivity, was correct while the dominator response is built up by a variety of different cones and rods. I think that his experiments on the effect of bleaching with red and green light on the dominator response proved his assumption. Now Brown's and Wald's recent experiments have definitively proved that Granit was right. For that reason I find it an act of justice to declare Granit's discovery of the modulator-dominator mechanism in the 1940s can now be better understood, which is in accordance with the regulations of Alfred Nobel's will…"

The committee asked Zotterman to make his fourth and as it turned out final and conclusive review. His review provided an excellent summary of the developments over the years, including the ups and downs in the case of Granit. He referred back to his 1951 review, which has already been discussed at length above. He further emphasized that the temporary labeling of Granit's contributions as "at the present time not worthy of a prize" in 1949 was lifted in 1956, when he was again considered worthy of a prize. Only Bernhard temporarily had a dissenting position. This was not adjusted until in 1960 when he abstained from making a reservation, as already mentioned. In Zotterman's concluding review, 12 out of the 31 pages presenting the three candidates were devoted to Granit. The emphasis was on the development of the conceptual understanding of the basis for color vision. He initially referred to Adrian's description of the situation as it had developed. He then cited extensively from all his previous evaluations and eventually raised the question why it has taken so long to recognize the existence of modulators in the human eye. Granit had posed this question in 1945 and consolidating data were published only twenty years later by Wald and collaborators. Hereafter Zotterman cited in full the excerpt from Heymans' nomination presented above. His conclusion was that it was self-evident that Granit was worthy of a prize. This was confirmed by the committee.

In 1967 Heymans repeated his nomination of the three candidates and also separately added to his list Christian de Duve, who however had to wait for seven more years for his shared prize. We will return briefly to his prize in the final chapter. There was also a separate and very extensive nomination for Granit alone. It was submitted by none other than his good friend and forthcoming co-recipient of the prize, Hartline. In this year the committee asked Bernhard to make a review of Granit and Hartline and Theorell to make another evaluation of Wald. Thus Bernhard had an opportunity to explain the emergence of his questioning of Granit's candidacy for a prize and to elaborate on his present supportive view on his contributions. Most of his report concerned Hartline, as we shall see, but over five pages he presented the transformation of his view on Granit. He first praised the impressive comprehensiveness of Granit's production and noted that he had acquired a remarkably authoritative position in the field. He referred to Granit's important reviews in the field, the two books already mentioned[11,12] and also a chapter entitled *The Visual Pathway* in the later 1962 book[13]. Besides Granit's work on the sense of seeing, Bernhard also mentioned with respect his more recent, partly pioneering contributions on the super spinal control of the muscular spindles and the influence of the small brain, cerebellum, on the coordination of motor activities of the body. He suggested that these contributions could be considered to have strengthened his position as a candidate to a prize. However, it serves to remind that a prize is given for a single discovery and nothing else. Bernhard appropriately remarked that Granit's later studies were a natural derivation from his early training with Sherrington. He then paid attention to his retinal studies that have been extensively commented on in the many reviews discussed above.

It was emphasized again that a seminal component in Granit's production concerned the principles of color discrimination. The publication in 1939 together with Svaetichin[7] was highlighted as particularly important. The follow-up studies of the retina from a range of different species consolidated the picture with emphasis on the postulated occurrence of neurons of a higher order towards which signals deriving from many light-sensitive cells converge.

It was deduced that the color-distinguishing mechanisms were trichromatic, a conclusion that was partly confirmatory in nature. In the 1950s two new findings caused a discussion of Granit's general conclusions. The first one was that Svaetichin had managed by use of improved delicate micro-electrodes to identify a new kind of slow potentials, which also were graded. These were referred to by other researchers, Motokava and collaborators, as S-potentials. The phenomenon observed seemed not to fit in with Granit's general conclusions. The other question raised was if the cat used in Granit's central studies really could distinguish colors. It soon became clear that none of these new reservations forced a reinterpretation of Granit's original general conclusions. The S-potentials did originate in the cells containing photo receptors but from other kinds of integrated cells and hence are of interest in a more general context. Further it was shown that cats indeed do see colors. As the technology advanced it finally became possible to determine the adsorption spectrum of individual cones. The experimental animal used in these studies originally was goldfish, but very soon the same experiments could be performed in monkeys and man by Brown and Wald. We will return to these studies in the next chapter. Thus in the end Granit's general hypothesis turned out to be a "discovery" that stood the test of time. Bernhard wholeheartedly supported a prize to be divided equally between the three candidates, Granit, Hartline and Wald as we shall see.

Altogether Granit's contributions had been dissected over more than 100 pages in ten different reviews before his candidacy could finally be moored in a safe harbor. Hartline and Wald had a shorter voyage, but in fact they also had some head-wind and nasty weather to manage. We will meet them in the next chapter and see how it all ended well. It turned out that in this case the saying that all good things come in threes — *omne trium perfectum* — could be applied.

Chapter 5
Visionary Contributions Gave a Happy Trio

THE ROLE OF SIGNALS
TO ACTIVATE OR SUPPRESS
NETWORK COMPLEXITY

The evolution of life on Earth is a magnificent story. The principles by which this had occurred did not become apparent until after the book *The Origin of Species* was published by Charles Darwin in 1859. Since then our insights into how different forms of life have emerged and developed have grown immensely. It is a fascinating story. As conditions on Earth have changed new forms of life have emerged. Originally development took place in the oceans and the existence of some kind of simple cellular life has been identified as far back as 3.8 billion years ago. Reading the books of life provide facts on genetic relationships as mentioned in my previous books on Nobel Prizes in physiology or medicine[1,2]. Comparison of differences between various species by use of genome characterizations has sometimes given surprising insights. New forms of life are continuously being unraveled and new unexpected relationships are identified. We start to get an insight into the early developments of life not least by studies of viruses and microorganism, which we shall briefly return to in the last chapter of this book.

It took a long time before multicellular organisms with separate functions allocated to different cells developed. During the so-called Cambrian explosion which took place some 550 million years ago a great diversity of shape and form of different animals emerged. And with time some kinds of animals started to adapt to a life on land instead of in the sea. The overall majority of these life forms then became extinct because of mostly unidentified dramatic changes in the conditions on Earth at different prehistoric times. The origins of these changes to some degree depended on the reshuffling of continents by plate tectonics and the associated volcanic activities introduced in the previous chapter. The major changes over long periods of time were connected to possible major earthquakes which could result in dramatic changes of the atmospheric conditions. But there could also be extra-terrestrial influences. After each extinction, which involved not only life forms on land but also to some, if lesser extent in the sea, the evolution of different forms of life started anew, departing from the limited numbers of survivors of the altered conditions. Two major extinctions occurred some 245 and 200 million years ago and, during this time, first the dinosaurs had started to emerge and later also the first mammals. The dinosaurs came to develop into many forms and sizes, but eventually they were also challenged by a major change in living conditions, which led to their extinction. This occurred 65 million years ago and in this case the catastrophic changes to the condition of their habitat were due to a major meteorite hitting Earth.

The source of the dramatic changes, also referred to as the K-T event, reflecting the transition from the Cretaceous (German *Kreide* for chalk) and Tertiary, the first main subdivision of the so-called post-Mesozoic era, for a long time remained mysterious. However, evidence accumulating towards the end of the previous century led to the conclusion that an asteroid (a large meteorite) postulated to have a diameter of about ten kilometers had hit the Yucatan peninsula in Mexico. The evidence provided was the increased concentration of the rare earth metal iridium in the area of the Chicxulub crater. In fact the general presence of this earth metal showed a major increase at the time of the postulated event. The impact dramatically changed the living conditions on Earth and only some selected species of animals survived. One of them was a form of color-blind shrew-like burrowing animal that became the ancestor of all the mammals living on Earth today. Although presumably a very rare event the challenge for humankind of a new asteroid hitting our world remains. However, an even more immediate challenge remains the possibility of major volcanic eruptions. In 1815 there was an extremely powerful volcanic eruption of the mountain Tambora on the Sumbawa island in Indonesia. This eruption had dramatic consequences for the climate in the Northern hemisphere. The year 1816 became the coldest summer since the year 1400. Crops failed around the world and there was a major shortage of food.

At Monticello the home of my "house-God" Thomas Jefferson the crops failed and he was forced into bankruptcy. As a means of stabilizing the disastrous financial situation Jefferson sold his unique collection of some 6,500 books to Congress. This became the major foundation stone of the Library of Congress, whose own smaller book collection to a large extent had been destroyed by the British during the 1812–14 war. But there were also other effects in the field of human culture. In 2010 there was a marriage between the Swedish Crown Princess Victoria and Prince Daniel in Stockholm. In my function as Lord Chamberlain-in-Waiting at the Royal Swedish Court I was the personal host of our guest of honor, Queen Margarete of Denmark. She is an impressive person who besides having a brilliant intellect is also a respectable artist in her own right. She is also a smoker and hence there were opportunities to personally discuss a range of matters of interest at intermissions to satisfy this vice. On one occasion I brought up the artist William Turner who painted some remarkable brightly colored sunsets. I referred to the fact that these spectacular motifs were due to the occurrence of ashes in the atmosphere as a result of a volcanic outbreak at the time. In this context I mentioned the 1883 eruption of Krakatoa, located in Indonesia. Then Queen Margarete said: "No, I think it was another volcanic outbreak." And of course she was right. I was not fully aware of the fact that Turner was such an early "impressionist," living between 1775 and 1851 and that hence it was the 1815 Tambora outbreak, also in Indonesia, that caused the atmospheric conditions giving the brilliant colors to his sunsets besides leading to world starvation due to failure of crops. However, I found it quite acceptable to be corrected by a Queen! It is now high time to return to our story.

The horseshoe crab.

The horseshoe crab, *Limulus polyphemus*, is a master of survival. It has populated Earth for some 450 million years managing the dramatically changing conditions, to some extent possibly dampened by the aquatic environment of its living. Because of its ancient origin it often is referred to as a living fossil, a contradiction in terms, Latin *contradictio in adjecto*. This blue-blooded invertebrate animal is kept together by an external skeleton and has a series of legs at its sides. Its carapace shell is dome-formed and it has a sharply spiked dagger tail. It is one among many crustaceans, a group also including crabs, lobsters, shrimps and others. However, despite its name genetically it is more closely related to spiders, ticks and scorpions than crabs. It is in fact not a crab.

The reason this exceptional animal is introduced here is that it has unique eyes. The animal even has several different kinds of eyes, but of particular interest in the present context are the relatively large compound eyes. This kind of eye is found in crustaceans and in insects. They are composed of thousands of independent photoreceptor units, so-called *ommatidia*, each one of which has a lens and photoreceptor cells, separately identifying brightness and color. The different ommatides point in different directions and hence in contrast to our eyes they allow integration of optical signals coming from many different directions. The ommatidia of the horseshoe crab are remarkably large, about a millimeter across. They are the largest receptors for light in the animal kingdom. In 1926 the large eyes of this unique animal attracted the interest of a young physiologist with a particular interest in the mechanisms of vision who at the time was visiting a marine biology station on the southern coast of Massachusetts. His name was Haldan Keffer Hartline. Let us look more closely at his background.

A Student Born to Become a Scientist

Haldan K. Hartline, recipient of one-third of the 1967 Nobel Prize in physiology or medicine. [From *Les Prix Nobel en 1967.*]

Haldan Keffer inherited an interesting mixture of genes. He was brought up by two enthusiastic naturalists. On his father's side he stemmed from Pennsylvania Dutch (in fact meaning German–Deutsch) stock. His paternal grandfather was a farmer and cabinet maker from Reading, Pennsylvania. One of his sons Daniel Schollenberger showed talents in school and studied to become a teacher. Unusually for that time he even pursued some studies at the universities in Bonn and in Heidelberg. It helped that German was one of his original languages and he also came to share this language with his son. In 1897 Daniel married Harriet Franklin Keffer. She came from a musical family and played the violin. However, professionally she, like her husband, was a teacher (of English) at Bloomsbury State Normal School. On the side she was an active gardener and shared a deep interest in nature with her husband. The couple had a son born in 1903, who remained their single child. He had been given two names, Haldan and Keffer, but came to be referred to by the latter name which originally was his mother's maiden name. Not surprisingly he was brought up to become a naturalist and his childhood in a large house in a very attractive setting in the Susquehanna Valley was harmonious. There were endless opportunities to learn from the well-read father — referred to as "my first and best teacher" — and also from his gardener mother. Besides a broad exposure to the diversity of flora and fauna he also acquired deep insights into astronomy and geology. When he had finished his schooling in Bloomsbury in 1920 his father sent him to Cold Spring Harbor, Long Island to learn comparative anatomy for six weeks.

It was then time to start College education and for this purpose Lafayette College Easton, Pennsylvania was chosen. He was inspired by an encouraging teacher of biology B.W. Kunkel to study land isopods, an order of crustaceans, including woodlice. He collected some of these in a culture only to ask Kunkel "What do I do now?" The answer was: "Well, that's research!" Hartline went back to his culture and observed that the animals seemed to avoid light. He started to search the literature for insights into avoidance responses to light in animals. A book by Loeb, the profiled scientist introduced in the first chapter, on *Forced Movements, Tropisms and Animal Conduct* gave him inspiration to start quantitative experiments researching visual responses to light. He examined the behavior of the wood lice Oniscus, when it was exposed to beams of light against a black background. Through this study he gathered his first experimental results. Kunkel had advised Hartline to apply to medical school, which he did, but apparently he did not have his heart set on a career to become a physician. He was a born experimenter.

Selig Hecht (1892–1947).

Before starting medical school he spent a summer at Woods Hole Marine Biology Station where he even met his research hero Loeb, the qualified contestant for a Nobel Prize who was introduced in Chapter 1. Loeb was delighted to learn about Hartline's data and shared them with Selig Hecht, whom we already met in the previous chapter. Hecht, a professor of biophysics

Visionary Contributions Gave a Happy Trio 185

at Columbia University in New York, was examining the light response of a clam, *Mya arenaria*, and was at the time one of the authorities in the field of visual research and color blindness. Hartline learnt a lot from this summer experience and even had his first scientific article published. However, he was very reluctant to start medical school and sought advice from the aging Loeb, who actually died the following winter. Loeb's response was that he should go ahead and study medicine since "there was no future in biology"! Hence Hartline started medical school in the autumn of 1923. He would have liked to expand his insights into mathematics and physics, but that had to wait. He did, however, appreciate that Johns Hopkins University lived up to its reputation as a school encouraging research. He established good contacts in the Department of Physiology and was introduced to the use of sophisticated instruments available at the time to measure fine electrical activities. He started to record retinal action potentials in frogs, rabbits and cats. The data obtained were published in 1925, but Hartline felt he had to try a new approach using eyes from species other than vertebrates.

Once when he had finished his experiment on a cat he noticed a big bluebottle fly buzzing around the carcass. He caught this insect, mounted it and placed electrodes on the eye and on the body. He then learnt that he could record a response which was even higher than the one he had recorded from the eye of the cat. This led to a switch of experimental objects towards use of insects and arthropods. Using the grasshopper he could confirm the validity of a concept called the Bunsen-Roscoe law. This law stated that the response of the eye to short flashes was a product of their duration and intensity. These results were published in 1928, but already the year before that he had received his medical degree from Johns Hopkins University. His own formulation was that he was granted his degree only under the promise that he would never practice medicine!

During the summers of his medical studies he continued to visit marine biology stations, an environment to which he had been introduced by his father. It was in the summer of 1926 that he had his first encounter with the horseshoe crab, introduced above. He found it hard to believe his father when he had pointed out that the nodule on the side of its head was an eye, but of course he was right. It was during summer visits to Woods Hole Marine Biological station that he developed his work on this exceptional animal and simultaneously he enjoyed the teachers giving the exciting summer courses. His work on *Limulus* focused on dark adaptation. He learnt a lot from the single-fiber preparations he could obtain from this animal. But Hartline was also an impressive hands-on experimentalist and became involved in improving the electronics of his readings. He built an amplifier for a small permanent magnet string galvanometer. Similar technical development work took place in other laboratories, not least in that of the forthcoming Nobel Prize recipient Edgar D. Adrian, introduced earlier[2] and also mentioned in the previous chapter.

Hartline was not satisfied with his basic education in mathematics and physics and therefore went to Europe. There he joined in the lectures and discourses at the two Meccas of these disciplines — the one managed by Arnold Sommerfeld, the mentor of many Nobel Prize recipients in physics, in Munich and the other by Werner K. Heisenberg in Leipzig, who three years later received this prize. Together with students of the latter group he traveled to Berlin to hear Einstein present his unified field theory, but again the physics was beyond what he could comprehend. To his disappointment it turned out that he did not have sufficient background to follow the advanced teaching. He noticed this with regrets but took the opportunities offered by the visit to Europe to expand his mountaineering experiences both on skis and climbing. During the one and a half year he was away from the U.S. he got to know many parts of Europe and could also do some studies on the retina at the well-known Zoological Station at Naples. In England he could devote himself to another of his many fields of interest, astronomy.

Hartline's Career Takes Off

Detlev Bronk was a scientist, educator and administrator of high caliber. He established the sub-discipline of biophysics. As a relatively young scientist he was offered the Directorship of the Eldridge R. Johnson Foundation. This was the foundation that Granit and also Rushton had already joined at its start in 1929, as mentioned in the previous chapter. In 1931 Bronk also recruited Hartline, who thrived in the environment of biophysics and with the collegiality of the other high-quality physiologists employed. Hereafter Hartline

Detlev Bronk (1897–1975).

followed Bronk wherever the latter's leadership talent took him. This was true both in 1949 when he became the president of Johns Hopkins University and in 1953 when he moved to the prestigious post of President of the Rockefeller Institute. Already from the start Bronk began to change the organization of this institute to embrace university characteristics. In 1954 the institute started to award Ph.D. degrees. The group leaders originally referred to as members were called professors. The introduction of these changes eventually led to the formal transformation of the Institute into a University in 1965. Bronk stayed on as president until 1968. It can be added that he was also, on the side, the President of the National Academy of Sciences, 1950–62. With Bronx at the helm Hartline always had the resources he needed for his original research. The other two scientists who Hartline got to know in 1931, Granit and Rushton, became friends for life. They all became deeply involved in analyzing the neurophysiological basis for the function of the retina, Granit up to the end of the 1940s, as we have seen, whereas Rushton got involved in the field of vision hereafter, spending the second half of his career making important contributions. It was at the Johnson Foundation that Hartline met his future wife, Elisabeth Kraus. She was a psychologist who came to work at the Foundation. Her father was a renowned chemist and she had a broad academic background also including biology.

In 1931 Hartline made a very fundamental observation. He managed to make a recording from a single nerve fiber of the eye of the horseshoe crab. Together with his collaborator he had attempted this throughout the summer, but consistently failed. Careful dissection of nerves was performed, but it was first in the last two days of experimentation during the summer that success was achieved. Single-fiber responses were identified, which a few years earlier had been observed by Adrian and Zotterman in work on the frog stretch receptor as described earlier[2]. The secret turned out to be to use large adult animals and this became the system that Hartline continued to refine and develop throughout his career. A number of fundamental discoveries were made. One of them was summarized by himself in 1974 stating "The second development in our studies of the *Limulus* eye turned my attention away from receptor mechanisms. Once alerted I could easily show that shading the regions of the eye neighboring the receptor whose nerve fibre I had isolated restored its activity." This phenomenon was referred to as *lateral inhibition*, a discovery first presented in 1949. In essence the expanded work showed that the retina was more complex than simply activation of light-sensitive cells and ensuing signaling in neurons. There were also other cells in the retina involved in the integrative processing of the information about light, under different conditions. It should be added that Hartridge also expanded his work to involve the mechanisms of function of a vertebrate retina. He used the bullfrog in these experiments.

Once when he was flying his own airplane it struck him that it might be possible to get hold of a single nerve fiber in the retina by focusing on the blind spot mentioned earlier. As the nerve fibers assemble to exit the retina by passing through its cell layers there should be an opportunity to identify individual nerve fibers. He was successful and could classify their responses into three categories; "on," "off" and "on-off" fibers. He then decided to return to an experimental system he had used earlier, the scallop, *Pecten irradians*. The retina of this animal has two layers and each layer has its own exiting nerves. The outcome of the experiments was exciting. The inner layer of the retina showed a response like that he had found in the horseshoe crab, an "on-response." The other outer layer, however, showed instead a pure "off-response," reflecting the "shadow reaction" mentioned above. It should be added that these delicate experiments became possible only because Hartline was a master dissector and he only used tools of the highest quality, often sharpened by himself. As emphasized, not infrequently advanced hands-on work is very important for the development of experimental science[2]. In contrast to the high quality precision instrument employed it needed to be acknowledged, according to Hartline's close collaborator for decades Ratliff that the laboratory was an untidy environment. In fact he even qualified it to be "a slightly disorganized, but extremely fertile, chaos." In the memoirs he wrote he summarized Hartline's four major achievements as follows[3]:

> "Hartline's four major accomplishments were all 'firsts' in their respective fields: With Clarence H. Graham he recorded the activity of single optic nerve fibers. He mapped the activity of the visual receptive field to reveal a system of many convergent pathways from many photoreceptors (the foundation for modern concepts of parallel processing by specialized channels). He recorded — with Wagner and MacNichol — intracellular generator potentials. And finally, he discovered lateral inhibition in the retina and described the integrative activity of neural networks with the Hartline–Ratliff equations."

We will return to these and other discoveries in the reviews by members of the Nobel Committee and also in Hartline's Nobel lecture.

A Rich Personality

The memoirs of Hartline by his close friend Granit together with his most important collaborator Ratliff[4] and also by Ratliff alone[3] give a picture of a very likable person. He thrived in sophisticated experimental work and he was a technical genius in designing experiments. His broad insights into biology at large were very helpful. Administration was not to his liking. He detested deadlines and once they had passed he simply noted that that was that. He seems to have lived following what I think was one of Mark Twain's maxims "Do not delay until tomorrow what you can do the day after tomorrow!" He had a special, no in fact two, filing boxes labeled "PROCRASTINATE" and they were just that. As to the etymology of this word it is worth remembering that Greek mythology tells about Poseidon's son Procrustes who had an iron bed into which he forced everyone regardless of their length. He used stretching or amputations. In summary Hartline seems to have searched throughout life to find the proper balance between virtues and vices. He had a good dry sense of humor and was a charming and personable host. Even when he had moved his science to the Rockefeller Institute the family remained in their home outside Philadelphia. It was located in charming natural surroundings and was called Turtlewood. His wife Betty and their three boys, who eventually all became biologists, continued their rich family life there. Hartline took long weekends to join them. He preferred to live in the country rather than in the large city of New York. Their home was an open house for friends and colleagues — there was no "lateral inhibition"! And still, at heart, Hartline has been described as a very private person.

Hartline's preferred recreations were climbing and flying, briefly mentioned above, but also sailing. During the 1930s he was to sail with Granit in the Baltic archipelagoes. He also participated in weekend sailings with the Bronk family along the coast of Maine. We know about the sailing friendship that developed between Hartline and Granit from their correspondence. Granit's archives, including his large correspondence, are cared for at the Centre for the History of Science, where I have my office. The richest part of this correspondence is with Hartline and with Eccles, his friends since the late 1920s and the early 1930s. The early handwritten and hence partly illegible letters from Hartline contain a lot of technical advice including hand-made drawings. They are less personal, but this changed after their joint sailing in 1938. Hereafter they addressed each other by their given names. It can be added that in Sweden in the 1930s titles were very important reflecting the major bourgeois influence in society. What may seem surprising today is that the telephone directory of course was sorted under family names, but then subdivided alphabetically according to the first letter of the professional title. In Sweden there was nothing similar to the English "you" or to "you, Mrs Brown." The need for a convenient term to address each other in the immigrant country of the U.S. must have been a historical necessity. By way of contrast the Swedish "Du" was used only between close friends and after mutual agreement. This changed in 1968 and the person of major influence in this reform was the recently appointed Director General of the Royal Board of Medicine Bror Rexed. He had a background in neurobiology working with the central actors of the previous chapter, Granit and Bernhard. Rexed presented a thesis on "Postnatal development of the nervous system" in 1945. It was mostly a morphological study. He then held different academic positions at the Karolinska Institute, after which he became Professor of Anatomy at Uppsala University. However, he is best remembered for his public service involvements during the 1960s at the National Medical Board (Medicinalstyrelsen). He introduced the use of "Du" firstly for collegial address among all his employees. This practice progressively later spread through society, but there was also considerable resistance to this reform. An alternative use of "Ni," previously serving only in a plural form, was introduced to allow a grading of proximity of the relation, but this never caught on. With time Du has become a general way of addressing people you know as well as those you have not encountered before. It is now high time to return to the joint sailing adventures of Granit and Hartline.

Bror Rexed (1914–2002).

In 1938 Hartline and his wife arrived by boat at Mariehamn in the Åland islands, after a challenging climbing trip to the Jotunheim mountains in Norway. At Mariehamn they met not only Granit and his wife, who had sailed there from their home island Korpu in their uncomfortable 6 m R-yacht.

One more boat joined them at Mariehamn. The skipper was the professor of physiology at Harvard, the electro physiologist Alexander Forbes. He and his family had arrived from Copenhagen in Denmark where they had picked up a newly-built ketch. This was of the kind referred to as wish-bone ketch, meaning the presence of a permanent splitting gaff between the two masts. These two kinds of boats for obvious reasons had very different performances. The small R-yacht was much faster in particular when tacking. When the sailing in the Finnish archipelago was over, Forbes' boat, named "Stormsvala" (a Scandinavian name meaning storm petrel) was transported by a steamer to the U.S. Hartline and his wife afterwards were often a part of the crew when this boat made weekend sailings from its home harbor, Naushon, opposite Woods Hole. But Hartline also had his own sailing boat which the family used when visiting their summer retreat nearby.

Way into his seventies Hartline, against the recommendation of his cardiologist, decided eventually to make the journey through Grand Canyon, which he had planned and of course postponed for many years. He and his wife took the trip and it was quite an endurance test naturally because of rafting under varying weather conditions. Still they enjoyed it and were very pleased to have managed it. Grand Canyon is a remarkable piece of nature. The age of the canyon is hotly debated and predictions vary from 6 million years to ten times this long, stretching perhaps even back to the ages of the dinosaurs. Whatever the truth is it is a remarkable piece of nature. In the late 1980s my wife and I and our two younger children, at the time in their early 20s had a fantastic experience of the valley taking the mule ride down to Phantom Ranch, staying overnight and riding back the next day. This was early November and hence we started on a frosty morning, but when we reached the first watering station at Indian Gardens the temperature had already become much milder. During our ride it increased further and was very comfortable when we reached the ranch at the end of the day, where we slept well in its rustic milieu. My horse-like mule, Mercedes, safely managed the narrow hairpin bends, including Jacob's Ladder. It was all a matter of forming "a nice compact group." Like the descent to the bottom of the canyon, climbing out of it the next day in pastoral light was an equally unforgettable experience. There is something magic about this piece of nature.

The multitalented Hartline also had many other intellectual interests. He was well-read in American history and enjoyed classical music. He was a frequent visitor to the classical music concerts that for a long time and into the present are offered on the Rockefeller University campus. The role of astronomy as a major obsession of studies seeded by his father has already been alluded to. Together they saw Halley's Comet in 1910. But since Hartline died of a heart attack in 1983, he did not have an opportunity to see the return of the comet in 1986, which he had hoped for. He had declined any official memorial service after his death. Instead there was a concert at Rockefeller University dedicated to him performed by the Stuttgart Chamber Orchestra, conducted by Karl Münchinger.

The Nobel Committee Reviews Hartline

Hartline was nominated for the first time in 1949. He was proposed together with Granit by none other than the 1932 Nobel laureate in physiology or medicine Adrian. The nomination highlighted the 1932 discovery together with Graham based on readings of action potentials in single optic nerve fibers which had made it possible "… to construct visibility curves relating to wavelength to simulating effect for different colors." As presented in the previous chapter von Euler was selected to be a reviewer from the committee and in addition the unusual step was taken of using an external reviewer, in this case the 1944 Nobel Laureate Gasser, to make a parallel analysis of the two candidates. Since von Euler had reviewed Granit before he only added information on him that was new. Still the major part of the review, nine out of 14 pages, was primarily devoted to Granit's work. Four pages specifically commented on Hartline's work. Hartline and Graham's successful registration of impulses from individual nerves of Limulus was praised and compared to preceding similar work in other systems by Adrian in collaboration with Bronk and with Zotterman. As in other nerve sense organs subjected to, for example, stretching or pressure, higher intensity of exposure correlated with a higher frequency of signaling. The use of short light exposures also demonstrated that a certain amount of light was needed to elicit a response. The visibility curve analyzed by use of light of different wavelengths had demonstrated a maximum at a specific wavelength, just like the characteristic response of the darkness-adapted human eye. Von Euler considered that the experiments examining the receptor elements in vertebrate eyes were of particularly importance. It was found that on-off effects were common in vertebrate eyes, but absent in the *Limulus* eye. The conclusion was that Hartline had made fundamental contributions to knowledge about the physiology of the retina. Somewhat surprisingly the summarized view was that Hartline, based on his contributions, could not alone be considered worthy of a prize, but when combined with Granit it might be justified to award him the prize. This proposal was in conflict with the rule of agreement that when the prize is given to more than one recipient, each one of them should be capable of carrying a prize on his own.

Gasser's three-page letter evaluating the two candidates was somewhat more positive about Hartline as compared to Granit. Hartline's discovery of "on-fibers," "on-off fibers" and pure "off-fibers (with the recognition that the latter depend on synaptic connections)" was brought to light. They had been confirmed by Granit and used extensively by him in his work. Hereafter Gasser credited Granit with having "…placed color-vision physiology on a new plane." In the second to last sentence he said "had I done what I have now done last winter, in conformity with the rules, I should have nominated Granit and Dr Hartline jointly." He had only nominated Granit, as presented in the previous chapter. It was in the year 1949 that Jorpes on his own initiative wrote a provocative analysis questioning Granit's candidacy. The consequence was that the committee considered that neither Granit's nor Hartline's contributions should be considered for a prize at the time. Three dissenters von Euler, Liljestrand and Zotterman were of a different opinion and found both candidates worthy of a prize.

Two years later Liljestrand, the secretary of the committee, strategically nominated Hartline to make him available for discussions. Again the committee decided to use a reviewer within their own circle, Zotterman, and also to seek advice externally, from Adrian, a previous nominator. Zotterman as always wrote a very comprehensive 18-page review focusing mostly on Granit. He wanted to reemphasize Granit as a strong candidate on his own. The last six pages of the review, however, discussed Hartline's contributions. Zotterman was moderately positive. He referred back to von Euler's extensive review two years earlier and then noted that there had been only a few additional publications from Hartline's laboratory. It was mentioned that a part of the reason for this might be that, during the late 1940s, he had moved his laboratory from Philadelphia to Baltimore. Zotterman gave Hartline credit for the pioneering registration of nerve impulses from individual nerve cells in *Limulus*, extending the earlier work by Adrian and himself. The new findings led to the important conclusion that the retina of the "crab" behaved like the dark-adapted, scoptic, mammalian eye. It was then emphasized that Hartline was the first to register the electrical signaling in individual nerve threads in the innermost layer of the retina in vertebrates and their division into the three categories as already mentioned. Zotterman referred to the unresolved observation at the time of finding all three categories of responses in the eye of the grass snake, which is only composed of cones. He then praised Hartline's studies of the two-layer retina structure, with contrasting signaling patterns, in clams. Zotterman's conclusions were somewhat contradictory, probably because he wanted to secure Granit's position as the dominating scientist in the field. He cited von Euler who in 1949 had stated "although Hartline's and his collaborators contributions are particularly beautiful it is difficult to let them alone form the basis for recognition by a Nobel Prize." Zotterman declared that he whole-heartedly agreed with this statement. He then went on to praise Granit's "large richness of ideas and his larger self-confidence as a scientist." Somewhat unexpectedly he concluded, after having cited Gasser's earlier review, that he wholeheartily supported a joint prize to Granit and Hartline. No wonder the committee was somewhat bewildered. As mentioned in the previous chapter the committee at large supported a prize to both Granit and Hartline but three members, Bernhard, Hammarsten and Jorpes disagreed and one member, Kristensson abstained from voting. Thus Hartline's candidacy was not off to a good start.

It would take nine years until Granit and Hartline again were nominated for a joint prize. In 1960 the Professor of Ophthalmology in Utrecht Jurriaan ten Doesschate nominated the two candidates together. A reference was given to Granit's two books summarizing the field and for Hartline some ten different key references were given. As mentioned in the previous chapter no review was made but the committee, including Bernhard who had previously objected to a prize for the two nominees, unanimously voted that they were worthy of a prize. After an intermission of six years a repeat nomination for a prize to Granit and Hartline was submitted by Adrian, but this time he also included Wald. In addition there was another nomination for all the three candidates together by Heuymans, as mentioned earlier. The committee asked Zotterman to make one more concluding review. This review provided broad analyses of the three candidates and Hartline's work was described over seven pages. Zotterman started out by citing from the nomination by Lord Adrian. It read:

"Keffer Hartline's name was put forward with Granit's on an earlier occasion. His records of visual discharges from the eye of *Limulus* have given precise information about the action of light as a stimulus and they have led to the recognition of an important fact in connection with the physiology of sensation generally. This is the effect of lateral inhibition as a means of heightening contrasts and providing a general 'editing' of the sensory message."

And further on

> "Hartline's records from the *Limulus*' eye have in fact provided the basis for much of the work on the visual regions of the brain which is beginning to show how visual information is organized into the temporal and spatial patterns which determine behavior."

Zotterman hereafter cited extensively from his earlier evaluation of Hartline, whereafter he focused on his more recent contributions. He supported that priority should be given to Hartline for his discovery of lateral inhibition. It serves to remind that discussions about priority to a discovery represent one of the most important parts of evaluations of candidates for a Nobel Prize. It can be noted that Granit had subjectively given a somewhat different interpretation of the priority in this case. In his biography[5] he argued that he observed this phenomenon in 1934 and that it aroused in him as translated from Swedish, "an incredible pleasure." And further on: "Eureka, I could exclaim: the lightening-up and turning-off processes were antagonists in the retina (translated from Swedish)." Later on he went on to say: "The blocking of impulses was confirmed the following year by Hartline." Sometimes it is very difficult, not to say impossible, to decide on priorities, when the critical experiments were made and how this relates to when they became published. Zotterman presented the sequence of events to encompass discoveries of Hartline of the already mentioned three elements "on," "off and on" and "off" in the early 1930s whereas Granit in studies of the electroretinogram reached his conclusions on this matter towards the end of the 1930s. Hereafter Zotterman gave some details of the *Limulus* eye as a background to the presentation of Hartline's results.

This organ as mentioned is a faceted (insect) eye including about one thousand ommatides, each of which functions as its own eye. The integration of the information accumulated by all these "eyes" takes place more centrally in the nervous system. Each ommatide contains about a dozen cells of two different kinds, wedge-formed retinal cells and one bipolar neuron, a so-called eccentric cell. Nerve fibers extend from both kinds of cells and they compose the thin bundle of nerve threads extending from an ommatidium which in its turn interconnects with nerve bundles from other ommatides to form a complex plexus. Confusingly, Zotterman then went on to say that the lateral inhibition was discovered and described by Hartline as late as in 1949. One wonders if there may be some confusion here in the use of a specific terminology. Be that as it may, the high-quality work resulted in the formulation of the Hartline-Ratliff equations already referred to above. They have turned out to be important in the later research on more advanced visual systems in different vertebrates, such as the later work by Barlow and by Kuffler in 1953 and the much later work by Torsten Wiesel and David Hubel that brought the latter a Nobel Prize in physiology or medicine in 1981[2]. Towards the end of the review the introduction by Hartline and collaborators of computers for neurophysiologic work was mentioned. This was pioneering work.

In the year of the prize, Heymans repeated his nomination of the three candidates. As already mentioned Bernhard was selected to review Granit and Hartline. He now had an opportunity to resurrect Granit, who he earlier, perhaps somewhat surprisingly, had been hesitant about. In addition he could, for the first time, give his views on Hartline. He first referred to the fact that Wald whom we will soon meet, had been declared worthy of a prize first in 1965 and that the year after he together with Granit and Hartline, were considered a very strong combination of candidates for a prize. Although no formal ranking had been made Bernhard remarked that they were second to Rous and Huggins. His introductory presentation of Hartline was full of praise. It read (translated from Swedish):

> "Hartline's discoveries concern the properties of the photoreceptors to become excited and the relations of the impulse patterns conveyed from the receptors to the quantity of the light stimulation and the principles of treatment of information in the neuron channels that lead from the intercommunicating sensory cells (some sentence!, my remark). His production is not extensive but each publication is a pearl and Hartline's results and conclusions represent cornerstones in sense physiology."

Bernhard then praised Hartline's exceptional technical skills and his choice of biological systems to examine. His description of the different discoveries made by Hartline and collaborators was full of enthusiasm. He praised the choice of the Limulus system because the unit of vision, the ommatidium, is so large that individually it can be exposed to light, that individual fibers from the receptor units can be isolated and put in contact with electrodes and finally that there is no layer of ganglion cells, between the stimulated receptors and the impulse carrying vision nerve fibers, as in the case of vertebrate retinas. It was these conditions that allowed the formulation of the basic principles of generation of impulses when light of different intensity and duration was used, and their further spread, a research field in which Bernhard and his collaborators themselves had been involved. It also allowed an analysis of the sensitivity to light of different wavelengths. The characteristics found could later be correlated with the occurrence of a specific photochemical substance in the Limulus compound eye, as we shall see. According to Bernhard, Hartline had presented "the code for the integration of in-put signals to central nervous systems."

Bernhard later in his review returned to work made by use of the Limulus eye in the late 1950s and which hence not had been commented on before. In this work it had been found that there were interconnections between neurons, which had not previously been recognized. Such interconnections could explain the fundamental phenomenon of lateral inhibition. This advance was used in further in-depth studies of the dynamic underpinning of visualization of fundamentals like form, contrast and movement. This field was later to be followed up by many other eye neurophysiologists. In the early phase of such studies Hartline and Ratliff defined the fundamentals of threshold conditions, the interaction between separate groups of neurons and the quantitative relations of the resulting inhibition and the distance to the units in which the inhibition was expressed. It was shown in the early 1960s that responses in the eyes of Limulus by complex integrative processes could approach perceptions managed by the vertebrate eye concerning identification of form and movement[6] and interpreting spatial contrast effects.

Bernhard also praised Hartline's work in the late 1930s with the vertebrate eye, using the frog retina as his experimental object. The already mentioned identification of three different reaction patterns to light exposure, the "on," the "off" and the "on-off" reactions, was highlighted. The results were referred to as milestones in eye physiological research. Not surprisingly Bernhard concluded that both Granit and Hartline were very strong candidates to a prize and that a combination with Wald was very attractive as discussed by the committee in 1966. Thus it is now time to see what additional qualities he could bring to the group of three wise men or were they three musketeers?

An Exceptional and Narcissistic Eyewitness

George Wald like many Nobel Prize colleagues had a Jewish background. His mother came from Germany and his father from Poland. It was in November 1906 that he first saw the light of day in Brooklyn, New York. He had his upbringing in a working-class neighborhood and his parents ensured that curiosity about mechanical things and learning about science were stimulated. His success in building a crystal detector radio made him popular among his friends. Together they could now listen to the 1919 World Series. He received manual training at the Brooklyn Technical High School and considered for a while becoming an engineer. However, he had an appetite for wider involvements. This was met when he entered Washington Square College of New York University, where he was exposed to art, classical music and literature. He retained interests in the broad aspects of humanities and in addition developed a deep involvement in the justice of human

George Wald, recipient of one-third of the 1967 Nobel Prize in physiology or medicine. [From *Les Prix Nobel en 1967*.]

affairs. He was in fact a pre law student which is said to have been due to his documented success as a performer in a vaudeville act at a neighboring Jewish community during his high-school years. During two summers of his College years he gained important experiences working on board a passenger ship en route between New York and Buenos Aires. This social experience stimulated him to change orientation and he became a pre-medical student. Still, like Hartline, he did not have his heart set on making a clinical career, he was too curious about science. Like many others he was stimulated by reading Sinclair Lewis's *Arrowsmith*, introduced in Chapter 1. He was accepted for graduate studies at Columbia University.

In this new environment he was exposed to some very influential teachers. The 1933 recipient of the Nobel Prize in physiology or medicine Thomas H. Morgan, gave the course in genetics and Hecht, whom we already met above, taught biophysics with an emphasis on his favorite object, the eye. As mentioned his main contribution to science was the documentation that visual mechanisms conform to photochemical laws. The exposure to his thinking sealed Wald's future. It became his mission to transform Hecht's concepts, that a photosensitive substance was decomposed by light into two subcomponents which became re-associated during an ensuing darkness. It was interactions of this kind that Wald and his collaborators were eventually able to explain in molecular terms.

Not surprisingly Wald received his first experience of science in studies of the fruit fly. He examined its visual performance. Interesting similarities to the properties of vision in animals and even man were found. Dark adaptation that was the target of studies by Hecht at the time also caught his interest. Since Wald wanted to go deeper into the molecular mechanisms of importance for seeing he looked for a reputable chemical laboratory to join. Fortunately he was awarded a National Research Council Fellowship for studies in Otto Warburg's laboratory. Warburg at the time had received his 1931 Nobel Prize in physiology or medicine and his laboratory was a very dynamic place. The existence of visual pigments had been identified already in 1876 by the German physiologist Franz C. Boll. He had noticed that it had a reddish-purple color and originally it was named visual purple. Later it was named *rhodopsine* from Greek *rhódon*, rose, and *opsin*, sight. The color of rhodopsine is changed to yellowish-orange and further on to colorless when exposed to light. Its chemical properties demonstrated that it was a protein.

During his stay with Warburg Wald analyzed the light absorption spectrum for rhodopsine and deduced that it was a carotenoid (organic pigments from plants having vitamin A activity)-linked protein. The discovery of vitamin A and its physiological role had already been recognized by a shared Nobel Prize in physiology or medicine to Frederick G. Hopkins in 1929. Since it was known at the time that vitamin A deficiency could lead to night-blindness (nyctalopia) and further to complete blindness, Wald speculated that the vitamin might play a direct role in the process of seeing. In order for Wald to consolidate his finding he was sent by Warburg to a laboratory in Zürich managed by Paul Karrer. Within a few years, in 1937, Karrer received a shared Nobel Prize in chemistry "for his investigations on carotenoids, flavins and vitamins A and B2." From Zürich, Wald continued to the laboratory of yet another Nobel Laureate in physiology or medicine. It was Otto F. Meyerhof in Heidelberg, who in 1922 had shared a prize with Archibald V. Hill.

This scientific odyssey well illustrates the importance of choosing the right mentors, viz. if one wants to become a Nobel Prize recipient one should choose to work in laboratories of other previous prize recipients. The situation that developed also illustrated the increasing threat to Jewish scientists in the advancing 1930s in Germany. Both Warburg and Meyerhof worked in non-governmental institutes, not immediately affected by the anti-Semitic legislation which heavily afflicted civil service institutions. Warburg was able to continue his tumor research throughout the war, apparently because Hitler was afraid of cancer, but Meyerhof finally had to leave the country for France in 1938 only to have to carry on what became an adventurous journey via Portugal to the U.S. A position had been arranged for him at the University of Pennsylvania by the Rockefeller Foundation[7]. The emerging situation also became a threat to Wald and the National Research Council called on him to come back home. However, before he did this he took an opportunity to conduct what turned out to be a very critical experiment in Meyerhof's laboratory.

Holiday time had begun when a shipment of some three hundred frogs arrived. The first reaction of the remaining staff was to release the animals, but then Wald requested them for an experiment. He dissected their retinas, used them unexposed and exposed to be bleached to the yellow color and hereafter subjected the different materials to various selected solvents. He then discovered the existence of a new kind of substance he called *retinene*. In other preparations he instead found rich quantities of vitamin A. This led him to a bold postulation of a vicious cycle. The proposal was made that retinene normally was bound to a protein to form rhodopsine but that exposure to light yielded the visual yellow product. He then speculated that there was a gradual conversion of retinene into vitamin A, reflected in the appearance of visual white. Reversion of the process was suggested to make rhodopsine again to become available. These groundbreaking findings were published in *Nature*. At the time of publication he had returned to the U.S. and used the second year of his fellowship doing physiological research at the University of Chicago. From there he moved on to his final academic home, Harvard University.

At Harvard he became instructor and tutor in biology in 1935 and faculty instructor in 1939. He then advanced to associate professor in 1944 and full professor in 1948, after which he remained at Harvard till the end of his career. The first experiments at Harvard obviously aimed at demonstrating whether the visual circle identified in the frog could also be identified in other vertebrates. Like many other successful biologists and physiologists Wald had a very close relationship with the Marine Biological Laboratory at Woods Hole, Mass. This is a private institution established already in 1888. Since 2013 it has become affiliated with the University of Chicago. As has been mentioned in previous chapters of this book and in my earlier books on Nobel Prizes[1,2] a large number of forthcoming Nobel Prize recipients have enjoyed the teaching and research at Woods Hole, generally in the summer. Wald was no exception. For many years he returned to the station as a researcher and he also participated as teacher in its reputable course in physiology. Later he also became a Trustee.

I have visited Woods Hole only once and my host was none other than Byron Waksman. He was the son of the 1952 Nobel laureate Selman Waksman.

Visionary Contributions Gave a Happy Trio 201

Since his father had a laboratory at the marine biology station he had spent all his summers there. He knew all the back alleys. Byron himself became a reputable scientist with particular interests in autoimmune diseases. This was the reason why our paths crossed. In my laboratory at the time we had a major interest in the restricted immune responses in the brains of patients with multiple sclerosis. This involvement of possible autoimmune responses in the disease led to our interaction. To this should be added that Byron at the time had become the vice president for research programs and medicine at the National Multiple Sclerosis Society in the U.S. The reason for our encounter at Woods Hole was that he had arranged a conference there for discussions of the state of the art of knowledge about autoimmunity and multiple sclerosis. It was a memorable visit to this historically important place, not least due to Byron's insightful and charming hosting. Let us now return to Wald.

He used his summers at Woods Hole to investigate the vision of different kinds of fishes. He could confirm the existence in salt water fishes of the cycle found in mammals; rhodopsine released retinene, later converted to vitamin A. However, when examining fishes from brackish water interesting differences were found. Earlier studies had demonstrated that this kind of fish had a different form of visual pigment, named pyrophyropsin, since it had a purple color. It took light of longer wavelengths to activate this visual pigment and the products identified were different. They were referred to as retinal$_2$ and vitamin A$_2$. Logically the next step was to examine fishes which alternated between fresh and salt water or animals like frogs which spawn in fresh water and then live on land. Interesting observations were made and for example it turned out that during metamorphosis of tadpoles to frogs a switch was made from the use of vitamin A$_2$ to A$_1$. These findings led to reflections on aspects on the evolution of life in general. Later in his career Wald enjoyed writing books on particular aspects of evolution for larger audiences.

It was an obvious temptation to examine if the phenomena in cones might differ from those found in rods. In early experiments using chicken retinas Wald could isolate a pigment that was activated at longer wavelengths, which was more sensitive to red light. It was referred to as iodopsin, but its break-down products could not be confirmed since it was always admixed with rhodopsine. It took time to fully interpret the prevailing situation and this was not clarified until by experiments in the 1950s. During the time of the Second World War Wald was involved in developing binoculars that could be used by soldiers at dawn. It turned out that it is possible for the human eye to see into the region of infrared. In peacetime these binoculars are of great value for use by night bird watchers and hunters. Once at dusk when we were approaching the main island of Fiji in Craig Venter's sailing boat Sorcerer 2 during the global circumnavigation to search for all forms of DNA in the oceans, as described earlier[1], I tried this kind of night binoculars. It was fascinating to observe how visibility under these conditions could be improved.

When Wald was able after the war to return to studies of cone pigments he built a strong research group including Ruth Hubbard and Paul Brown. Later Ruth became Wald's second wife. At this time another research group had demonstrated that retinene was an aldehyde of vitamin A. It could now be demonstrated that the cone pigment iodopsin used vitamin A_1 and retinene and it was postulated to use a different opsin protein. New terms were introduced; retinene was given the name *retinal* and vitamin A was referred to as *retinol*, designations used into the present time. As knowledge advanced it became possible to examine in the test tube the interaction of retinol, opsins and the appropriate enzymes. To cut a long story short it turned out that retinol derived from natural sources like cod liver oil, which was used as a source of fat-soluble vitamins in my childhood, could appear in two different chemical forms. These were referred to as cis-trans isomerization forms. These terms are defined by the form in which the chemical double-bonds are present in the molecule. Wald and his collaborators were able to show that the precursor to all visual pigments was one particular isomer, a so-called 11-cis isomer. This finding has been considered as the first example of a biological importance of a *cis-trans* formation. In order to give a background to explanations of further developments it may be appropriate to cite a section from Theorell's evaluation in 1965, which aimed at providing background information to his fellow committee members. It read (translated from Swedish):

> "Both rods and cones contain light sensitive receptors connected to carotenoids as a chromophore group and proteins, opsins. The chromophores derive either from vitamin A_1, with a double-bond, or vitamin A_2. Animals with a retina containing A_1 usually do not have A_2 and vice versa, but there are interesting transitions connected with the metamorphosis of certain animals. Rods and cones have different opsins, and hence one can distinguish between four forms of chromo proteins: Rhodopsine with A_1-aldehyde = retinal[1] and porphyrosine, with A_2-aldehyde = retinal[2], in rods; iodopsine with retinal 1, cynopsin, with retinal[2] in cones. These forms appear in the dark-adapted eye. It has been known since long that upon exposure

to light a bleaching of these chromo-proteins occurs, after which they are rebuilt in darkness."

After this summary intermission let us go back to a description of further elegant experiments in which it was possible to answer the central question of what effect light has on rhodopsine. By complicated experimentation it was found that it was a matter of a single critical chemical effect. This was that light isomerized 11-*cis* to the all-*trans* form. However, this was only a part of the postulated cycle. It remained to understand how the molecule could return to the original 11-cis form. It could be speculated that this involved various conformational changes of the protein. This part of the story was not resolved until two decades later. It represents a sequence of reactions involving the G proteins mentioned in Chapter 3 and cascades of enzyme reactions of the kind found to be prevalent in biology. It can finally be mentioned that for periods Wald was also involved in studies of vitamin A deficiency and night blindness and also in presence of cones of different spectral sensitivity in normal and color-blind humans. Vitamin A deficiency may lead to night blindness and in more severe cases to complete blindness and other severe symptoms causing death in several hundred thousand individuals each year in the world. There is an urgent need to prevent these major problems of vitamin A deficiency and the best way forward appears to be to use the so-called Golden Rice. This is a form of rice which by use of gene technology has been enriched to also contain vitamin A. Regrettably there has been strong resistance by certain civic groups, in particular environmental protection organizations, to accepting the registration of this much needed product. As already mentioned in my previous book on Nobel Prizes[1] there is an appeal arguing for the formal release of the Golden Rice for global use signed by 121 Nobel laureates (as of 2016).

Finally it can be added that Wald was also involved in studies of color blindness. Red-green color blindness, *deuteranopia*, is a common disease in men affecting about 6–8% of that part of the population. The difference in colors identified by a normal eye and a deuteranoptic eye can be seen from the picture. Both of our two sons have this kind of color blindness, which their mother carried over in her genes from her father. Wald registered a pattern of absorption of rod pigment (black curve) and of absorption of cone pigments (red, green and blue curves) as already illustrated on page 150.

The memoirs about Wald[8] also highlight his many other rich qualities including those required to be an educator. He was a superb lecturer and also an excellent and engaged writer. He took on the large topics of the origin of life or of the origin of death. In his actions he represented a politically committed humanist. He heavily involved in discussions on the Vietnam War, had strong views on the military-industrial complex and of course also on the arms race, including the development and use of nuclear weapons. When he finally retired from laboratory work he fully involved himself in these large political matters. But, like many other great scientists such as Crick, Edelman and, as we shall see in the next chapter, Marshall Nirenberg, he also remained interested in the most tantalizing topic of all, human self-consciousness. Thus he did not refrain from discussing all-encompassing topics like "Life and Mind in the Universe." He was married twice and had two children by each marriage. He died at the age of 91, in 1997, having seen nine grandchildren and even three great-grandchildren.

Wald's Discoveries Catalyze the Prize Discussions

Among the three candidates to the 1967 Nobel Prize in physiology or medicine Wald had the shortest career. He was nominated for the first time in 1955 by the giant Albert Szent-Györgyi. This vitamin researcher could appreciate the central role of vitamin A that Wald had demonstrated in the function of the retina. The nomination is very complete and highlighted the major findings made in Wald's laboratory. The committee apparently considered this nomination to be of great importance and selected two reviewers, Hammarsten to cover the biochemical aspects and Granit to focus on the eye physiological aspects. The choice of Granit may seem a bit surprising. Could he really provide an unbiased opinion about a field that he himself had pioneered and in which he had been a candidate for such a long time? Even if prudent objectivity prevailed in the committee it would have been better to let another neurophysiologist do the job. It may be appropriate to cite from the subjective comparison of these two candidates by Bernhard in his biography[9]. He wrote (translated from Swedish):

> "Wald and Granit were intellectually and emotionally each other's opposites and in addition they were not on good terms. Both of them wanted to show off. Wald by displaying verbal clarity, almost accentuated by a certain snobbishness, whereas Granit mumbled and had greater difficulty in making himself understood, in particular when speaking English, which he in fact spoke perfectly, but tainted by an

excessive Oxford intonation ever since, in the 1920s, he had worked with his master, Sir Charles Sherrington. He (Granit) was at his best when he functioned in a group where he did not feel compelled to take the lead. On those occasions he was a charming presenter of stories. Both were very energetic both in the laboratory as well as at their desk. Their route to the prize was characterized by a large number of scientific publications."

Granit, it would seem, attempted an objective and extensive evaluation of Wald over 18 pages. However, he did not present a conclusion! Granit started by criticizing Szent-Györgyi's nomination in that it said Wald has extracted the cone pigment cyanopsin. According to Granit it had only been synthesized. The evaluation was divided into two parts: the first one dealing with vitamin A and the cycle of visual purple and the other with the substances present in rods. He then went on to criticize the nominator's description of events arguing that he had been too much influenced by semi-popular presentations in the American literature and hence had been remiss in giving credit to contributions from English researchers "who may be less prone to use this form of (self-promoting) activities." According to Granit there were a number of studies prior to Wald concerning the relationship of the presence of vitamin A and night blindness. However he gave credit to Wald's presentation in 1934 that the previously described intermediate stage "visual yellow" at the bleaching of visual purple could be of carotinoid nature. In spite of insufficient experimental evidence Wald postulated that its nature was a kind of carotinoid and he named it retinin. He assumed it to be the chromophoric part of the visual purple. Parallel work in the laboratory of Hans von Euler in Stockholm, the 1929 Chemistry Nobel Prize recipient, had failed to give satisfactory chemical evidence of the postulated reaction. Thus it was accepted that Wald was on the right track, but in the next paragraph he was on the wrong track. There was a long discussion of "transient orange" in studies of which many groups have been involved. The conclusion was that it was not correct to say with Szent-Györgyi that Wald was the discoverer of "the chemistry of visual excitation."

Then Granit referred to his own studies of the correlation between the amount of visual purple and excitation. He took notice that Wald in a 1954 publication in *Science* had given credit to, at the time 15 years old, the investigation by Granit *et al*! The conclusion after this first section of the evaluation was that Wald could hardly be considered as a candidate based on his contributions to elucidation of the problem of "visual excitation," and which he had no claim to have been the first one to understand. Thus he should be judged on the basis of his biochemical studies. Granit then in fact went on to give credit to the work in 1937–39 on the visual violet, porfyropsin, in fishes, mentioned above. This work led to the identification of the carotinoids 1 and 2, a finding Granit praised. He then discussed the possible priority of Wald's contribution to the identification of vitamin A as the final product in the bleaching of rhodopsine. It was considered to be confirmatory. Hereafter there was a discussion of studies of the identification of three photochemically active substances at wavelengths 5100, 5330 and 5500 Å. Granit argued that physiological experiments were lacking. He presented a number of similar studies by others. He even in one context used the following formulation (translated from Swedish) "Wald, who has always been unwilling to accept other results than those by himself or his group, has for a long time neglected…"; The polemical style continued when in one place Granit stated "This acknowledgement he could have allowed himself 12 years earlier." Hereafter followed an analysis of the visual pigment in the cones of birds and the polemics went on. Maybe it would be best to cite the final paragraph of the exceptional review, which, as already mentioned, did not present a conclusive recommendation. It read (translated from Swedish):

> "It is documented by this report that Wald's most important results are the synthesis with vitamin A1 – and A2 – aldehyde plus receptor egg white from animal eyes of four of the photochemically active substances, which are responsible for the four known dominators, identified by the undersigned by use of electrophysiological methods in several cases using the same kind of experimental animals. Concerning the isolated photochemical problems it was the Lythgoe group which made the pioneering contributions. These have not been discussed except as concerns the question of transient orange. Also Morton and collaborators have made important contributions and at a later time this has also been true of Ruth Hubbard in Wald's laboratory where she, however, has an independent position. Many of Wald's best contributions have been made together with her. However, Wald's strength has been the continued living contact (that he has kept) with the physiological aspect of vision, a field within which he as a collaborator of Hecht at the time received a solid training."

This was the end of the review, but it was obvious that Granit was not enthusiastic about seeing Wald as a recipient of a prize in the field that was his home turf!

The second review was made by Hammarsten, who had been a member of the committee for most years since 1928 and until his retirement 1957. During the last three years he was the chairman of the committee, including the year 1955 when he made his review of Wald. This review was very critical throughout and reflected the attitude of a scholarly biochemist towards a biologist. He was very meticulous in giving historical references, about ten per page and it is obvious that he had carefully perused the relevant literature. The review was full of critical remarks like "his experiments do not justify him in proposing," "the data presented have a comparatively low value due to the properties of the preparation," "the last proposal cannot be defended on account of the impure preparations used," "rests on insufficient data," "the characterization of the "opsin" does not even fulfill the most elementary requirements from a biochemical point of view," etc.

The two final paragraphs of the review read (translated from Swedish):

"Wald's contributions, which have been reviewed above, as concerns the results presented, without doubt may be considered a coherent discovery. This might possibly be described as the discovery of the turnover of carotenoids in rods and cones. The priority of this discovery is, however, markedly circumscribed by contributions from preceding and concomitant researchers within this field of research. In addition there are missing links in the knowledge about the course of events of this turnover. Wald is exceptionally broad in his presentation and makes up for the absence of sufficient knowledge by what, for a biochemist, appears as a biological whitewash (sic, my remark).

With reference to my review above I want to express as my opinion that Wald's contributions at present are not worthy of a prize."

Unsurprisingly the committee expressed a joint conclusion that Wald at the time should not be considered for a prize.

As already mentioned nominations of Granit continued to be submitted during the end of the 1950s and in parallel Wald was nominated in 1957–59. The 1957 nomination by the truly senior colleague Otto Loewi, who received his shared Nobel Prize in physiology or medicine in 1936[1], was very comprehensive.

Similarly the 1959 proposal by K. K. Chen was relatively detailed. No further reviews were made and there were no comments on these nominations in the protocols by the committee, except that Granit was referred to in 1959. The matter of Wald's candidacy was not returned to until in 1965 when he was nominated by his Harvard University colleague John T. Edsall, an authority on proteins. A belatedly mailed extensive supplementary material for the nomination, which Jim Watson had co-signed, was added to this proposal. The nomination was also endorsed by the 1953 (shared prize) laureate in physiology or medicine Fritz A. Lipmann. The materials of the proposal were addressed to Theorell, who in this year was the chairman of the committee. He was also selected to make a new review. The first page of the seven-page-long report summarized the very critical comments made in 1955 by Granit and Hammarsten by citing from their evaluations. Theorell then noted that a number of important new data had been added during the recent ten years. Some critical problems relevant to the mechanisms of vision had been solved. He devoted his review to these advances.

The first important advance made concerned the above-mentioned molecular variants referred to as "cis" and "trans." It was known from earlier data that a certain isomer, named "neo-b" present in the aldehyde of vitamin A, called retinal, played a central role in the influence of light on rhodopsine, the visual purple. Together with an important collaborator at the Ortho Research Foundation, William Oroshnik, Brown, Hubbard and Wald were able to show that the "neo-b" vitamin A was identical with the 11-cis isomer of the molecule. Theorell then described as a representative example the reactivation of light bleached rhodopsine, the most examined chromophore. Hubbard (alias Mrs Wald) and Kropf discovered in 1958 that exposure to light over a long time led to the presence of a mixture of all different cis and trans isomers. This work was extended by Japanese researchers Yoshikawa and Kito who exposed rhodopsine cooled to −195°C which led to the appearance of another form of end product. Yoshikawa then joined Hubbard and collaborators in Wald's laboratory at Harvard University. Three different kinds of products were identified; rhodopsine 11-cis isomer, absorption max. 505 μ: prelumirhodopsine, all-trans isomer, absorption max. 543 μ; and isorhodopsin, isomer 9-cis, absorption max 401 μ. Further experiments led to the conclusion that the primary reaction to light was rhodopsine to prelumirhodopsine which took place in the chromophore group as a cis to all-trans isomerization.

The regeneration of rhodopsine took place by a mechanism examined by Hubbart using an extract of a retinine-isomerase prepared from bovine eyes. Theorell tried to explain this to his committee members by referring to a crumpling of an originally flat molecule, reducing its further capacity to absorb light. He further described how the karotinoids served as photoreceptors wherever these were found in the animal or plant kingdom. Studies of the lobster, *Homarus Americanus* had revealed that 90% of all vitamin A found in the animal was located to its eyes. Theorell described that recent findings of the importance of 1-cis retinal were of a high quality. His concluding paragraph read (translated from Swedish):

> "As described above it is my view that many of Wald's (recent) publications have (considerable) merits and in part serve the function of strengthening Wald's candidature, but the question must be raised what importance his collaborators may have had, as for example his wife Ruth Hubbard. As concerns the color-sensitive receptors in the retina there are a number of researchers who have employed irradiation and light absorption techniques — Granit, Rushton, Stiles, Lythgoe, Weale and others. The nominators also highlight "perhaps it should go to more than one man." This possibility is not available this year, since no nomination for splitting the prize has been submitted (Granit but not Hartline was nominated this year, my remark). In conclusion I therefore recommend, with a certain hesitation, that Wald is declared worthy of a prize for his discovery of the primary photo receptor of the eye, but that the committee, without awarding Wald the prize this year, follows the development of this field (of research) with attention."

The committee simply wrote that Wald's studies of the chemistry of the visual processes was worthy of a prize. As already mentioned the same protocol concluded that Granit's studies of mechanisms of muscle control were not at the present time worthy of a prize.

In 1966 the three candidates for the first time were nominated simultaneously as already described. Zotterman's review of all three of them therefore also included Wald. It covered ten pages and started by citing the nominator Lord Adrian. He wrote.

> "George Wald is the greatest living authority on the chemistry of the retinal pigments. He (with his wife and other collaborators) showed the importance of "retinal," the aldehyde of Vitamin A in the reaction

of light on rhodopsine and his recent studies have identified 4 types of chromoproteins which are present in the dark-adapted eye. He has found that the carotinoides are the photoreceptors in a great variety of eyes and has related this to their molecular arrangements, a chain with double bonds rather than a ring."

Hereafter Zotterman cited the reservation in previous reviews by Granit and Hammarsten and noted that it had been emphasized in Theorell's later review that important new data had been presented during the last ten years. The discovery of 1-cis retinal was considered to be of high quality even though there were some questions about who had made the leading contribution to this work. Zotterman then stated that not much had been added during the year that had passed concerning knowledge about the structure of the chromophores. However, data of a related but different kind had been presented during the recent year in studies of the receptor mechanisms of the human eye. He emphasized that in humans and other primates two particular related phenomena had been observed. This was the occurrence of the yellow spot, *macula lutea*, Latin spot and yellow, and its importance for color vision unique to the primates. The spot is 2–3 mm in diameter and 80% of the structures it contains are cones, the elements responsible for color vision. The yellow spot contains antioxidants like lutein and zeaxantin. In the recent work a method had been developed by Brown and Wald to examine it spectroscopically. Measuring circular areas with a diameter of 0.2 mm at first in darkness and then after a powerful bleaching first with red light and then even further using shorter wavelengths it was possible to identify a pigment sensitive to red at λ max of 565–570 μ and a pigment sensitive to green at λ max 350–540 μ. Very soon after this discovery had been made another research group could identify one more cone substance, which had sensitivity to the blue color at λ max 450 and 457 μ (p.150). Within each maximum there was a certain differentiation between cones for activation of light of different wavelengths. This work was followed up by the use of a relatively simple psycho-physical method, showing similarities to an approach used by Granit already in the early 1940s. The technique was applied for measurements on two well-trained subjects, one of whom was Wald's wife. The maxima for absorption of lights of different wavelengths were finally determined to be 430, 540, and 575 μ.

Following these new insights it became possible to examine humans with different kinds of color blindness. Such blindness is rare in women but occurs in as many as 6–8 percent of all men. The reason is that the genes for the color pigment of two of the three kinds of rods are located to the X chromosome. Since there is no allele to compensate for the defective gene in the X chromosome in men, the defect becomes apparent as discussed earlier. Wald's conclusion from the work on color blindness which was continued at the time of evaluation was that in essence it was only a single syndrome with certain variation in the different fields of light wavelengths. Zotterman summarized that there was no question about the fact that Wald was worthy of a Nobel Prize. He emphasized that Granit's discovery of the dominators and modulators combined with Wald's identification of chromatophores provided a clarified and simple explanation of the sense of color vision. Zotterman finally praised all the three candidates arguing that they supplemented each other excellently and should receive a combined prize. However they had to wait one more year.

In 1967 Wald was again nominated alone by Szent-Györgyi and together with Granit and Harline by Heymans, like the year before. As already mentioned the committee let Bernhard make a final review of Granit and Hartline, whereas it let Theorell make a supplementary review of Wald. Since he had made a review just two years earlier he was very brief and summarized his comments over three pages. Theorell focused on two recent publications from Wald's laboratory ("…to those who during the previous year…"?). One was the publication from 1966 about the wavelength maxima for the three receptors, discussed in the previous paragraph, and then a publication from 1967 by Yoshikawa and Wald available for only two months at the time of writing. The work presented the situation concerning the photoreceptors in the cones, the iodopsins. It was documented that they showed considerable similarities with, but also a certain difference from the rhodopsine present in rods. These differences became apparent at very low temperatures (−195°C). It was found that if, under these conditions, a small quantity or 11-cis-retinal was added only the iodopsin product was synthesized, interpreted to mean that cone opsins have a much higher affinity for the 11-cis compound than the opsin in rods have. The final paragraph in Theorell's review read (translated from Swedish):

> "As is apparent from what has been brought forward above, Wald during the last years has strengthened his candidacy for a Nobel Prize, in particularly by his work on the iodopsine and color blindness. His foremost contribution, however, from the point of a Nobel Prize remains his discovery of the primary molecular mechanism of the sense of seeing, the one for the whole part of the world of animals that

can see, joint "trigger"-mechanism by which a steric rearrangements in retinal leads to that light energy being converted into a nerve impulse."

Interestingly Wald had also been nominated for a Nobel Prize in chemistry in 1967 by Szent-György. The nomination also included another candidate Daniel I. Arnon from the University of California, Berkeley, who studied photosynthesis. The committee responsible for this prize asked Theorell to make a review of Wald. He did not make a separate new evaluation but composed a verbatim amalgam of his 1965 and the 1967 reviews requested by the committee for physiology or medicine. Only the concluding paragraph was newly written. It read (translated from Swedish):

"George Wald's scientific production is presently not of the uncommon kind for which recognition by a Nobel Prize either in chemistry or in physiology or medicine may well be considered. An argument for a prize in chemistry is the fact that his most original and most unexpected discovery is of a chemical nature. Then I am thinking of his documentation that nature uses a relative steric restriction in a molecule very suitable for this purpose, 11-cis retinal, as a trigger to convert the energy of the photons to nerve impulses. The mechanism employed is used universally throughout the life forms which use eyes for seeing. It is therefore my view that George Wald is worthy of a Nobel Prize in chemistry, for example for the discovery of the primary molecular mechanism for seeing colors."

The committee noted in its summary that Wald had been nominated for the first time. It was concluded that he was a strong candidate and he was listed among the eight runners-up, but it was noted that a prize in physiology or medicine might be more appropriate. The prize in chemistry this year recognized with one half to Manfred Eigen and the other half jointly to Ronald G. W. Norrish and George Porter with the extended motivation "for their studies of extremely fast chemical reactions, effected by disturbing the equilibrium by means of very short pulses of energy."

Finally the Nobel Committee at the Karolinska Institute reached its conclusion, but it took extended and repeated discussions. As mentioned and discussed in Chapter 2 in 1966 there were 35 contributions by different scientists, often in combination with others that were considered worthy of a prize. In 1967 this number had increased to 37, which obviously did not make the decision process simpler. We will see in Chapter 7 that in 1968 there was a significant further increase of prize-worthy candidates to 52. Thus it seems that the 1960s was a decade when many important discoveries were made. A part of the explanation may have been the introduction of an impressive new armory of techniques. They allowed separation of large molecules, comparing their biophysical, biochemical and functional properties. The era of molecular biology had come of age.

The first report by the 1967 Nobel Committee to the Faculty of Teachers was submitted on September 26, but it did not include a firm proposal for prize recipients that year. After further deliberations the committee came back on October 18 and proposed that the prize should jointly recognize Granit, Hartline and Wald "for their discoveries concerning the primary physiological and chemical visual processes in the eye." It had been a long journey on a rocky road for the three candidates, not least Granit. In the end it was judged that they had provided support to each other's findings leading to a broad insight into the processes of seeing. It was not a matter of a single momentous discovery but of a progressively improved understanding of the mechanisms of seeing. But what remains debatable was the fact that some findings were accumulative, meaning that each one of the three candidates would appear possibly not to be capable of carrying a prize on his own. This does not agree well with the rule of guidance, repeatedly referred to, which states that in the case a prize is given to more than one candidate, each one of the recipients should be capable to carry a prize on his own. The deviation from this rule will remain a question for future discussions. But let us leave this matter and now turn to the prize events.

The Festivities and a Charming Mishap

When the Hartline and Wald families arrived at Arlanda Airport outside Stockholm they were met by Granit and his wife Daisy. It should have been a memorable reunion; in particular between the close friends Granit and Hartline. At the time, the ages of the three prize recipients were 67, 64 and 61 years, with Granit being the oldest and Wald the youngest. Among other prize recipients the same year there was one who was older than Granit and that was the recipient of the literature award, the Guatemalan Miguel A. Asturias, who was 68. There was also an exceptionally young prize recipient.

Ragnar Granit and his wife Daisy (in the rear line to the right) greet fellow Nobel laureates Wald and Hartline (rear left and right) and their families welcome at the Arlanda Airport in Stockholm. [From Ref. 17.]

That was the 40-year-old Manfred Eigen from Germany, one of the three prize recipients in chemistry already referred to. I have mentioned him in my first two books on Nobel Prizes[1,10], because I have had the privilege of befriending him during many visits to the Scripps Research Institute in La Jolla throughout the 1980s and 1990s as well as later. Eigen was an impressive person, a skilled and visionary scientist and also a great humanist in the early 20th century German tradition.

Eigen passed away on February 6, 2019, while the text of the present chapter was being finalized. Following the rules which state that the archives of prize recipients in physics and chemistry only become available after they have died, it then became possible to review the history of his nominations. Without going into detail it can be summarized that he was already nominated for the first time in 1960, when he was only 33 years old and then increasingly over the years. In the year preceding his prize and the one when he received his prize there were, respectively, eight and seven nominations of him. He was reviewed four times, in 1960, 1963, 1966 and finally in 1967 by the professor

Manfred Eigen (1927–2019), the recipient of half a Nobel Prize in Chemistry in 1967, receives his prize from the hands of His Majesty the King.

of physical chemistry at Stockholm University, Arne Ölander. At the time the chemistry committee was restrictive as to sharing of prizes in this discipline, but in this particular year they decided to award Eigen half the prize and to divide the other half of the prize between Ronald G. W. Norrish and George Porter with the joint motivation "for their studies of extremely fast chemical reactions, effected by disturbing the equilibrium by means of very short pulses of energy." Ölander gave the presentation at the prize ceremony. In his concluding 1967 review he emphasized that "Eigen in later years had started to focus his interest on information and memory problems in biological systems viewed from the perspective of a chemist."

Over decennia Eigen was a very central figure in this border line field of advancing science. He was the author of several influential books on science, the latest of which is named *From Strange Simplicity to Complex Familiarity*[11]. This book provides a number of reflections on a wide range of scientific topics by a thoughtful, erudite scientist. At the time of writing of the later texts for the present book the Nobel prizes of 2018 were announced. The prize in chemistry recognized with one half Frances H. Arnold "for the directed evolution of enzymes" and the other half George P. Smith and Gregory P.

Winter "for the phage display of peptides and antibodies." The possibilities to exploit evolutionary phenomena in the laboratory for development of useful macromolecules was first introduced as a concept by Eigen.

The evening before the prize ceremony Bernard and his wife, whose nickname was Gullan, had a lively dinner in their very special home at Lidingö, an island city north of Stockholm. There is a vivid description of this evening in his autobiography[9]. The 1934 Nobel laureate Lord Adrian was the guest of honor. He was a good storyteller and referred to the fact that six of his predecessors as Chancellor of Cambridge University had been decapitated. Zotterman apparently was an entertainer of class talking to everyone and it was only by locking the grand piano that he was prevented from taking over the evening playing dance music. The hostess Gullan entered into an intense debate with Wald about religion. She had converted from the Lutheran Swedish State Church at the time to Catholicism whereas Wald with his background was a believing Jew. When Wald said "I would not surrender Jesus to Goyim (non-Jews or heathens in general)" he was rebuked by his wife. A good-quality class red wine, Chateauneuf du Pape, was enjoyed in generous quantities and it was way beyond midnight before the coffee and cognac were finally finished and the guests could return to the Grand Hotel in the limousines waiting in the wintry weather. It was time for Bernhard to concentrate on his Nobel address to the three prize recipients later that same day.

Lord Adrian (1889–1977).

The prize ceremony in the Stockholm Concert Hall was ceremonious as expected and the events can be followed in the words of a deeply involved eyewitness, Bernhard, who has described what happened in the above-mentioned autobiography. After the introductory piece of music the prize recipients enter in the order physics, chemistry, physiology or medicine and literature, following the listing in Alfred Nobel's will. Each prize recipient was accompanied by a member of the responsible committee. Granit was accompanied by Bernhard, Hartline by Zotterman and Wald by Gustafsson, the new secretary of the committee. When the procession had lined up hidden in the entrance-way, Bernhard discovered that there was a misprint in the program and that Wald's name was not mentioned. So of course Bernhard had to inform him that he could go home, dressed in tails and everything. However, there was the problem of his accompanying committee member, so Wald was allowed to stay in the procession. But then something surprising happened.

The first piece of music had finished and now von Euler's voice was to be heard. Being the Chairman of the Board of the Nobel Foundation at the time he was to give the welcoming opening address. His voice carried on for a while, but then it grew quiet. Influenced by the stressful situation von Euler had overlooked the fact that he should not start his speech until the procession had entered and the prize recipients had taken their seats. He was alerted to the embarrassment of the calamitous situation by Nils Ståhle, the Director of the Nobel Foundation and quietly returned from the podium to his place. However, human mistakes like this in a uniquely stately event such as the Nobel Prize Award ceremony provide an outlet for the audience to spontaneously share their reaction of positive emotions. The whole audience, including His Majesty the King shared in a discreet laughter. The members of the procession could

His Majesty the King and the audience in the auditorium laughing. [From Ref. 17.]

audibly follow the unexpected events and at some stage Bernhard carried his joke even further and said that maybe it was time for all of them to go home. However, eventually the procession accompanied by music was able to enter the stage and the program continued uneventfully as planned.

Bernhard gave a very stimulating introduction[12], of which the second paragraph has already been cited in my previous Nobel book[2], but the introductory section deserves also to be cited. It reads (translated from Swedish):

"Light, shadows and colors do not exist in the world around us. What we perceive visually and call light is the result of the action of a certain portion of the electromagnetic radiation of the sensory cells in the retina of the eye. Our awareness of the play of light in nature, the multiplicity of the forms and the richness of the colors is ultimately dependent on the pattern of this radiation with respect to frequency and intensity. The light is composed of packets of energy, which combine the properties of waves and particles. When these particles — the quanta — strike the retina of the eye they are caught by the specialized sense cells — rods and cones. It is known that one quantum, which represents the least possible amount of light, is sufficient to initiate a reaction in a single rod. The excitation of the sensory cells results in messages (an electric signal, my remark) directed towards the brain. As there are no direct connections from the eye to the brain the messages must be transmitted through several relays which combine signals from many sensory cells and translate the message into a language which can be understood by the brain. The primary relay is in the retina itself, represented by an intricate nerve net, the structural beauty of which was originally revealed by the neuroanatomist Ramon y Cajal, Nobel laureate in 1906. In this complex structure messages from a great number of sensory cells converge on a far smaller number of optic nerve fibers and this results in a transformation of the pattern of signals."

By reference to the way Picasso had described how he painted Bernhard then emphasized that the eye does not function like a camera. It dissociates the pattern of incoming signals and by reassembling this information in the brain the perception of a picture can be obtained, but with "a sharpening of contrast so that forms stand out more clearly, colors are exaggerated and movements accentuated." Bernhard then introduced the importance of Wald's finding of the

Granit (a), Hartline (b) and Wald (c) receive their Nobel Prizes from the hands of His Majesty the King [From Ref. 17.]

role of cis-trans changes in the vitamin A-opsin complex, a discovery to which his wife as mentioned had made major contributions. Hereafter Bernhard turned to color vision reminiscent of the early historical contributions of Isaac Newton, Thomas Young and Hermann von Helmholtz. He then saluted Lord Adrian in the audience and referred to his, together with Zotterman, pioneering first recordings of signals in nerve cells. Hereafter Granit's pioneering 1939 findings together with Svaetichin of rules for color vision, consolidated by Wald's later discoveries, were mentioned. Finally the fundamental contributions by Hartline were mentioned with an emphasis on integration of signals, not least in the form of lateral inhibition already mentioned above. When Bernhard finally turned to English to ask the prize recipients to collect their insignia he took the opportunity to cite the formulation by Hartline's collaborator Ratliff[3] in describing their laboratory as "a slightly disorganized but extremely fertile chaos." This led to a second wave of laughter in the audience. Finally all the three scientist could receive their Nobel Prize medal and the diploma from the hands of His Majesty the King.

At the ensuing banquet the three laureates had a good time. They all gave short speeches, with Wald in particular addressing the students in attendance. Granit emotionally shared not only the joy of the moment but also referred back to his home islands and the turmoil of his early life. He also cited Darwin by quoting "I remember well the time when the thought of the eye made me cold all over." It was the daunting complexity of this sense organ that the three prize recipients had unraveled. Hartline referred to the contrasts of experiences by saying "One works in one's laboratory — one's chaotic laboratory — with students and colleagues, doing what one most wants to do — then all this happens!" He also emphasized that practical benefits would flow most freely when the fundamentals of the function of an organ are understood. Finally Wald emphasized to the students that science is a serious business but generally allows happiness in its pursuit. He referred to the scientist as a learned small boy. Finally he commented that the honor of the prize found its way into the hearts of the public at large. New knowledge is available for everyone to access, a valuable shared property.

The day after the festivities it was time for the Nobel lectures. The rule that a Nobel Prize recipient should receive his prize before giving his lecture was upheld until 1976, the year when I chaired the presentation by Baruch Blumberg and Carleton Gajdusek. Hereafter the committee changed the arrangement and allowed a presentation of lectures prior to the prize ceremony and the banquet. In 1967 Granit was the first among the three presenters in the field of physiology or medicine[13]. He returned to a field of experimental science that he had left almost twenty years earlier and finally summarized in 1962 in a review in *The Eye, volume II*[14]. He started by referring to the introduction of techniques to record impulses in a single fiber and to the early studies of the electroretinogram. The major steps in his contribution to the understanding of retina physiology were recapitulated, including the lateral inhibition — first observed by Hartline — as a mean to enhance contrasts. The importance of minor fluctuations of eyeballs as a means of accentuating contrasts was also presented. Finally most attention was paid to the basis for color vision. Departing from Hecht's (whose widow was present during the events) fundamental observation of color psychophysics, Granit described his findings of three individual modulators responding to filtered light with wavelengths corresponding to the blue, green and red colors. In a figure from 1945 he summarized the pioneering results.

Hartline in his Nobel lecture[15] mainly focused on his favorite "the Xiphosuran arachnoid, *Limulus polyphemus*, commonly called "Horseshoe crab"...." Many pictures of electrical recordings, so-called oscillograms, were shown as was the anatomy of the ommatidia of the lateral eye after removal of the chitinous corneas and crystalline cones. The progressive technical refinement and proficiency of the experimentation was apparent from the presentation. Towards the end of the lecture the overriding importance of "on-off" phenomena in the repeatedly recapitulated concept of lateral inhibition was described.

Wald was the last to speak[16]. His in parts eloquent lecture was off to a good start. It read:

> "I have often had cause to feel that my hands are cleverer than my head. That is a crude way of characterizing the dialectics of experimentation. When it is going well, it is like a quiet conversation with Nature. One asks a question and gets an answer; then one asks the next question, and gets the next answer. An experiment is a device to make Nature speak intelligibly. After that one only has to listen."

Except that I was always taught never to start a letter with "I" this is very well formulated. Biological science is all about reading the books of life. Against the background of evolutionary capriciousness the way nature has ingeniously solved the management of challenges of the environment and endowed the organism with a competitive advantage towards surrounding other organisms represents the story of life. To this can be added that we have come a long way in interpreting the books of Nature, because of our impressive capacity to read and even write the genetic language, but it should be emphasized that at present most secrets of nature still remain unresolved. Wald's initial historical account described how his critical teacher Hecht in his turn was inspired by the renowned Swedish physical chemist Svante Arrhenius, the profiled 1903 Nobel Prize recipient in chemistry.

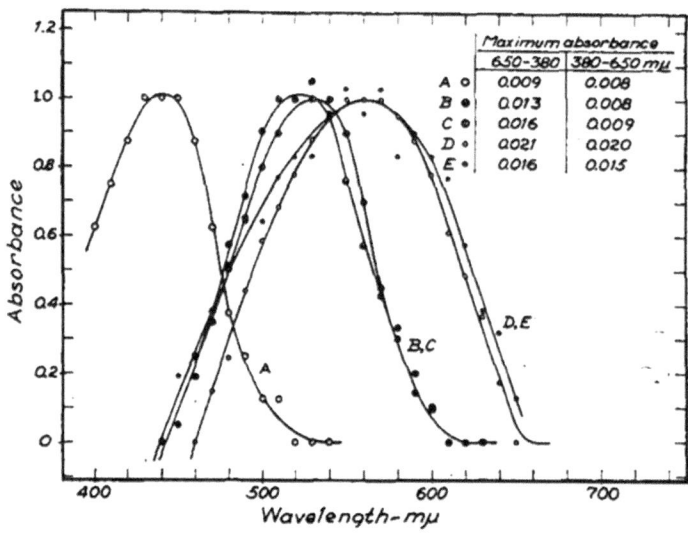

The wavelength sensitivity of the three kinds of cones in the human eye. [From Wald's Nobel lecture[16].]

Wald then went on to present his early "Wanderjahr" in three different laboratories managed by Nobel laureates — Warburg, Karrer and Meyerhof. This experience steered him into a life with molecules. Appropriately he described that an essential characteristic of the scientific enterprise was its relentless development. He said "it is an organic growth, to which each worker in his time brings what he can; like Chartres or Hagia Sophia, to which over centuries a buttress was added here, a tower there." He then described the story of unraveling the nature of visual pigments. It was described how, in vertebrates, the two retinals 1 and 2, were joined with two families of opsins, those of rods and those of the cones. Four major pigments were represented with different light sensitivities. Models of the chromophoric sites were shown and the mechanisms of cis-trans molecular changes were described. Hereafter stages in the bleaching and regeneration of rhodopsine were illustrated. Finally it was described how insights into the paths of vision chemistry and electrical physiology of the eye could be combined to provide a general interpretation of critical mechanisms. The extensive written version of the Nobel Lecture presented the accumulated experimental evidence for the final conclusions. The second last paragraph read:

> "Looking back over this account I am struck with the thought that some of the most significant aspects of the photoreceptor process come from its being laid out in two dimensions: on the molecular level, in two-dimensional arrays of oriented molecules, the membrane that compose the photoreceptor organelles; and on the cellular level, in the single layer of receptor cells that composes the retinal mosaic. In these arrangements each molecule of visual pigment and each receptor cell can report on its own. The absorption of a single photon by any molecule of rhodopsine among the many millions that a dark-adapted rod contains can excite it (reference); and the early receptor potential signals in detail the synchronized reactions of populations of visual pigment molecules. Similarly, quite apart from the skilful electrophysiological procedures that measure directly the responses of single retinal units, we have no great difficulty in sampling, through their differences in spectral absorption and sensitivity, the properties and behavior of each receptor type over the surface of such highly variegated retinas as that of man."

It is now time to leave the field of eye physiology and dramatically change themes of presentation. The following concluding chapters will concern one of the most dramatic developments in biological science. It is the discovery of the genetic language used by Nature since the first emerging self-replicating chemical systems later providing the basis for the origin of cellular life on Earth and in all forms of life that since then have appeared on Earth by the magic of aimless evolution. Most of these diverse forms of life have now become a part of history due to cataclysmic events in the history of Earth. However, the richness of those that exist on Earth is immense and only characterized to a limited extent. One of those forms of life is a particular primate with a unique self-consciousness and intellectual capacity – us. It is we humans who have learnt to read and write the books of life and we can do this because we have managed to decipher the language of evolution. The way this came about is a remarkably adventurous story. It all started with the discovery of the double-helix nature of DNA[1].

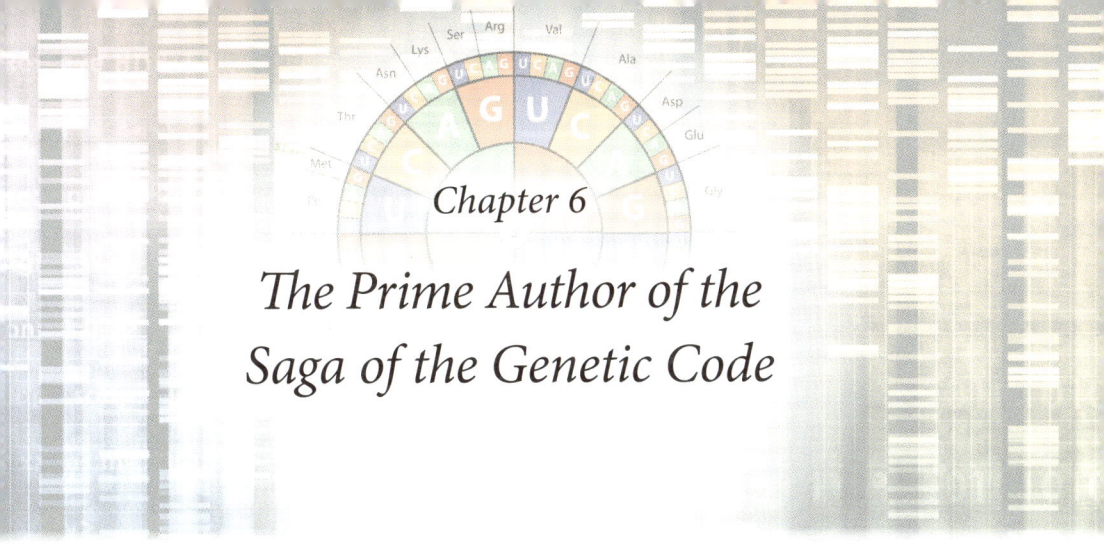

Chapter 6

The Prime Author of the Saga of the Genetic Code

THE ROSETTA STONE
UNDERSTANDING A UNIQUE CODE
THE LANGUAGE OF LIFE

The revolutionary discovery of the unique equidistant pairing between the nucleotide bases guanine and cytosine and adenine and thymine led Watson and Crick to propose the double-strand structure of DNA in 1953. This momentous finding had two major consequences. The first one was that the structure allowed a semi-conservative replication preserving the complementary sequence of nucleotides for succeeding generations of cells. Separating the two strands and building a new complementary strand provided two copies of the double-helix, identical to the parent molecule. The second major implication was that the existence of four different bases offered a potential insight into the information storage and the language used by nature to control protein synthesis and metabolism in general. The question of coding and the possible access to a Rosetta Stone for deciphering the language used by nature was raised soon after the discovery. However, several fundamental questions first needed to be answered.

There were many critical steps in the growing understanding of the relationships between DNA, RNA and protein during the 1950s. These questions have been repeatedly returned to in my previous books on Nobel Prizes in the life sciences[1-3]. The central question obviously was the chemical nature of the genetic material. DNA, already at an early stage, was considered to be of importance in carrying the genetic information since it was associated with the chromosomes. However, Phebus Levene's conclusion that the relative proportions of the four bases in diverse unpurified materials was constant led to the assumption that DNA was some kind of support material. It was considered a boring and uninteresting molecule. Two major observations during the later 1940s challenged this assumption, which for a considerable time severely held back further developments.

One was Oswald Avery's experimental demonstration that it was DNA that carried the genetic properties demonstrated in transformation experiments with pneumococcal bacteria and the other was two major discoveries by Erwin Chargaff. The latter scientist by use of purified fractions of DNA was able to prove that its base composition in fact varied extensively between different forms of life. The other seminal finding he made was that the representation of the four kinds of nucleotide bases was always of such a nature that the concentration of thymidine (T) and adenosine (A) was the same as that of guanosine (G) and cytosine (C). We will return to Chargaff later in this chapter, but it can be noted already here that he himself did not find an explanation for this so-called equimolar representation. This remained for Watson and Crick to do in 1953. Although it would seem that the description of the double-helix structure of DNA could provide answers to many critical questions it took time for a full delivery of the momentous impact of its identification. There were only a few references to the original finding of the structure of DNA during the mid 1950s and surprisingly there were no nominations for a Nobel Prize in physiology or medicine to Watson and Crick until 1960[2]. There was no campaign among nominators in 1962 for the momentous discovery by Watson and Crick of the kind that we will see for the nominations to the 1968 prize, recognizing the description of the genetic code. However, there were a number of unexpected cumulative original findings in the early 1950s which provided a momentum for an understanding of the central role of DNA in heredity.

The results of one such important experiment were published by Alfred Hershey and Martha Chase in 1952. They had labeled the nucleic acid and the proteins of a bacterial virus, a bacteriophage, with two separate isotopes. To their great surprise it appeared that upon infection only the labeled nucleic acid penetrated into the bacteria, whereas the protein shell of the virus remained at the surface of the infected microorganism. Other concept-breaking publications appeared in 1956 when Alfred Gierer and Gerhard Schramm and separately Frankel-Conrad demonstrated that RNA extracted from TMV on its own carried a full infectious capacity. Hence the nucleic acid appeared to be the sole origin of genetic information[1]. Hershey was recognized by a shared Nobel Prize in physiology or medicine in 1969, which means that the major role of studies of bacteriophages in understanding fundamentals of genetics can soon be reviewed in detail. In this chapter we will have to restrict ourselves to the prize the preceding year, recognizing the discovery of the genetic code.

In parallel with these early developments there was a renewed interest in the chemistry of nucleic acids and their components. These developments were pioneered by Alexander Todd, who was to receive the 1957 Nobel Prize in chemistry "for his work on nucleotides and nucleotide coenzymes." As discussed earlier[2] in reference to Arne Fredga's reviews of Todd it was finally accepted in 1955 that it had "passed beyond the state of conjecture" that DNA was the carrier of the genetic information. Not surprisingly nucleic acids became a very hot target for recognition by Nobel Prizes in physiology or medicine at the end of the 1950s. In 1958 the prize was divided between George W. Beadle and Edward L. Tatum "for their discovery that genes act by regulating definite chemical events" and Joshua Lederberg "for his discoveries concerning genetic recombination and the organization of the genetic material of bacteria" and in 1959 the prize recognized Severo Ochoa and Arthur Kornberg "for their discovery of the mechanism in the biological synthesis of ribonucleic acid and deoxyribonucleic acid." It was found later that in fact none of the latter two scientists had found an enzyme that could replicate RNA or DNA. It deserves to review this prize in more detail, not least since later on Ochoa forcefully entered the race to decipher the genetic code.

Severo Ochoa (1905–1993), recipient of the shared Nobel Prize in physiology or medicine 1959.

Arthur Kornberg (1918–2007), co-recipient of the 1959 Nobel Prize in physiology or medicine.

An Important Nobel Prize Given for Experiments Later Shown to be Flawed

The basis for the 1959 prize to Ochoa and Kornberg was the identification of two different enzymes involved in nucleic acid processing. Ochoa and his colleagues used an extract from the microorganism *Azotobacter vinelandii* and Kornberg an extract from the common gut bacterium *E. coli*. It was later discovered that the enzymes assumed to be responsible for replications of the two forms of nucleic acids in fact did not have the proposed activities. Before reviewing these flaws in experimentation it would be appropriate to discuss the different principles in replication of these two functionally and structurally different kinds of nucleic acids in the light of knowledge emerging in the late 1950s and early 1960s.

The replication of DNA is semi-conservative which means that upon separation of the two strands, two new copies of the original double-helix structure identical to the mother molecule are formed. However, it should be noted that the two original strands are what is called anti-parallel. Hence a DNA polymerase enzyme may run along one of the strands and continuously make a copy leading to a full double helix. However, this is not possible in the case of the complementary strand, since it runs in the opposite direction. The way nature has solved this is that pieces of the complementary strand are being consecutively produced — so-called Okazaki fragments. These fragments are then stitched together by use of yet another enzyme to form a complete strand of DNA, which can combine with its sister DNA strand to form a double-helix molecule. In spite of the complexity of this replication its fidelity is impressive. The error of base insertion in newly formed DNA is only one in a million. The problem with the system used by Kornberg was that it was later found that mutants of *E. coli* lacking the specific enzyme he studied could replicate efficiently. Hence there must exist some additional DNA polymerase.

Disparaging remarks in a biology class taken by Kornberg's middle son Tom made him look into the possible occurrence of other polymerases in the bacterium. Tom originally did not intend to become a natural scientist. He was involved in learning how to play the cello at the Juilliard School in New York. An unfortunate hand injury forced him to change the orientation of his academic career. In some early experiments with *E. coli* he was lucky to find two previously unidentified DNA polymerases, named II and III. Their existence explained the disposability of the enzyme originally identified and used by his father. DNA replication has been found to be very complex and two additional polymerases were later found in the bacterium and in eukaryote cells at least 15 different polymerases have been identified. The enzyme originally studied by the elder generation Kornberg was later found to extend one end of the DNA in a partial duplex. It could in fact under certain conditions produce some double-stranded DNA. It can be added that the original manuscripts describing the replication of DNA submitted by Kornberg were rejected by *J. Biol. Chem.*, but later rescued by a separate editor John Edsall[4]. The situation in Ochoa's laboratory was more serious.

Whereas DNA represents a stable information carrier it has been progressively understood that RNA may serve an impressive diversity of different functions. This was of course not known in the middle of the 1950s and hence it was fully acceptable to believe, as Ochoa did, that there might exist a specific separate enzyme capable of replicating RNA using preexisting RNA as a template. The early insights into the existence of different kinds of RNA were related to their role in protein synthesis. The developments in understanding the existence of different forms of RNA can be found in many textbooks, but in fact Watson's Nobel lecture in 1962[5] provided a good description of the early developments. Also the Nobel Prize in physiology or medicine to Albert Claude, Christian de Duve and George E. Palade in 1974, to be briefly discussed in the concluding chapter, provided a useful historical background. It was Claude who first observed particles in the cellular sap that he called microsomes and later Palade would show that they played a role in protein synthesis and could also be membrane-attached. Their name became *ribosomes* when it was documented that they contained RNA, in the following referred to as *ribosomal RNA*. Watson compared these RNA-protein complexes to virus particles but later came to the conclusion that they were much more complex. Later on it was found that they differed in size in nucleated cells and in bacteria and also that they contained two populations of RNA of different constant sizes. The size difference was originally determined by ultracentrifugation techniques and size determined from the varying sedimentation characteristics was expressed in Svedberg (S) units, from the 1926 Swedish Nobel Laureate in chemistry The Svedberg. Ribosomes from prokaryotes have one 50S and one 30S subunit and they contain 30S and 16S RNA, respectively, the rest being protein. In eukaryotes there are 60S and 40S subunits containing 28S and 18S RNA, respectively.

One additional form of RNA was discovered by Mahlon Hoagland in Paul Zamecnik's laboratory at Massachusetts General Laboratory in Boston in 1956. It was found that amino acids chemically activated by a certain enzyme were connected to a form of low molecular RNA, originally referred to as soluble RNA (because of its low molecular weight it stayed in the supernatant when a material was subjected to ultracentrifugation) but later called *transfer RNA*. Two other scientists were also in the lead of identifying this form of RNA. One of them was Paul Berg, who was later to receive half of the 1980 Nobel Prize in chemistry for his work on recombinant-DNA. In his biography[6] it is stated that he preferred to highlight not the discoveries recognized by his Nobel Prize but his contributions in the 1950s when working in the laboratories of the star biochemists, both of them eventually Nobel laureates, Fritz Lipmann and Arthur Kornberg. The other scientist involved in the identification of transfer RNA was Robert W. Holley. In the next chapter we will learn a lot about the structure of this kind of RNA and its role in transporting amino acids to their proper place in the growing protein chain, since he was added to the 1968 prize honoring Nirenberg and Khorana. He provided the first chemical characterization of this form of RNA. It can be noted that Crick in the fall of 1956 circulated a manuscript among the RNA tie club members[7]. Its title was "On Degenerative Templates and the Adaptor Hypothesis." Thus Crick by his intellectual analysis preempted the experimental discovery of transfer RNA. But there remained one more, a third critical category of RNA to be discovered. This turned out to be a short-lived form of the molecule, which is copied from DNA and transfers the information for synthesis of proteins in ribosomes, in bacteria in an all connected machinery but in eukaryotes after transport from the nucleus to the cytoplasm.

The apparent simultaneous discovery of this form of RNA in bacteria in two laboratories was described in my previous book on Nobel Prizes[3], since François Jacob, in collaboration with Brenner, was one of the main contributors to this advance. It was Jacob who together with Monod named this form of RNA, *messenger RNA*. This discovery had already been presented in Watson's Nobel lecture[5], since a somewhat similar observation was made also in his laboratory at Harvard University presumably by inspiration from the Pasteur group. It was base pairing mechanisms that allowed the transfer of information — the sequence of nucleotides — from DNA to RNA, with the modification that T was substituted for by uracil (U) in RNA. It should be added that it took a longer time to identify messenger RNA mechanisms in eukaryotes. Large molecular RNA precursors were identified at an early stage by James Darnell's group and others but it remained to identify the processing of these molecules into functional molecules and also to map out the mechanisms of their transport from the nucleus to the cytoplasm where the protein synthesis took place. We will return to a special aspect on processing of messenger RNA in the final chapter.

The new fundamental insights led to Crick being able to formulate the so-called *central dogma* introduced already in Chapter 2. One may ask why he proposed the term "dogma," which is normally used in a theological context to formulate a system of belief or a doctrine. According to Sydney Brenner, Crick, who of course was an agnostic, used the term in a teasing way to provoke representatives of different belief systems. The truth of the matter is that the term "dogma" is not appropriate. The flow of information from DNA to RNA and to protein represents the core products of the early evolution leading to the emergence of cellular life. This insight has been deduced from facts that have been established by qualified and imaginative scientific pursuits. It is not a concept to be formulated by man in a conjectural way. In spite of this the concept "dogma" has come to stay and it is seen by many to explain phenomena appearing to be set in stone. Still, as we shall see towards the end of the concluding chapter also the nature of the information system used and the role of different actors also represent a unique reflection of accidental events during the very early evolution. Let us leave these conceptual speculations and return to the generally accepted highly consolidated concept of the central dogma. It simply states that the information carried in the sequence of nucleotides in DNA can be copied into a newly synthesized messenger-RNA by a process of *transcription*. This messenger-RNA can then find its way to the protein-synthesizing machinery, the ribosome, where the sequence of nucleotides is used to determine the sequence of amino acids in a polypeptide chain representing a specific protein. The latter process is called *translation*. In later studies using crystallography as we shall see it has been possible to identify details of the highly complex molecular machineries for transcription and translation, respectively.

The problem with Ochoa's enzyme was the fact that it was later found that in real life there is *no* replication of RNA in normal cells using RNA as a template. The template used is always DNA. This replication leads to the appearance of all the different kinds of populations of RNA introduced above. The only exception to this, as discussed in Chapter 2, is when a cell is infected with certain RNA-containing viruses. In this case the virus may either direct the synthesis of an RNA-dependent RNA polymerase or carry this kind of enzyme along in the virus particle. There is also a third alternative as we have learnt in Chapter 2. Certain viruses carry along a reverse transcriptase which allows a copying of RNA into DNA. The latter molecule can then become

Karl Myrbäck (1900–1985).

integrated into the host cell genome and by use of the existing cellular machinery copied into messenger RNA

It was later found that what Ochoa and collaborators had been examining was an enzyme the function of which was to add phosphorous groups to RNA as a prerequisite for degradation of the molecule. Under certain conditions it was possible to drive the reaction backwards and to achieve some synthesis of RNA. This is an artifact but as it turned out also a reaction that could be exploited by experimentalists. It took a long time before Ochoa appreciated the limitation of the enzyme he was studying. At the time of his Nobel lecture[8] he still believed that he had found the key enzyme for RNA-dependent RNA synthesis in cells. The question may be raised if it would have been possible for the reviewers at the Royal Swedish Academy of Sciences and at the Karolinska Institute to have sensed the existence of possible defects in Kornberg's and Ochoa's experiments.

The reviewers were, respectively, Karl Myrbäck and Hugo Theorell. Their reviews were briefly alluded to in my first book on Nobel Prizes[1]. Myrbäck evaluated Ochoa's work for a prize in chemistry in 1956, 1958 and 1959 during which years there were a number of proposals for a prize to him. The first two reviews noted the high quality of the work and found them promising. However, because of certain uncertainties regarding interpretations of the biological significance of the observations made Myrbäck recommended the committee to wait. However, in 1959 he concluded that he now interpreted that Ochoa (translated from Swedish) "for the first time had clarified the mechanism of the biosynthesis of nucleic acid." He added that the observations "had exceptional importance for biology, genetics and medicine." The fact

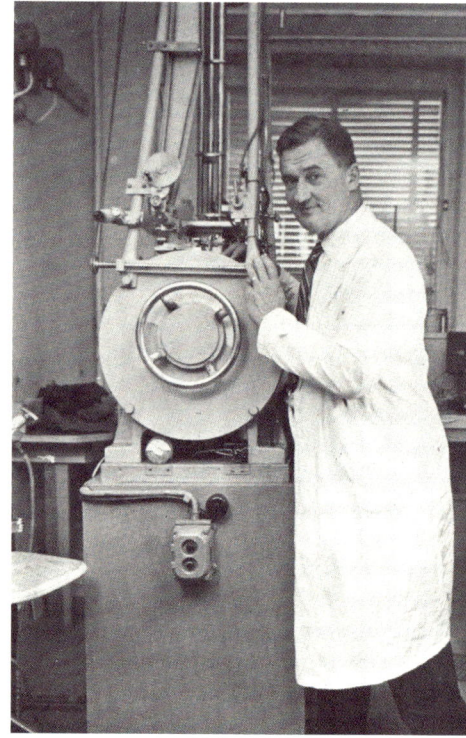

Hugo Theorell (1903–1982), the recipient of the 1955 Nobel Prize in physiology or medicine.

that, at the time, nucleic acids were taking the center stage most likely provided an incentive to a benevolent interpretation of the data. Kornberg was nominated for the first time for a prize in chemistry, not in physiology or medicine, in 1959 by John Runnström, a Swedish cell biologist of high standing. Myrbäck again served as a reviewer and evaluated Kornberg over 12 pages whereas his already mentioned follow-up review of Ochoa covered 8 pages. He noted that Ochoa's enzyme could be activated even by a single nucleotide as a "primer," whereas Kornberg's enzyme required a mixture of all four nucleotides. In the end he recommended a shared prize for their achievements "for the first time having clarified the mechanisms of the synthesis of a nucleic acid." However, the chemistry prize in this year went to the Czech scientist Jaroslav Heyrovsky for his discovery of the polarographic method.

At the Karolinska Institute the first nomination for a prize to Ochoa was submitted already in 1954. New nominations were submitted over the ensuing years and in 1956 there was a campaign with 10 nominations from Valencia in Spain. Two brief preliminary evaluations were made in 1954 and 1956 by Hammarsten, but the first comprehensive review also including Kornberg was made in 1958 by Theorell. It is clear from his summary in 1958 that he was more convinced by the experimental documentation provided by Kornberg, evaluated in parallel, than the one presented by Ochoa. He wrote (translated from Swedish):

> "Even though Ochoa had a certain lead in time with his RNA synthesis it is quite obvious that Kornberg independently of this has pursued his analysis of the synthesis of DNA, which has been found to be a highly different mechanism. In addition one gets the impression that the system used by Kornberg is closer to the processes used by nature, since the synthesis requires the presence of all four nucleotides that are constituents of DNA and in addition the presence of a primer is needed. By comparison, in Ochoa's system the necessity for a primer is uncertain and a synthesis can be initiated by the presence of only one of the nucleotides. Ochoa's enzyme, like Kornberg's, can form the natural 3'-5'-phosphodiester bonds, but Ochoa does not seem to care which nucleotides are connected with other nucleotides. With consideration taken to the biological specificity of nucleic acids such a mechanism of action of an enzyme system cannot explain the complete biological RNA synthesis. This is however possible although not proven in the case of Kornberg."

In 1959, the year of Kornberg's and Ochoa's prize at the Karolinska Institute, there was no external nomination of Kornberg. Hence this was secured by Theorell, who also again made a review of the two candidates. He argued in his final review for a recognition of both nominees, but voiced a higher appreciation of Kornberg's work by a statement in the last paragraph (translated from Swedish): "Ochoa's lead in time of the project in initiation of his experiments is counterbalanced by Kornberg's recent publications which, if possible, display a greater stringency (of work) and interest than (those published by) the Ochoa group." Finally, it can be added that Theorell also stated in his summary "The definitive, chemical proof of identity, in the form of a sequence analysis of polynucleotide chains, and a determination of their (spiral-) structures of a higher order (? My question mark) should be expected to have to wait a long time." It was of course difficult to foresee the explosive developments into present day speedy sequencing and mathematical information management of essentially unlimited amounts of information. In addition it should finally be emphasized again that the role of the two kinds of nucleic acids started to be clarified first towards the end of the 1950s. As already described there are three different kinds of RNAs serving distinct functions in the processes of formation of proteins.

The Lady Is a Trump

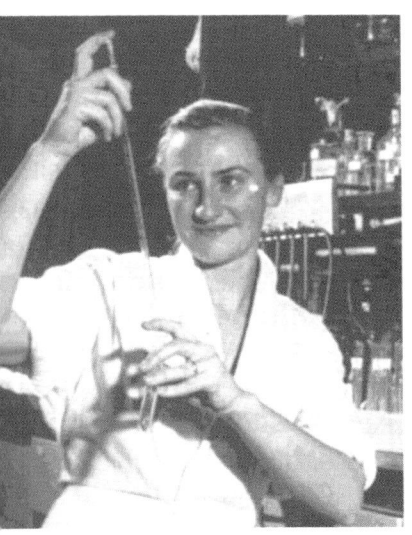

Marianne Grunberg-Manago (1921–2002).

It was not Ochoa who made the original identification of the critical enzyme he used in his RNA synthesis studies. It was his collaborator Marianne Grunberg-Manago. At the time of discovery of the enzyme in 1956 she was according to one historian[9]: "… a young French biochemist of gypsy handsomeness with a vibrant deep voice." In fact she was of Russian and not French origin, but she moved with her family from St. Petersburg when she was less than one year old. Most members of her family had artistic inclinations, but her mother was an architect, an unusual profession for women at the time. Marianne herself had difficulties choosing her own professional direction. Eventually she settled for the natural sciences, a choice consolidated during her Ph.D. work at the Marine Biology Laboratory at Roscoff on the northern coast of Brittany in France. Her curiosity about science led her to join Ochoa's laboratory at New York University in 1953. It was during the mid 1950s while working in his laboratory, that she discovered an enzyme that could phosphorylate nucleotides in RNA. It was originally believed that it was an enzyme that could copy RNA from RNA, an RNA polymerase, but this turned out to be wrong as already emphasized. Instead it was later found that the primary function of this enzyme was to facilitate break down of RNA. However, under certain conditions the enzyme could run the reaction backwards and in the presence of a high concentration of nucleotides link a few of them together. It was this reaction that was originally examined in Ochoa's laboratory providing results that were recognized by the Nobel Prize awarded to him.

Although the true nature of the enzyme had started to become apparent at the time it was still argued in Ochoa's Nobel Lecture[8] that he and his collaborators were studying a RNA-dependent polymerase enzyme. Incorrect data indicating that there might be some correlation between the DNA composition of a host organism and the base composition of an RNA product generated from the organism used to prepare the enzyme were presented. Disregarding these data, now hidden by oblivion, it should be emphasized that Grunberg-Manago's enzyme came to critical use in later attempts to identify the genetic code as we shall see. Later on in 1960 the true RNA polymerase, an enzyme that needed a DNA template, requiring the presence of all four nucleotides and using energy-rich tri-phosphates, was discovered by Samuel D. Weiss, University of Chicago, the aforementioned Berg, Stanford university and Jerard Hurwitz, at the Sloan-Kettering Institute, New York. And to reiterate, RNA synthesis using RNA as a template only occurs in cells infected with different kinds of RNA viruses.

When Grunberg-Manago had finished her time with Ochoa she returned to Paris, and developed a successful independent career. She continued to work on the enzyme she had discovered but then turned into studies of protein synthesis and also into the regulation of translation and expression of the critical RNA components involved in protein synthesis. She came to play an independent important role as a "passionate and energetic woman" in the development of the emerging field of molecular biology and in the early 1960s she temporarily joined in the concerted efforts in Nirenberg's laboratory to crack the genetic code[10], as we shall see. In 1982 she became a Foreign Associate member of the U.S. National Academy of Sciences. She pioneered lectures by a female professor at Harvard University; she became the first female president of the French Academy of Sciences in 1995–96 and also the first woman to head the International Union of Biochemistry.

Grunberg-Manago came from a historically well established family having strong connections in the culture of Moscow. At the time of the 1961 congress she invited Berg and a number of other American scientists to dine at the Moscow home of her family and also invited them to visit their dacha in the surrounding of the city. This was at the time of height of the Cold War and care had to be taken only to exchange information of private and politically sensitive matters when it was certain that the discussion could not be overheard. Berg was to remember this visit vividly[6]. It was at this congress that Nirenberg presented his breakthrough experiment, which we will soon return to. Berg himself had planned to get involved in the analysis of, at the time enigmatic, genetic code. He decided, hearing Nirenberg, not to do this. There was however another major research group that reacted differently!

At the time of Ochoa's Nobel Prize Grunberg-Manago wrote him a letter of congratulations. All she received in return was a printed card with his and his wife's names expressing their thanks for the congratulations. This hurt and from an eyewitness, a molecular biologist who had been taught by her as a part of his Ph.D. studies on how to purify the enzyme she had discovered, I have learnt that she had said with tears in her eyes — origin uncertain, since she was preparing onion soup — "He could have sent me a rose."

The Chagrin of Chargaff

Next to Avery, Chargaff, carried the main responsibility for turning the focus on DNA as the carrier of genetic information. Like many Nobel laureates he also had a Jewish background. The origins of his family were in present-day Ukraine, at the time, the city of Czernowitz, in the Duchy of Bukovina, in the Austro-Hungarian Empire. When the First World War broke out the family moved to Vienna, where Erwin received his early education and obtained a Ph.D. in chemistry.

A young Erwin Chargaff (1905–2002).

He, together with the family he had formed at the time, then tried the United States, but returned to Europe for work at the University of Berlin. When Hitler came to power his situation as a Jew became impossible and the family then moved to Paris, where he worked at the Pasteur Institute, but within two years he returned to the U.S. and in 1935 settled at Columbia University where he pursued his life-long career. In the Second World War he lost his mother in the Holocaust.

Chargaff was a highly opinionated traditional chemist to whom the adjective "irascible" has been attached. He was not given to speculation and conjecture. Hence his personal traits had no congruence with those of Watson and Crick, as witnessed in many books. And still he made two absolutely essential experimental findings in studies of DNA from different species. By use of high quality experimental studies starting in 1944 in which he employed the newly introduced paper chromatography technique and a clever way of identifying the nucleotides by their pronounced absorption of ultraviolet light, he was able to identify the relative concentration of the four nucleotides of DNA in purified materials. The first seminal finding made was that the relative frequency of occurrence of the four nucleotides varied extensively in different kinds of biological materials. As already mentioned this led to the demolition of a dogma introduced by Levene in the early part of the previous century that the four kinds of nucleotides (in unpurified material!) occurred in matching predictable proportions. This conclusion, referred to as the tetranucleotide rule, had led to DNA being labeled as a molecule of little capacity to show variations in its structure, as mentioned. Hence this monotonous molecule was speculated to represent some uninteresting support structure in cells. However Chargaff's findings proved that there was a major relative difference in occurrence of the four separate nucleotides depending upon the origin of the DNA; animals, insects, plants or bacteria.

The other momentous discovery was that the nucleotide bases adenosine and thymidine, respectively cytosine and guanosine always occurred in equimolar amounts. Why this was the case became obvious when Watson and Crick identified the double-helix structure of DNA. Surprisingly their vague knowledge of the Chargaff "rules," as they came to be called, appears to have had a rather small influence on the model building. However, the model immediately explained the reason for these rules. In an interview with Judson[9] Chargaff admitted "Nonetheless, I must say that I myself underestimated the role of base pairing, the multitude of functions it performs in the cell." Visionary speculations were not his strong suit.

Chargaff was highly upset when he was not included in the 1962 Nobel Prize in physiology or medicine to Crick, Watson and Wilkins, but he had in fact never been nominated for a prize at that time. He wrote to a number of colleagues to relieve his frustrations. It was not for nothing that he was referred to as "an irascible authority." However, later he was nominated both for a prize in physiology or medicine in 1964 and for a prize in chemistry in 1965 and also in 1967. In 1964 there were two nominations for Chargaff at the Karolinska Institute. One was by the Nobel laureate Szent-Györgyi. He proposed Chargaff "for his discovery of base pairing in nucleic acid" and Nirenberg "for his discovery of the first shown interrelation between the constitution of RNA and the synthesized protein." There was a reference to one publication for each of the two nominees, but it is not clear if this was a joint nomination, although this was most likely the case. The second nomination was by another Nobel laureate, a major authority in the field, Fritz Lipmann. The first half of his two-page letter was a nomination of Lwoff and Monod, but the second page was entitled "Proposal to include Erwin Chargaff with Monod and Lwoff," briefly mentioned in my previous book on Nobel Prizes[3]. It is worth returning to this second page in full. It read:

"It was to the considerable regret of myself and others, for example Professor Sonneborn on the University of Indiana, that Chargaff's contribution to the recent award for the elucidation of DNA structure was not included."

Several years ago, when Crick gave a series of lectures here, he more or less started by stating that he had always been surprised that Chargaff had not already reached a conclusion similar to that eventually gained by him and Watson in their analysis of the X-ray diffraction pattern. At a later stage, Watson, on being questioned as to what influence Chargaff's work had had on their conclusions, remarked a bit defensively, that he thought they probably would have reached these same conclusions without having Chargaff's data available.

These statements recognized the potential importance and the impact of these data. It is, I feel, very difficult to judge *a posteriori* because, when the X-ray diffractions patterns appeared, the Chargaff data were available and, presumably, did guide the thoughts of the investigators although possibly unconsciously into the base-pairing pattern. After all, Chargaff had recognized the base correspondence in

DNA and the equality between the sum of the matching pyrimidines and purine derivates long before anything was known about the double helix.

These data speak for themselves as a feat of analytical acuity which guided others to the important discoveries years later. I think it somewhat unjust to ask that Chargaff should have realized the meaning of the finding of pyridine to purine correspondence, although, in retrospect, one might think that if he had consulted with Pauling they might together have approached the presently accepted structure. However, if I remember right, at that time Pauling had in mind a nucleic acid structure with phosphate inside rather than outside as it appears to be now. Nevertheless, a sharp look at the Chargaff data may have sparked the idea of a hydrogen bonding between the purines and pyrimidines and that, in consequence, may have suggested the double helical structure.

But this is speculation; yet I feel that Chargaff made an immense contribution which, to my mind, greatly influenced what eventually happened."

The committee took these *post festum* nominations seriously and let Hugo Theorell make a preliminary evaluation. He concluded that Chargaff's contribution was of such a magnitude that it deserved to be thoroughly looked into. A full evaluation was performed by Peter Reichard who we will encounter repeatedly as a skilled critical reviewer in the following. Over four pages he carefully described Chargaff's important contributions in the late 1940s which finally wiped out the tetranucleotide theory paralyzing the field of DNA research and paved the way for a fuller appreciation of Avery's advances in allocating a central role to DNA as the carrier of genetic information. Chargaff developed an efficient method for breaking DNA into its components, still used into the present time, but he was less successful in characterizing the composition of RNA. Reichard of course emphasized the critical finding of equal amounts of purines and pyrimidines in DNA, but stressed the fact that although he described the phenomenon, Chargaff could not explain it. This remained for Watson and Crick to do. Interestingly Reichard agreed with the nominator Lipmann that it would have been equally justified to divide the prize between Watson, Crick and Chargaff as between Watson, Crick and Wilkins, which was the choice by the committee in 1962. In the summary report of the committee in 1964 Chargaff was listed as not worthy of a prize at the present time. The events of time had passed him by.

At the Royal Swedish Academy of Sciences there was in 1965 and again in 1967, a nomination by Warren M. Sperry, professor of biochemistry at Columbia University, hence a department colleague of Chargaff. The nominations were rather extensive and in 1967 it began by mentioning: "… that the current awards of Nobel Prizes to Mulliken, Kastler, and Rous represent examples of the distinction being conferred on research work the importance of which had not been fully appreciated when it first appeared." In his nomination Sperry initially referred to the conclusion of Maurice H. F. Wilkin's Nobel lecture[11]. It read:

"More generally, I thank:

Francis Crick and Jim Watson for stimulating discussions; ….; and especially, Erwin Chargaff for laying foundations for nucleic acid structural studies by his analytical work and his discovery of the equality of base contents in DNA and for generously helping us newcomers in the field of nucleic acids."

The latter remark may appear somewhat out of context. There are a number of references to the fact that the contacts between Chargaff and Watson and Crick prior to the presentation of the double helix were not of a productive kind. There was a complete lack of personal chemistry. Sperry finally noted that Chargaff in 1965 had been elected to the National Academy of Sciences and that he had received a number of important prizes.

The committee at the Royal Swedish Academy of Sciences in 1965 asked Einar Hammarsten, the frequently cited medical chemist with some thirty years experience of work in the committee at the Karolinska Institute, to make a review. He has been introduced at length previously[3]. His review is of considerable interest from a number of both historical and actual perspectives. The research field of nucleic acids had been pioneered by Hammarsten's uncle Olof. Einar followed in his tradition and spent his whole scientific career studying nucleic acids. Together with Caspersson he was able to demonstrate already in the 1930s by use of the newly constructed ultracentrifuge that they had molecular weights in excess of millions. A surprising, but, as was first understood much later, an important observation. Finally he was the critical reviewer of Avery and carried the prime responsibility for the fact that he never received the prize which was due to him. Hammarsten's review of Chargaff was brief but rich in important statements.

He first gave credit to Chargaff for managing, by use of the simple and reproducible techniques developed by him, to determine the content of individual nucleotides in biological materials of many different sources, including a number of tissues from humans. As already emphasized, this work led to the important conclusion of the existence of a *species-specific* relative presence of the frequency of purins (adenine, guanidine) and pyrimidins (cytosine, thymidine). This was the nail in the coffin of the generally accepted rule that nucleotides were regularly arranged in fixed groups of four. Hammarsten then cited from one of Chargaff's own 1950 publications. The text read:

> "As the number of examples of such regularity increases, the question will become pertinent whether it is merely accidental or whether it is an expression of certain structural principles that are shared by many desoxipentose nucleic acids, despite far-reaching differences in their individual composition and the absence of recognizable periodicity in their nucleotide sequence. It is believed that the time has not yet come to attempt an answer."

This is a formulation by a scientist who does not like to speculate.

Hammarsten then briefly mentioned that Chargaff in the late 1950s and early 1960s attempted to develop methods to determine the sequence of nucleotides in DNA and in RNA. However, it would take some time before such methods had been developed by Walter Gilbert and Frederick Sanger, who received half a Nobel Prize in chemistry in 1980[3].

The second to last paragraph of the review expressed some important reflections. It reads (translated from Swedish):

> "There is an apparent connection of ideas between Avery's discovery of transformation and Chargaff's discovery hereafter of the connection of the frequency of purine and pyrimidine content reflecting the specificity of a certain species. It seems to me reasonable to assume a similar connection between J. H. Matthaei's and M. W. Nirenberg's finding (1961) and the joint discovery of the importance of purine/pyrimidine sequences in DNA for genetic information."

Then came a very interesting statement: "…I feel it as a burden that the Nobel Committee never proposed Avery for a prize since it was of the view that the priorities were not clarified."

This is an important statement from the person who carried the main responsibility for the fact that Avery never received a prize. The background to this has been reviewed in my first book on Nobel Prizes[1] and it has also been referred to in the 2003 book by Watson[12] and also the more recent 2016 book by Siddhartha Mukherjee[13]. To this can be added that Watson and Crick for unknown reasons omitted to make a reference to Avery in their two seminal Letters in *Nature* in 1953. They did refer to Chargaff, however. Hammarsten had reviewed the first nomination of Avery for his discovery of the transforming principle in 1946. He hesitated to accept the conclusion. Six years later the professor of bacteriology Berndt Malmgren made a 19-page review of Avery with a very vacillating conclusion. The committee persisted in its hesitancy. Two years further on Hammarsten returned as a reviewer. He now accepted the proposed phenomenon, but suggested a deferral of decision since the mechanism was not known. In 1955 there were four nominations of a prize for Avery. Interestingly one of them came from Chargaff. He proposed a prize divided between Hammarsten (+Caspersson?) for showing the large size of nucleic acids in the 1930s and Avery for his discovery of the transforming principle. What a stroke of irony! Of course no new reviews were made since Avery died in this year. In an exceptional posthumous review made by Tiselius for the chemistry committee in 1957 it was concluded that a major mistake had been made. One may wonder about Hammarsten's thoughts when he was writing the 1965 review of Chargaff. Firstly, why did he write as an attempted excuse "that priorities were not clarified" which obviously was not true and secondly what were his reflections on his own career. This had over decades been devoted to nucleic acids and still he had never fully understood that there were grains of gold running between his fingers!

It may now be time to close the books on Avery, who has figured in all my three previous books on Nobel Prizes[1-3]. In March 2018 I visited Vanderbilt University at Nashville, Tenn. to give a lecture on topics in my third book on Nobel Prizes. Assisted by my host and recent friend Jacek Hawiger I had the opportunity to visit Avery's grave. It is hard to conceptualize that the very humble tombstone belongs to a man who by his discovery laid the foundation for the whole field of molecular biology.

Finally it is time to return to the conclusion of Hammarsten's review of Chargaff. He summarized (translated from Swedish):

"Even though Chargaff's discovery of the connection of purine-pyrimidine frequencies relation to species specificity in its importance is

The author at Oswald Avery's discreet grave.

secondary to Avery's discovery, it is my view that the discovery made by Chargaff is worthy of a prize. It is my view however that Chargaff should not be considered for a prize alone, but on a later occasion he might be considered together with Matthaei and Nirenberg."

The committee cited Hammarsten's conclusions in its summarizing report. It then wrote that the candidates nominated for a Nobel Prize in chemistry for the first time, Chargaff, Nirenberg and possibly the candidate Taylor should be retained for continued considerations, although for the reasons given one should await further developments. This was as close as Chargaff ever came to be considered for a prize. In 1967 the committee took no further action following the repeat nomination by Sperry. No additional review was made and in its summary report Chargaff was mentioned together with 23 other potentially prize-worthy candidates, noting that nothing new had been added and that no action was motivated.

Premature Discoveries Revisited

In discussions about Nobel Prizes in physiology or medicine the term "discovery" cited in Nobel's will is frequently returned to. It is a decisive and critical concept. Over and over again in the present and also in the preceding books on Nobel Prizes[1-3] we have taken notice of how theoretical and predominantly experimental data, managed as a proper ferment for the intellect of an anointed individual, has led to the taking of a quantum leap — the epiphany of creation. The phenomenon is readily identified because of the avalanche of confirmatory research it releases. But there are a number of examples when a truly visionary proposal, in many cases even underpinned by solid experimental data, has been presented without having any impact on the course of events. A number of examples of such situations were referred to throughout this book.

There is no question about the fact that Ellerman and Bang recovered a filterable agent from a tumor of avian origin prior to the similar finding by Rous. Ignorance about the fact that leukemia, in its very nature, represents a cancer corresponding to that of tumors originating in solid organs held back the full appreciation of the discovery and they were never nominated for a Nobel Prize. Rous introduced the concept of progression in a description of tumor development but this did not come to represent a full discovery because he was hostile to accepting the role of accumulated genetic changes. It was Nordlund who first proposed that the accumulated changes in the development of cancer had a genetic nature. But his intelligent speculations had no impact because the time was not ripe to interpret the genetic underpinning of the proposed consecutive events. Similarly the possible existence of a central dogma of transitions from DNA to RNA to protein was suggested by Alexander Dounce (to be further mentioned in this chapter), but again the time was not ripe to understand the consequences of this in molecular terms. The proposal did not influence the course of events; one had to await the accumulation of solid experimental facts.

A final example is the unraveling of the nature of the genetic material. Chargaff's identification of the equimolar presence of the nucleotides adenosine and thymine and guanine and cytosine, respectively, was a major contribution. Still it appears only to have had minor importance for the final correct description of the double-helix structure of DNA based on the critical base-pairing discovered by Watson and Crick. Chargaff was cited in their seminal 1953 publication in *Nature*, but Avery and collaborators were not. This is indeed surprising. What Avery and collaborators presented in 1944 were solid experiments showing that purified DNA extracted from pneumococci could lead to a change of properties of the recipient cells. Their experimentation could readily have been confirmed by other scientists, but the group at the Rockefeller Institute was so much ahead of their time that the unique message was not accepted by the scientific community. In this case it was not flaws in the proposals, but limitations in the receptivity of the scientific community that caused the lack of impact. Hence Avery's momentous discovery should have been recognized by a Nobel Prize, but the Nobel Committees became aware of this too late as repeatedly noted.

Crick and the Early Speculations on the Code

Judging from the title of one of his books, *Francis Crick: Discoverer of the genetic code*[14], there is no doubt in Matt Ridley's mind about who solved the genetic code. The origin of the word "code" is interesting since it derives from the very first material used for writing, to create a book. The material originally used was the stock or stem of a tree, the pith, on which symbols were scratched. The Latin word for this material originally was *caudex*, which was later modified to the term "code." At the library of Uppsala University there is a unique book, the Codex Argenteus, the silver bible that the Swedes took as a trophy in Prague in connection with the Thirty Years' War. It is a remarkable incunabulum, written in the now extinct Gothic language in the early sixth century. In France the word "code" was later used to depict rules of law, as in the Code Napoleon.

Ever since the time of the discovery of the double-helix structure of DNA there has been a discussion of how sequences of the nucleotides could store genetic information which eventually would decide the sequence of amino acids in the protein products. But at this dawn of molecular biology there were, as discussed above, also a number of related questions that had already been discussed, such as what was the role of RNA; were there different populations of RNA serving different functions; how could such potential populations of RNA interact and where in the nucleated cell did this occur. Feelings were running high and without doubt Watson and in particular Crick had a central role. In the early debates about the nature of the code Sydney Brenner also made seminal contributions. These developments have been reviewed in part in my second book on Nobel Prizes, describing the background to the prize in physiology or medicine in 1962[2]. On March 11 and 12, 2019, I had two as it turned out

Sydney Brenner (1927–2019).

precious meetings with Brenner in Singapore. He died three weeks later after a uniquely active life in science. It is worth remembering that the Nobel lectures in 1962 started with Wilkins[11], who did in fact talk about DNA and how its structure had been derived from crystallographic analysis. The ensuing lectures by Watson[5] and Crick[15] presented the developments of molecular biological insights into the functions of RNA, as mentioned, and the genetic code, respectively. They hardly mentioned the double-helix DNA structure. This illustrates well the rapid development of the field of molecular biology at the time.

Speculations on the mechanisms for transferring information on how the information hidden in DNA could determine the sequence of amino acids in a polypeptide chain representing a protein started immediately after the discovery of the structure of DNA in 1953. A colorful individual in these early developments was George Gamow, introduced earlier[2] and presented in many books on the history of molecular biology[9,16,17] and by the central actors Watson[12] and Crick[7]. Although a theoretical physicist who had made major contributions already in the 1940s to the development of the concept of the Big Bang at the birth of our universe, Gamow was also very curious about the developments of biology catalyzed by the revelation of the double-helix structure of DNA as published and discussed in the two Letters in *Nature* in the spring of 1953. He had also briefed himself about Sanger's discovery of the amino acid sequence of insulin[3]. Combining the two fundamental discoveries he speculated on how DNA could provide information for the synthesis of the chain of amino acids.

He outlined his ideas on how the different amino acids might attach to twenty different cavities in DNA formed by different combinations of nucleotides in a letter to Watson and Crick in the summer of 1953. The first question concerned the number of amino acids for which code words were needed. Gamow had settled for twenty and when Watson and Crick in response to his letter sat down to determine the most likely number they also ended up with twenty, but several of the amino acids they included differed from those proposed by Gamow. Early on they also picked an argument with Gamow on using DNA as a matrix for assembly of amino acids to form a future protein.

It soon became clear that there needed to be an intermediary and it was deduced that RNA was a strong candidate — but what kind of RNA? It would take until the early 1960s before all three different principal forms of RNA involved in protein synthesis had been identified, as already introduced above. Let us first consider the speculation on the structure of the code. To be semantically correct, it is in fact not a matter of a code but of a cipher, as pointed out by Crick[7]. The individual code elements are used in different combinations, hence it is a cipher. However, the term *genetic code*, which has a better ring, caught on early and will be used in the following.

George Gamow (1904–1968).

Since there were only four letters for use in the code it apparently would not suffice to use two-letter combinations. A bi-nucleotide code could only identify a maximum of 16 amino acids. Hence at least three letters had to be used, but that gave 64 different code words, much more than needed. Another central question concerned if the code words would be overlapping or separate, the latter referred to as comma-less. In the mid 1950s the debate became intense and on Gamow's initiative the RNA tie club was formed as presented earlier[2]. It had 20 ordinary members, just like the number of amino acids, and also four honorary members, one for each nucleotide. It included authoritative biologists as well as physicists, but it should be noted that none of the central actors in this or the last chapter, like Nirenberg, Ochoa, Khorana and Holley were members. In unofficial circulars the members shared ideas and debated upcoming thoughts. Brenner and Crick provided the leadership of this group.

Gamow's original proposal stated that there should be an overlapping code, including two nucleotides of the triplet used. This immediately put a restriction of the amino acids that potentially could be included in a polypeptide chain. Supplementary theorizing made it clear that the code must be comma-less, which raised two major questions. One was how the system managed to put the first comma correctly, viz. to know where to start. The other question obviously was which triplet was responsible for which amino acid. Was it only one triplet for each amino acid or could there be multiple ones, meaning that the code would be redundant. This seemed like a problem that required the involvement of biochemists to solve it. But before this decoding started Crick had accumulated experimental evidence that the code was based on triplets.

For a short period in his long life as a dominating theoretician Crick in the early 1960s involved himself in practical experimental work. He used viruses (phages) infecting bacteria earlier studied by Seymor Benzer (a strong contestant for a Nobel Prize, who eventually received a Crafoord prize[3]) and a strain of phages selected by Brenner to cause a frame-shift in the reading of DNA by the deletion or addition of a single nucleotide, referred to as − or + mutants. Without going into any details of the evidence provided it could be demonstrated that if there was an accumulation of three insertions or alternatively three deletions in a single gene it was possible for it to become expressed. Crick was proud of his hands-on work and described it briefly in his Nobel lecture[15], which by the way used the word "triplet" more than thirty times. It was Brenner who coined the term *codon* for the triplets. Eventually the exact triplet code was deciphered by professional or self-taught chemists. As we shall see the first breakthrough was made by Marshall Nirenberg's group at the National Institutes of Health (NIH) in Bethesda, Maryland but in the continued work a major competition came from Ochoa's group at New York University, School of Medicine. Later, there was also a separate attack on the code by Gobind Khorana at the Institute for Enzyme Research, University of Wisconsin to be discussed in the next chapter.

The Development of a Humble and Unassuming Scientist

Nirenberg's life has been described in 2015 in a biography entitled *The Least Likely Man. Marshall Nirenberg and the Discovery of the Genetic Code*[18]. As the title indicates Nirenberg's developments in his early life did not foretell that he would became a highly successful scientist. He was a very modest person and perhaps because of that underrated. However, once he felt he was on to something important he displayed

Marshall Nirenberg (1927–2010) the co-recipient of the 1968 Nobel Prize in physiology or medicine. [From *Les Prix Nobel en 1968.*]

a pronounced endurance, tenacity and creativity. He set his goals high and stayed with the problem he had selected.

Marshall was of Jewish origin. His ancestors on his father's side came from Odessa, the seaport on the Black Sea coast of Ukraine. Because of a recent pogrom his great-grandfather left for the United States and arrived in Philadelphia in 1888, later settling in New York. He did well in a shirt business which was taken over by some of his six children, including Harry, who was to become Marshall's father. In 1924 Harry married Minerva Bykowsky, whose parents had also arrived from Russia. Marshall was born in 1927, a year after his older sister. He grew up in Brooklyn and appears to have been materially well provided for. The family managed to sustain a good standard of living even during the depression of the 1930s. As a young boy Marshall developed rheumatic fever which was a major cause of concern to the family. To improve his situation the family moved from New York to the city of Mount Vernon and he received tutoring at home. A neighbor of theirs in the new place of living told them that there was a State of "milk and honey," Florida. The family then decided to move to a more attractive and milder climate. They took off for Orlando. They soon discovered that there were good things and bad things. The father, who changed trade completely and became a dairy farmer, came to realize that this profession had its own challenges and his success in the trade varied. Hence frequent supplementary entrepreneurial initiatives often involving the whole family were launched. On the plus side Marshall eventually recovered from his disease. He came to love the country life and the closeness to nature that it provided. It was this newly awakened curiosity that took him into science. As described[18]:

> "For Marshall, Orlando was a natural paradise. The miles and miles of space gave him an opportunity to wade and hunt, view a profusion of birds and plants and even catch an occasional cottonmouth moccasin, a poisonous water snake. Marshall would put the snake in a bag and bring it home."

He soon got into contact with professional biologists. The reason was that because of the developing Second World War, a local airport was constructed to train pilots going into the South Pacific war arena. One of them was the ornithologist Lieutenant Frank McAmey, who brought him along on birding expeditions. On one occasion they went to a place called Meril's Island, today better known as Cape Canaveral. They were fortunate enough to see the dusky seaside sparrow. Already at the time this was a very rare bird and since December 1990 sadly it has been declared extinct.

During this time his family's struggles also appear to have become a concern to Marshall and they were to influence his personal characteristics. He developed into a somewhat withdrawn and private person, a trait he retained throughout life. At the time there were also racial tensions in the area with activity by the Ku Klux Klan movements. Furthermore the developing war led to massive conscription and Marshall wanted to join the Merchant Marine. However, he was rejected because of his earlier disease. He therefore went off to college. In contrast to many other successful scientists-to-be, Marshall did not impress his teachers in school and college. When he entered the University of Florida at Gainesville, this institution of learning did not admit women or African-Americans. However, it did not have a quota for Jews. It took him only three years to graduate and he was able to improve his finances by working as a teaching assistant in comparative anatomy. His grades were not impressive but he was admitted for a master's program in zoology. Thanks to this he gained some important experience in using radioactive isotopes, literally a hot topic at the time. However, his thesis work was conventional and did not include front-line use of technology. The goal of the project was to catalogue flies, more specifically the species of caddis flies in a certain region. He then started graduate studies to enter medical school. Possibly he wanted to polish the family's tarnished honor. His father had left medical school after two years at Cornell University. Marshall applied to the University of Michigan.

It was a mixed group of students at the biochemistry department, including many veterans from the war. Marshall befriended one of them in particular, Conrad Wagner. They worked closely in the laboratory projects. Wagner's comments on his laboratory mate were "a thoughtful and caring person; very focused but not brilliant; full of enthusiasm." Marshall was very good-looking and hence popular not least among dental hygienists, all female students. Hence he had "numerous girlfriends and very clean teeth." Approaching his Ph.D. studies he showed a capacity suggesting an interesting potential for his future scientific career. He designed his own Ph.D. project which focused on the processing of sugars in cancer cells. He did not display pronounced manual skill in the laboratory but "he labored quietly, constantly and intently." His acquaintance with working with isotopes was of good help in the development of his chosen project. He had selected a relatively new faculty member to be his mentor, James F. Hogg. When the work was finished in 1957 Hogg gave him the advice to apply to the NIH in Bethesda, Maryland.

An important new chapter in Marshall's life was about to start. A fellowship from the American Cancer Society paid his way.

NIH Provides an Important Home to Nirenberg

At the time of Nirenberg's arrival at NIH discussions about the genetic code were already in full swing. They had become more focused, when it was appreciated that it should be a redundant triplet code. A biochemical approach was needed to determine the specificity of the 64 different triplets that should exist. Mapping out the three different forms of RNA and their functions took up all the time of the molecular biological chemists at this juncture. It seemed that the next step in the deciphering of the code needed to be an innovative experimental approach. Nirenberg's original plans were to deepen his studies of the sugar metabolism of cancer cells but he had also started to reflect on the possibilities of directing his research towards the genetic code. He wanted to search for a field of major importance where he could mark out his independence. However, he still had a time of learning ahead of him and his mentor at NIH, W. DeWitt Stetten, Jr. was to become an important person for his early postdoctoral development as a scientist. Immediately after Nirenberg's arrival, Stetten took off for Woods Hole Oceanographic Institute, where he spent his summers, like many other frontline scientists already introduced. Hence Nirenberg had a slow start.

Nirenberg still had a lot to learn. He was assigned to collaborate with William Jakoby, a qualified microbiologist. From him he gained important knowledge about enzymes and how they should be studied. These insights into the function of proteins would prove to be important for the forthcoming development of his work. The large question of the chemical basis of heredity loomed in his mind and he consumed the large amount of information on this matter. But how could he start to approach this problem on his own initiative? One senior colleague who influenced him was Alexander Dounce. He had done his Ph.D. in the laboratory of James B. Sumner, the recipient (shared) of the 1946 Nobel Prize in chemistry. In his subsequent work Dounce speculated on how proteins could make other proteins, a question asked of him in connection with his 1935 thesis presentation. Much later, in 1952 Dounce published a very useful method to isolate DNA by use of gently homogenized cells employing a device that is still referred to today as Dounce homogenizer. A year later, the famous year of the identification of the structure of DNA, he speculated in an article: "…it could conceivably happen that the deoxyribonucleic acid gene molecules would act as templates for ribonucleic acid synthesis, and that the ribonucleic acid synthesis on the gene templates would then in turn become templates for protein synthesis." This was thinking ahead of its time, as already mentioned. Dounce became a member of the RNA tie club and stimulated Crick in his thinking. He also provided Nirenberg with ideas in his forthcoming work on the genetic code.

In retrospect it might appear as if there would have been a straight way forward, but that certainly was not the case. There was a litany of ideas and in essence confusion prevailed. One is reminded of a formulation by the 1960 Nobel laureate Peter Medawar "science is the art of the soluble." It appeared that in this case the code had to be resolved by a chemical approach. When it came to the choice of RNA to be used, there were two proposals. One was to use simple synthetic RNA molecules and the other to use virus nucleic acid, since it had been demonstrated that in the case of TMV the extracted RNA was fully infectious. This important discovery by Alfred Gierer, Gerhard Schramm and Heinz Fraenkel-Conrat from the mid 1950s was very close to becoming recognized by a Nobel Prize, but eventually never made it[2]. But if Nirenberg was going to test the use of virus RNA for examining the detailed genetics of protein synthesis he needed to have techniques not only to examine the amino acid sequence of the protein(s) synthesized, which had recently become available, but also the sequence of nucleotides in the nucleic acid. Since the latter could not be determined at the time the challenge was insurmountable. It would be a long time before nucleic acid sequencing technology was developed.

Still, regardless of the choice of approach to be taken to evaluate the functional specificity of the RNA used in the experiment there was a need to select a protein synthesizing method for use in the laboratory. This could be managed either by use of animal cells or of bacteria. Nirenberg selected the latter simpler kind of cells and was able to demonstrate that using TMV RNA material it was possible to demonstrate protein synthesis, an important positive control. He took advantage of pioneering work in this area by Alfred Tissières and co-workers at Harvard Biological Laboratories. The synthesis of proteins was identified by using radio-labeled amino acids of different kinds and by identifying if the radioactivity was incorporated in the polypeptides synthesized under the laboratory conditions used. It should be emphasized that in order to be able to test the effect of added RNA of predetermined characteristics, the system first had to be emptied of indigenous protein synthesis resulting from remaining RNA present in the original material. Such a material by itself had a short half-life, but more RNA could be produced by copying the DNA also present in the crude material. To remove this from the sample it was treated with an enzyme breaking down DNA. Finally the system was adjusted to allow identification of protein synthesis from added RNA as mentioned.

In order to allow a possible advance it was decided to use synthetic RNA as an alternative to TMV RNA. Techniques to produce such molecules started to become available. In fact as mentioned Ochoa received his 1959 Nobel Prize for the identification of an enzyme that he believed could synthesize RNA, but which later was demonstrated to be a degrading enzyme. In spite of this the enzyme could under certain conditions lead to the production of a short chain of nucleotides. The enzyme therefore came into a very critical use in a daring step taken by Nirenberg in collaboration with a German postdoctoral scientist, the plant physiologist Johannes H. Matthaei. The latter had received a NATO Research Fellowship. He started at Cornell University, but the proposed project eventually did not fit into the research environment made available. He therefore tried to relocate to the Nobel laureate Fritz Lipmann in New York but in his laboratory there were no places left. After a final unsuccessful attempt to join a near Nobel Prize candidate[3] Britton Chance's also over-filled laboratory in Philadelphia, he ended up with his new driving license and old Cadillac at NIH. At this time there were interesting developments in Nirenberg's laboratory, which became the temporary scientific home of Matthaei.

Over two years Nirenberg had accumulated a number of ideas about how to retrieve solid information about the existence of an information flow from RNA to protein. Stettin had handed him over to another mentor who came to have a major influence on his future development. It was Gordon Tomkins, who in 1953 had come to NIH with a medical degree and biochemistry Ph.D. from Harvard University and the University of California at Berkeley, respectively. Tomkins had a special interest in original research talents and apparently was intrigued by Nirenberg's personality. Although he worked in a different laboratory he regularly joined in the weekly seminars in Stetten's laboratory. In 1959 when Tomkins was appointed as head of the laboratory of Molecular Diseases at the Institute of

Gordon Tomkins (1926–1975).

The Prime Author of the Saga of the Genetic Code 253

Arthritis, Metabolism and Digestive Diseases he offered Nirenberg a position as independent investigator. This was important for his development. Tomkins somewhat later left NIH to develop powerful laboratories in other settings. It can be added that when Tomkins died prematurely at the age of 49 years from a brain tumor, his peers remembered him in the following way "… he tackled fundamental questions…by temperament and conscious esthetic choice, Gordon sought an underlying simplicity and unity in biology.…"

A Major Breakthrough

At the time when Nirenberg had become a group leader he started to get a grip on the protein-synthesizing technique in his laboratory, but he had also devised means to synthesize molecules of RNA. It should be remembered that at the time the concept of messenger-RNA had not as yet been conceptualized. A critical observation catalyzing the establishment of this concept was an experiment by Elliot Volkin and Lazarus Astrachan at Oak Ridge National Laboratory. They had shown that T2 bacteriophage infection of E. coli cells led to a shift in RNA characteristics from mimicking cellular DNA to instead have a base composition similar to phage DNA. Nirenberg and Matthaei had received a preparation of RNA containing only uridine, - UUUUUUUU (poly-U) from a colleague at NIH, Don Bradley. This was tried in their system. Other research groups earlier had tried to use poly-A, essentially as a control in their laboratory protein synthesizing experiments. No protein product had been found by the methods used to extract them. The protein synthesis in the laboratory was measured by use of radio-labeled individual amino acids. In fact they prepared labeled samples of all the 20 different L amino acids. In principle there are two forms of all kinds of amino acids, an L(levo) and a D(dextro) form. Evolution fortuitously has selected to use only L amino acids, but in a chemical laboratory it is possible to produce mirror image protein molecules only containing D amino acids or a mixture of both. They have sometimes been found to be of value for use as drugs. The aim in Nirenberg's experiments was to test if the labeled L amino acids could become incorporated into proteins in his bacterial protein synthesizing system. The hope was that radioactive material would appear in the newly synthesized polypeptides.

When Nirenberg was away for an intended one month at University of Berkeley to learn from Fraenkel-Conrat how to work with infectious TMV RNA he received a telephone call. Matthaei had made an earth-shaking

Heinrich Matthaei and Nirenberg, the discoverers of the first codon. [From Ref. 23.]

observation in recording the results of the experiment they had jointly planned. The poly-U stimulated a major accumulation of radioactive phenylalanine (Phe), but *only* this amino acid into the newly synthesized protein product. Nirenberg promptly returned home. This is how Matthaei remembers his initial observation[18]:

> "I was total(ly) excited foreseeing the expected result, enjoing (*sic*) working through the nights. The unusual activity of ... (UUUUUUUUUUUUUU) ... I knew from the first ...experiment...with the C^{14} amino acid mix on May 21. So I spent the days May 22-27 testing the amino acids in groups (to save ... (UUUUUUUUUUUUUU) .. of which I only had 1 milligram and had used much more than necessary in the first pilot experiment.. doing everything in duplicate). The final alternative for decision on May 27 at 3:00 a.m. was either tyrosine or phenylalanine."

It was then found that the polypeptide chain only contained phenylalanine. What Matthaei had experienced as the first eye witness was the Eureka moment identifying the first triplet of the genetic code, namely that UUU means phenylalanine. His experience was very similar to that of Dominique Stehelin who was the first to concretely see the results demonstrating that the src gene in Rous virus was of cellular origin (p. 100). Stehelin later argued that he should have been included in the 1989 Nobel Prize in physiology or medicine because of this contribution, a view not shared by the Nobel Committee. In the same vein Matthaei might have argued that he should share a prize with Nirenberg. We will return to this issue later because Matthaei was in fact repeatedly nominated for a prize. On the Saturday morning following the original discovery, Tomkins, who had been informed about the planned experiment, out of curiosity called around and became the first to be informed about the momentous advance made. He kept it to himself until Nirenberg and Matthaei had presented it to the scientific community.

As mentioned by Nirenberg himself much later[10] it turned out that he and Matthaei had the luck that applies to anointed scientists. They were not the first to test the protein-synthesizing activity of a polynucleotide RNA containing only one kind of nucleotide as we shall see. In his work with poly-U Nirenberg learnt a lot about the special properties of particular proteins enriched for a certain kind of amino acid from a scientific colleague on an adjacent floor at the institute at NIH. It was Michael Sela.

Michael Sela.

Sela is a world famous Israeli immunologist of Polish origin, who at the time was a visiting scientist in Christian Anfinsen's laboratory. Anfinsen was later, in 1972, to receive half of a Nobel Prize in chemistry "for his work on ribonuclease, especially concerning the connection between the amino acid sequence and the biologically active conformation." Most of his career Sela has been active at the Weissman Institute of Science in Rehovot, Israel, which he also has successfully headed for extended periods[19]. Throughout his career he has had a particular interest in synthetic polypeptide antigens and as a consequence had learnt a lot about chemical characteristics of different kinds of such products. The aim in part had been to search for means to treat autoimmune diseases and a certain polymer of selected amino acids has been successfully used to mitigate the development of the disease multiple sclerosis in some patients. It is in this context I have learnt to know him and his close collaborator Ruth Arnon. Soon after his and Matthaei's discovery Nirenberg went to the laboratory a flight below to contact Sela. He asked him if knew anything about the properties of poly-phenylalanine. Sela obviously became curious and wanted to know the reason for the question. He then became one of the first to learn that Nirenberg believed that UUU coded for this amino acid. He also later became an important nominator of Nirenberg for a Nobel Prize, as we shall see. Sela himself described the consequences of this first encounter in the following way[19]:

> "While I was somewhat skeptical of the story, I immediately looked for and found, hidden somewhere in an experimental section of a paper in the *Journal of the American Chemical Society*, that poly-L-phenylalanine was insoluble in all the solvents we had tested, with the exception of a saturated solution of anhydrous hydrogen bromide in glacial acetic acid (reference). Because on that very day I was preparing just such a solution (used to remove carbobenoxy groups) in the lab, I gladly gave the reagent to Nirenberg and was touched and surprised when he acknowledged this in the classical paper that resulted in his receiving the Nobel Prize."

There is in fact an extension of the story, which illustrates in what a capricious way science sometimes works. The text continued:

> "But the real point of the story lies elsewhere. Why did we try to use such a peculiar solvent? The truth of the matter is that years earlier, together with the late Arieh Berger in Rehovot, we were investigating the mechanism of polymerization leading to linear and multichain polyamino acids. One day I had two test tubes: one with polyphenylalanine, and one with polycarbobenzoxylysine — stuck in an ash tray on my desk. Arieh came to decarbobenzoxylate the lysine polymer, a reaction with hydrogen bromide in glacial acetic acid during which carbon dioxide is released. He took the wrong test tube away with him and returned, puzzled because the material had dissolved and he could not see any evolution of carbon dioxide. At once we realized the mistake, and I noted in my lab book that, at last, we had found a solvent for poly-L-phenylalanine."

This experience illustrates once more the role of serendipity in scientific research, in this case an accidental finding that provided Nirenberg with a critical solvent for poly-lysine from the only person in the world who had information about this.

It was time to publish the revolutionary results and also to announce them at a scientific conference. At the time of writing the manuscript, an additional supportive important finding had been made. An RNA containing exclusively cytosine was found to result in a polypeptide composed only of the amino acid proline. However, the first communication to the scientific community included only the poly-U data. Joseph Smadel, a former associate director for NIH introduced the manuscript for publication in *Proceedings of the National Academy of Sciences*[20] and the first public exposure of the data occurred in a special context. In August 1961 there was an International Congress of Biochemistry to be held at Moscow, as already mentioned. This was at the height of the Cold War, but indicated the beginning of a return to an exchange of scientific information between the two world powers. Thanks to actions by influential Soviet physicists the paralyzing effect of Lysenkoism had started to wane[3].

When Nirenberg was going to give his presentation he was disappointed to note that he had been allocated to a minor session with only a few attendances. However, one of them was Matthew Meselson from Harvard University, who together with Frank Stahl by ultracentrifugation studies of isotope labeled DNA in different phases of replication had experimentally documented the existence of semi-conservative replication[2]. He appreciated the impact of Nirenberg's discovery and in fact rushed up to him after the presentation and gave him a big hug. Nirenberg would remember this for many years. Of additional practical importance, however, was that Meselson actively alerted Crick to the new findings. The latter immediately ensured that Nirenberg could make a second presentation at a much larger plenary session, which Crick himself chaired. According to Crick, using different verbs in different presentations, the audience was "startled" or "electrified." The discovery served as a start signal to a race to identify the functions of all 64 different codons. One of the most powerful competitors was Ochoa. Perhaps he wanted to secure a second Nobel Prize, since he had become aware at the time that his first prize had been awarded on the wrong premises.

On his way home from the Moscow congress Nirenberg did in fact make a stopover in Paris for a brief honeymoon. Just before departing for Moscow he had married a fellow biochemist from Brazil, whom he had met in the apartment building where he lived on the NIH campus. Her name was Perola Zaltsman. When Nirenberg had returned to NIH, Lipmann kindly shared with him a preparation of partly purified enzyme critical for the transfer RNA association with amino acids. It was shown that in the presence of poly-U transfer RNA was charged with phenylalanine. It may also be added that Nirenberg's group had not been the only one contemplating using an mRNA containing only a single kind of nucleotide. Ochoa had a coworker, Mirko Beljansky, who for a year tried to see if poly-A could direct the synthesis of a protein. However, the poly-lysine product to be identified is a basic protein which was not precipitated by the chemical, 10% trichloracetic acid, conventionally used to isolate protein products. In fact it dissolved it. The same problem had been encountered by Tissières in Watsons laboratory. Later it became possible to demonstrate, by use of modified conditions, that poly-A led to production of a protein only containing lysine and that poly-G directed the synthesis of a polypeptide only containing glycine. In summary to be a successful scientist you also need some luck!

After the discovery of the first coding triplet there remained 63 more codons, the specificity of which needed to be determined. The fact that Ochoa's powerful and qualified laboratory concentrated all its effort to crack the code by use of unnatural synthetic "messenger-RNA" was an obvious threat to Nirenberg's operation at NIH. In spite of his mild personality Nirenberg later admitted that he enjoyed the competition[10]. After all, these new synthetic polymers could be produced by use of Ochoa's (Grunberg-Manago's) phosphorylating enzyme. NIH came to Marshall's support and expanded his resources and opened up for participation by a number of qualified scientists. For example Leon Heppel and one of his top students, Maxine Singer joined him to support his efforts and another temporary collaborator was a visitor with her, the already mentioned Marianne Grunberg-Manago. Heppel was one of the towering figures in physiological and biochemical research and his knowledge was crucial to the advance in Nirenberg's laboratory.

Heppel received his doctorate in biochemistry from the University of California, Berkeley. In this early work he made the fundamental discovery that sodium and potassium could pass through the cellular membrane. It was not such a closed structure as was originally believed. He was a medical school friend of Arthur Kornberg and the two of them ended up at NIH (in Kornberg's case he was reassigned from naval duty, not least due to efforts made by Heppel). Heppel focused on enzymes that could hydrolyze RNA. When Ochoa and Grunberg-Manago had discovered the enzyme discussed above they wanted to collaborate with Heppel to define its nature. It was by this collaboration that it was found to be a polynucleotide phosphorylase, which under certain conditions could cause formation of polymers of nucleotides. Singer (see p. 374) was an important post-doctoral fellow in this work. She later became one of the most influential female scientists of her generation in the U.S. She was to make major contributions to understanding the structure of chromatin and the genetic recombination between viruses. She also came to play a major role in refining science policy, not least in relation to the use of recombinant DNA. Heppel–Singer's contributions had a major importance in the synthesis of the first polynucleotides in Nirenberg's laboratory.

Nirenberg's group was also strengthened by an inflow of fresh and young intellectual manpower in the form of talented students who, for ideological reasons, did not want to participate in the Vietnam War. Fortunately they had an option, instead of becoming a "Green Beret" joining in the war efforts, to be exclusively selected to become a "Yellow Beret" at NIH. The temperature in Nirenberg's laboratory was high and the spirit good, but as always in science there were more uphill than downhill events. Another breakthrough was needed, but it was delayed. At an early stage Matthaei decided to return to Germany. He had an invitation to join the 1969 Nobel laureate-to-be Max Delbrück, in his, at the time, newly-established Institute of Genetics in Cologne. However, their first encounter did not go well and Matthaei therefore opted for the Max Planck Institute of Biology at Tübingen.

At the beginning of this high-charged pursuit Marshall had an offer to leave NIH and to go to the University of Michigan at Ann Arbor. Ham Smith, a good friend at the J. Craig Venter Institute has provided me with the background to this event. It was in 1961 that James V. Neel at the Department of Human Genetics at the University of Michigan (incidentally the first department of this kind at a medical school in the U.S.) wanted to recruit Nirenberg, who had received his Ph.D. at the University. It was agreed that Nirenberg should join the department in 1962. Smith, who was already there on an NIH post-doctoral position and a phage geneticist Mike Levine looked forward very much to the powerful development of the group. However, it never came about since Nirenberg backed off the agreement, something it took years for Neel to forgive him for. One additional reason that NIH did not want to let go of Nirenberg included discussion also involving Marshall's wife Perole. She was a skilled biochemist working in another section at NIH and they wanted to retain her in this position. However, most importantly, the Directory at NIH understood that a golden egg was incubating in their premises and decided to

THE GENETIC CODE

UUU △ ○ PHE UUC △ ○	UCU △ ○ UCC △ ○ SER	UAU △ ○ TYR UAC △ ○	UGU △ ○ CYS UGC △				
UUA ○ LEU UUG △ ○	UCA △ UCG △ ○	UAA △ TERM UAG △	UGA △ TERM UGG △ ○ TRP				
CUU △ ○ CUC △ ○ LEU CUA △ CUG △	CCU △ ○ CCC △ ○ PRO CCA △ ○ CCG △ ○	CAU △ ○ HIS CAC △ ○ CAA △ ○ GLN CAG △	CGU △ ○ CGC △ ○ ARG CGA △ ○ CGG △				
AUU △ ○ AUC △ ○ ILE AUA △ ○ AUG △ ○ MET	ACU △ ○ ACC △ ○ THR ACA △ ○ ACG △	AAU △ ○ ASN AAC △ ○ AAA △ ○ LYS AAG △	AGU △ SER AGC △ ○ AGA △ ○ ARG AGG △				
GUU △ ○ GUC △ VAL GUA △ ○ GUG △	GCU △ ○ GCC △ ○ ALA GCA △ ○ GCG △	GAU △ ○ ASP GAC △ ○ GAA △ ○ GLU CAG △	GGU △ ○ GGC △ ○ GLY GGA △ ○ GGG △				

△ BASE SEQUENCE. [AA-tRNA-TRINUCLEOTIDE-RIBOSOME] COMPLEX
○ BASE COMPOSITION. RNA TEMPLATES FOR PROTEIN SYNTHESIS

The genetic code. From Nirenberg's Nobel lecture[21].

markedly improve the conditions for Marshall's work. His resources in terms of space and size of his research group were expanded as already referred to. Thus he stayed at NIH.

Once the coding specificity of poly-nucleotides containing only one kind of nucleotide had been tried it was time to test various combinations of them. However, since it remained to access techniques by which they could be placed in a predetermined order one had to try random combinations. Hence various simple synthetic random forms of di-, tri- and even tetra-nucleotides were synthesized and tested in the protein-synthesizing system. Certain conclusions on the nature of triplets could be drawn. As summarized in Nirenberg's Nobel lecture[21] it was deduced firstly that the mono-nucleotide RNA chains containing U, C, A, and G resulted in poly-phenylalanine, as already presented, poly-proline, poly-lysine and no product, respectively. It was found that poly-G formed a secondary structure blocking its capacity to be active in the protein expression system. The use of combinations of two, three or four different nucleotides allowed the identification of the base composition of some 46 different triplet codons (circular marks in figure), but the exact position of nucleotides in some codons remained to be determined. There were some challenging and intense years 1961–64.

The Second Major Breakthrough

In the midst of all the turmoil Marshall remained "modest, soft-spoken and kind"[18] but it can be seen from his personal notes that he was under stress. How should they continue to develop their attack on the code? There were too many options and a fair number of codons still remained to be identified. On the practical side rigorous discipline was introduced to secure the purity of the chemicals used. One of those who had a peripheral contact with the tense scientific environment was a young Robert Gallo. He did not finally get involved personally in the hunt for the code, but learned important lessons for his own forthcoming ground-breaking research. Of course Nirenberg sought advice from the most outstanding chemists in the field, such as Todd and Gobind Khorana, whom we will get to know in more detail in the next chapter. However, within the group at NIH, the aforementioned Heppel had a central role. He was a highly qualified chemist, shy and meticulous to the extreme in his work. Eventually a goal was set to try to chemically add one nucleotide to another in a controlled way. Under Heppel's supervision another young chemist, Brian Clark, became committed to make the synthesis. Other close collaborators were Bill Jones and Robert Martin. A considerable fraction of the codons still remained to be categorically defined. There was a need for another breakthrough. Short synthetic poly-nucleotides with a predefined sequence might give the answer.

Philip Leder.

A number of talented young scientists were recruited and one of them who came to stand out was Philip Leder. Leder's father had planned for him to enter the U.S. Naval Academy, but the son had other thoughts, perhaps for moral reasons. Leder was aiming for Harvard and between his second and third year of college he fortunately spent his university vacation at Nirenberg's laboratory. He was allocated for an interview with Nirenberg. The potential mentor he met was full of enthusiasm and excitement and Leder was caught up in the whirlwind. After having graduated from Harvard Medical School and finished his residency at the University of Minnesota he returned to Nirenberg's laboratory. Another important recruit was Sidney Pestka, who came from Princeton University and the School of Medicine at the University of Pennsylvania. Like Leder, Petska was a perfectionist and he was highly proficient in producing clean preparations of ribosomes. The laboratory was active around the clock. Nirenberg was a very focused and demanding leader, caring for every detail of the experiments. He drove himself hard, but wisely he also gave himself some time to reflect. On Thanksgiving Day 1963 a new approach was conjectured. Why not focus on the shortest possible functional message. Could it suffice to use only a few nucleotides? And further, would it be possible to determine the potential linking effects of oligo-nucleotides on the binding between transfer RNA, which now had started to become available in a purified form, and ribosomes as a proof of their specificity? At this time it had been demonstrated by Akira Kaji and Hideko Kaji of Japanese origin but working at U.S. universities that the presence of polyribonucleic acid could lead to a linking of tRNA to ribosomes.

Leder was selected to be the leader of this new experimental approach. He found a company in Germany which could provide defined pairs of nucleotides which simplified the preparation of material to use. Encouraging early experiments did in fact demonstrate that even short combinations of nucleotides could link transfer RNA to ribosomes. In the beginning short nucleotide chains of varying length were tried. Leder recalls[18] a situation in April of 1964 when he, brimming with enthusiasm, walked into Nirenberg's small office to ask a very particular question. It was "How long do you think a nucleotide chain needs to be to become recognized by the ribosome and initiate protein synthesis?" Nirenberg's cautious answer was "I don't know exactly, but some number bigger than six, maybe around nine or ten…." Then Leder said, "Would you believe three?." Nirenberg became speechless. If it would suffice to use a combination of only three nucleotides it would be possible to go straight for the code using all 64 different combinations of the four nucleotides. A selected trinucleotide like UUU had turned out to connect the labeled selected transfer RNA carrying phenylalanine to ribosomes. This was a major breakthrough.

Thus the goal was set to synthesize all the possible trinucleotides. Each trinucleotide would then be experimentally used in a mixture of a purified ribosome-containing preparation and samples of aminoacylated transfer RNA labeled with each one of the twenty amino acids, one at a time. This work was started in 1964 and it took until about 1966 before Nirenberg and some ten of his hardworking and highly competent collaborators had managed to settle the definite code meaning for 63 out of the 64 triplet codons (triangular marks in figure, p. 261). Many challenges were encountered in this foray into chemistry. These data were confirmed and supplemented by the contributions by Gobind Khorana and collaborators, which we will meet in the next chapter. In the final chapter we will discuss the magic complete 64 nucleotide triplet genetic code and the particular features of start and stop codons as well as some other unique features of the code.

As the full picture of the genetic code started to emerge Nirenberg reflected on whether it would be general or not. At first he intuited that it might not be universal. In 1964 he was joined by a research associate, Tom Caskey, who was to make important experiments bearing on this problem. At the time competition was mounting not only from Ochoa's powerful group in New York, but also from the rising star of nucleic acid chemistry, Khorana, at the University of Wisconsin. It was felt that his superior chemical approach might provide an advantage. Many of the coworkers in Nirenberg's group were biologists turned biochemists by experience, to quote Monod. Experiments by Caskey and collaborators demonstrated that in fact the genetic code was universal, but this does not mean as we shall see that the use of single code words cannot vary in very special situations of biological contexts. Caskey's group also was interested in how polypeptide chain synthesis was terminated. As we shall see later there are a few selected stop codons, but the surprise in these experiments was that it did not suffice for the translation machinery to sense the presence of a stop codon. There was also a need for a special protein which could recognize the stop codon and secure a stop of peptide synthesis. This was a surprise finding.

During the intense period when Nirenberg's laboratory managed to crack the genetic code it came to attract a number of young scientists who were later to have impressive careers on their own. In 1968 Leder became the head of the comprehensive graduate program at NIH and four years later he was appointed director of the Laboratory for Molecular Genetics at the same institution. In 1980 he returned to his alma mater Harvard University and became chairman of the Department of Genetics. Leder's group was the first to define the complete nucleotide sequence of a mammalian gene, the one coding for betaglobin. This allowed an identification of a number of important control signals for the expression of the gene. He also did pioneering work on the diversity of genes coding for antibody molecules. Finally he and Timothy Stewart were the first ones to be granted a patent on a genetically engineered animal. These animals had genes added, which made them increasingly sensitive to oncogenic factors. Still one more collaborator was Ed Scolnick who has vividly described how he learned the basics of scientific pursuits from Nirenberg. Scolnick later became the president of Merck Research Laboratories. Another was Joseph Goldstein, described at length in my previous book[3] and as already mentioned Gallo. Goldstein only worked for a short time with Nirenberg, but he has summarized the atmosphere at NIH during 1964–71, when eventually as many as nine Nobel Prize recipients emerged from the general creative environment established[22].

Gallo has kindly shared with me reflections on Nirenberg's personality and his way of running a laboratory. Like everyone else he testified to Nirenberg's basic popularity as a very much liked, nice and warm human being. He had high ambitions and wanted to approach fundamental problems. He very much acted as a conductor in relation to his collaborators and wanted to make absolutely sure that the notes used were proper — that the chemical reagents used really were pure and contained what they were labeled as doing — that the instrument was carefully tuned and that the score and its interpretation was carefully attended to.

Since Gallo did not have the personality to serve as an individual specialist in a group and was already attracted to studies of blood cells and viruses he did not work directly with Nirenberg. Instead he chose to work with Nirenberg's close collaborator Pestka. Together with Pestka Gallo did in fact do in-depth work on properties of transfer RNA. In summary although Nirenberg's laboratory was a great breeding ground for biological and biochemical experimentalists, it might have had a limitation when it came to stirring the scientific fantasy and imagination of the scientists trained there. Nirenberg wanted to attack the big problems, after the genetic code he wanted to create a synapse in a test tube, but he was not attracted to speculate in an imaginative way in the spheres of heterodoxy. Probably therefore the groundbreaking work on the genetic code suited him well. In the later part of the next chapter we will return to the development of Nirenberg's professional and private life after he had received his Nobel Prize.

The Chemistry Committee Reviews Nirenberg

The forthcoming recipients of the 1968 Nobel Prize in physiology or medicine were also nominated for a prize in chemistry. The first nomination of Nirenberg for a prize in this discipline was submitted in 1965. Nominators were Y. Avi-Dor from Haifa and R. E. Davies and J. B. March, professors of biochemistry at the University of Pennsylvania, Philadelphia. The committee logically let Tiselius do a review. He was, as described earlier[2,3], the central figure in furthering the appreciation of the remarkable developments in molecular biology and he had himself taken initiatives to foster developments in this subdiscipline in Sweden. Over six pages he praised the work by Nirenberg and noted that he well deserved to receive a Nobel Prize in chemistry. As a preamble he took notice that (translated from Swedish):

The Svedberg (1884–1971) and Arne Tiselius (1902–1971), recipients of Nobel Prizes in chemistry in 1926 and 1948.

> "The chain of biochemical processes that lead to the formation of specific proteins in cells and which hence determines the biological and chemical character of organisms is at the present time one of the most intensively processed fields, within which a very rapid development is taking place because of the enormous investment in manpower and material resources."

Tiselius then introduced the major advances in the identification of messenger RNA, identification of the role of transfer RNA and progress in the deciphering of the genetic code. In the case of expanding knowledge about transfer RNA Tiselius referred back to his earlier investigations of Lipmann and Zamecnik. Nirenberg's pioneering contributions to the development of a catalogue or a dictionary highlighting the language of genes were identified as a major breakthrough. Tiselius gave an historical background to speculations about the nature of the code, including its postulated triplet nature, highlighted already in his 1962 review of Crick and Watson[2]. Hereafter he noted that at the time it was not possible to make a direct attack on the coding problem by a parallel characterization of the nucleotide sequence of a specific gene and the amino acid sequence of its gene product. Tiselius then summarized the developments of Nirenberg's work to crack the code, and the contributions of others like Matthaei and Ochoa. Finally the state of the art of insights into the genetic code at the time was presented with comments on the peculiarities of the code — the different importance of the first, the second and the third nucleotide in code words — and on the missing information.

Although Tiselius accepted that Nirenberg was a very strong candidate for a prize he recommended expectation. Thus he cited a review from 1963 by Carl R. Woese, the giant of molecular biology, who was later to authoritatively reflect on the conceptual dimensions of the unraveling of the genetic code and further discovered the existence of the Kingdom of Archaebacteria. It reads (translated from Swedish)

> "In one sense the state of understanding of the biological code has reversed itself in the last two years. Where we formerly had few facts and abundant theoretical explanations, we now have many facts and no acceptable theory to explain them: our present understanding rally does not exceed the simple cataloguing of the codon composition derived from the *in vitro* system."

As we have seen the situation changed rapidly and in a historical perspective Tiselius on this occasion appeared to have been overly cautious. Thus he recommended the committee to wait. The committee agreed with this opinion and concluded that there was a need for "a further clarification about the complex machinery of protein synthesis." At the present time it was "incalculable." Thus for example François Jacob's and Jacques Monod's discoveries, including the identification of messenger-RNA remained to be recognized by a prize. This happened the very same year at the Karolinska Institute[3].

In 1966 Nirenberg again received multiple nominations. They were submitted by R. Archer, Paris, R. Monier, Marseille and V. M. Ingram, MIT, Cambridge, Mass. There was also a proposal by the Nobel Prize recipient in chemistry Max Perutz that the prize should be divided between Nirenberg and Khorana. Myrbäck was asked to make a supplementary evaluation to the one made the previous year by Tiselius. This review was short and concluded that Nirenberg had further strengthened his candidacy, but that there were still uncertainties about the field at large. The committee accepted the recommendation to again delay a decision on a prize. In 1967 there was a detailed nomination of Nirenberg by Stephan Zamenhof, University of California, Los Angeles. He stated that Nirenberg's discoveries undeniably represented "the first breakthrough, that made the scientific world understand the important and earlier unknown phases of the mechanism(s?, my question mark) whereby genetic information, carried by DNA, finds its final expression as a sequence of amino acids; the discovery also made it possible to clarify the code for translation from mRNA to a polypeptide." In conclusion Zamenhof stated "no other discovery in this decade has had a greater impact on the field of chemistry (biochemistry, my remark)." This year, in addition, Perutz repeated his nomination for a shared prize between Nirenberg and Khorana. Myrbäck was asked to make a supplementary evaluation including both these candidates. His comments on Nirenberg covered seven pages. There were a number of new publications to review.

He took note that by now the coding role of all possible 64 triplets of nucleotides had been defined, although there might be a few uncertainties. He then dissected the further studies by Nirenberg highlighting the relative role of nucleotides in a certain triplet depending upon their position. He took notice of the relatively unimportant role, or even the dispensability, of the third nucleotide in many coding triplets. In some sense the first base in addition seemed to play a lesser role then the middle one. Nirenberg had speculated that this in some way reflected how the code itself had evolved before it reached the present "frozen" state. Myrbäck also discussed the detailed or general issues such as the role of the concentration of Mg^{2+} on binding efficacy of different codons, the identification of start and termination codons and also the universality of the code. In summary it was noted that it had been repeatedly concluded that Nirenberg's pioneering contributions represent a major breakthrough and that his continued work has further consolidated his very strong position as a candidate for a prize. Still, somewhat surprisingly the conclusion of the review was to recommend a delay. In a vague way it was stated "On the other hand it seems to me that the later developments in the field have not made the situation (regarding a prize) much clearer and that I now should be prone to consider time mature for awarding Nirenberg or someone else for their work on the genetic code." The reason for this hesitation is enigmatic. The committee supported the attitude of expectation. However, nominations continued the following year.

In 1968 the chemistry committee had distributed 725 letters of invitation to nominate for the prize and they had received 132 responses. These included seven nominations that proposed Nirenberg. One nomination intended for 1967, but arriving too late, was from professor V. Caglioti, at Rome University, and there were two more nominations from the same university by colleagues of his, A. M. Liquori and G. Montalenti. Then there were three independent nominations for Nirenberg together with Khorana from the Pasteur group, François Jacob, Jacques Monod and André Lwoff, and also a repeat nomination for the same two candidates by Max Perutz. This year the committee decided to use a reviewer outside the committee. It was none other than Peter Reichard, who as we shall see had already the preceding year reviewed the candidates for a prize at the Karolinska Institute. His total report covered 26 pages. We will return to his comments on Khorana and Holley in the concluding chapter and here only give some selected remarks on how he judged Nirenberg as a candidate for a prize in chemistry over five pages. Reichard first cited the somewhat lukewarm reception of Nirenberg's candidacy for a prize in chemistry by the previous reviewers Tiselius and Myrbäck. He then described the progress of Nirenberg's work under the following headings (translated from Swedish); General background; Statistical polynucleotides; Synthetic triplet codes; The degeneracy of the code; and The universality of the code. Impressed by Reichard's review the chemistry committee had now become much more convinced that it was time for a prize. It wrote (translated from Swedish)

> "Great strides have been made within this field (the nature of the genetic code, my remark) and it appears as if — not least thanks to the contributions by Khorana and Holley — the situation had now become much more clear and that the questions which were the cause of hesitation by the committee have now been answered."

The committee then summarized in more detail the major contributions by the three candidates. In the final stage there were four more highly qualified candidates, two of them forming a pair. Eventually the chemistry committee did not settle for the discoverers of the genetic code and the reason was the following (translated from Swedish):

> "The discovery of the genetic code without doubt represents one of the greatest advances in the natural sciences at the time being and it is self-evident that it should be recognized by a Nobel Prize. At the present time it is the view of the committee that Nirenberg, as well as

The Prime Author of the Saga of the Genetic Code

Khorana and Holley, all clearly deserve a prize. However, the committee has learnt that the Nobel Committee of medicine seriously considers proposing these three researchers for a prize in medicine this year."

It was against this background that the chemistry committee with two adjunct members, including Reichard, proposed as its candidate for 1968 Lars Onsager. He received his prize "for the discovery of the reciprocal relations bearing his name, which are fundamental for the thermodynamics of irreversible processes."

Evaluations of Nirenberg for a Nobel Prize in Physiology or Medicine

As we shall see the way the committee at the Karolinska Institute handled the growing insight into the nature of the genetic code in the 1960s was that they waited to see all the different code words securely identified both as concerned the specification of each one of the twenty amino acids and also as potentially serving other functions. There was a rapid development, not to say a race, during the early part of the 1960s, and in the end the committee had a choice to recognize one, two or three prize recipients. They settled for three, which allowed them to include findings on how the nascent polypeptide chain might develop. Until the year of the prize, the reviewers at the Karolinska Institute generally were much more enthusiastic than their colleagues at the Royal Swedish Academy of Sciences about a prize to Nirenberg.

The first nomination for a Nobel Prize to Nirenberg was submitted to the Karolinska Institute in 1964. The following year he was nominated for a prize in both chemistry, as already discussed, and in physiology or medicine. The nominations of himself alone or in combination with various different outstanding scientists continued until 1968 when he received his joint prize in physiology or medicine. He was reviewed repeatedly at the Karolinska Institute. One nomination already mentioned was the proposal by Szent-Györgyi in 1964 to jointly recognize Chargaff and Nirenberg. The latter was nominated "for his discovery of the first known interrelation between the constitution of RNA and the synthesized protein." No additional information was presented. There were two additional nominations, one from Sela introduced above and the other from Vincent G. Allfrey, professor of molecular biology at the Rockefeller Institute, both of whom provided a wider documentation. Sela gave some general background and summarized the main features of the deciphered genetic code, mentioning also the (mostly) confirmatory findings by Ochoa and collaborators, in the following way "1) high coding efficiency by synthetic polynucleotides; 2) marked code word specificity; 3) degenerative code words; 4) RNA with a high degree of secondary structure has little ability to code; and 5) almost all amino acids tested can be coded by polynucleotides containing only two bases." He also listed 12 major references to the work by Nirenberg and collaborators. Allfrey's nomination was also very insightful and full of praise for the discovery. It had "revolutionized the science of molecular biology" and was "analogous to the finding of the Rosetta Stone." It was noted that Ochoa and his collaborators "were able to verify and expand the coding possibilities of the Nirenberg system." The letter ended with praise for Nirenberg's personality stating that it was a special pleasure to recommend him for the prize since "He is a fine, sensitive and modest individual…." Interestingly, Khorana's work, discussed for a prize in chemistry already the previous year, was not referred to in parallel.

Theorell was asked to make a preliminary evaluation. He referred to the rapid progress in developing insights into the structure of the genetic code. He wrote (translated from Swedish) "In particular Ochoa's group has made important contributions, but Nirenberg was the first to break the field open and his priority seems to be clear. With consideration to the elegance of the work and the boundless importance it seems to me obvious that an in-depth review, together with Chargaff, should be undertaken." It was decided to let Peter Reichard, make a full review. He was introduced in my previous three books on Nobel Prizes, in particular the third one[3] since he played a major role in the 1965 prize to Lwoff, Jacob and Monod.

The main discovery in science by Reichard himself was the identification of the enzyme ribonucleotide reductase. This is a catalyst that can convert the nucleotide building stones for RNA into components that can be used to build DNA, hence a very important enzyme. Reichard had returned to the Karolinska Institute to a chair in medical chemistry the year before from Uppsala University, where he had held a chair in the same discipline. In 1964 he was for the first time an adjunct member of the Nobel Committee. This was the start of an involvement for almost two decades in the Nobel Prize work at the Karolinska Institute. Because of his wide and in-depth knowledge of the rapidly developing field of molecular biology he came to play a major role in the work by the committee. He was to be speaker at the Nobel Prize ceremony no fewer than four times, starting with the 1968 prize for the genetic code as we shall see in the next chapter. As already described in Chapter 4 there are only three professors at the Karolinska Institute who have given a larger number of presentations. Reichard finished his involvement in the Nobel Prize work

Peter Reichard (1925–2018) in the center in conversation with Torgny T. Segerstedt, vice-chancellor of Uppsala University (1955–78). On the left is Torvald Laurent (1930–2009).

in 1982 after having served for six years as the chairman of the committee. I can testify from my own experience that he meant a lot for the work by the committee through all his years of involvement. He was very pedagogic in his presentation to colleagues in the committee and it was a challenge to give balance to his argumentative persistence and to match the sharpness of his intellect.

In 1964 Reichard made two separate reviews, one of Chargaff, already discussed above, and one of Nirenberg. The evaluation of Nirenberg is exemplary in its simplicity and availability also to those of the committee with many different other specialties. It was comprehensive and covered nine pages. The requirement on the laboratory (*in vitro*) system established to allow measurements of protein synthesis was described. Since this was an extract from cells continuously active in transcription and translation, the system had to be emptied of the originally occurring molecules initiating these specific activities. The quenched translation capacity of the system could be reactivated by offering it TMV RNA as mentioned, but of particular importance was that it responded to synthetic RNA molecules. The great breakthrough was when it was found that offering it poly-U led to a production of a protein chain only containing phenylalanine. Later it was found that poly-A gave a chain containing only lysine and poly-C a chain of only proline, but alternate means for extraction of the latter kinds of newly synthesized peptide had to be developed, as described above. Nirenberg also found that a synthetic DNA molecule only containing T (thymidine) could lead to appearance of RNA with only A, hence giving a polypeptide only containing lysine. Reichard then described how a synthesized RNA chain containing a random representation of two nucleotides could be used to identify additional code words. Within one year of work code triplets for 14 amino acids had been identified, and the data demonstrated that the code as anticipated was of triplet nature and degenerate — more than one code term for a single amino acid. In further studies the number of randomly positioned nucleotides was increased to three and this led identification of still more codons, although in some cases the allocation remained tentative.

The review also mentioned that evidence had been found supporting speculation regarding universality of the code in experiments using an *in vitro* protein synthesizing system not only in bacteria but also the much more complex nucleated eukaryote cells. It concluded by mentioning that Nirenberg at the time was developing a technique by which small RNA molecules with a predetermined combination of only three nucleotides had been produced and tested for their capacity to connect specific amino acid labeled transfer RNA to ribosomes. Nirenberg's pioneering contributions were then summarized. It was mentioned that Ochoa and his collaborators independently had identified a number of code words, but that there was no question about who had made the original experiment breaking the field open and who had then led the further developments. The review ended (translated from Swedish) "The discovery belongs within a very central field of biology, most closely associated with biochemistry and genetics, but also of an explicit theoretical medical interest. It appears to me to a high degree to deserve (to be recognized by) a prize in medicine and physiology."

The committee agreed that Nirenberg's discovery was documented to be worthy of a Nobel Prize.

In 1965 a German professor E. Bauereisen from Würzburg nominated Nirenberg together with Matthaei and separately Ernst Ruska, as will be discussed in the last chapter. Another nomination of Nirenberg was submitted by G. D. Preston, from Dundee, Scotland. Matthaei's candidacy was evaluated in a preliminary review by Reichard, but it was concluded that his contributions were clearly secondary to Nirenberg's and that hence it was not justified to make a full review of him. Reichard evaluated Nirenberg one more time and noted that major progress had been made during the preceding year. This had become possible by use of the briefly mentioned new technology. Isolated and defined trinucleotides of different combinations were employed to measure their capacity to bring about an association of ribosomes and specific amino-acid-charged transfer RNA. In this work with Leder it was discovered that the effect was optimized when three nucleotides were used. Two or more than three did not have any effect. The fact that an optimal effect was obtained by the use of triplet nucleotides was interpreted to represent chemical evidence for the existence of a triplet code. At the time when Reichard wrote his report the coding capacity of as many as 45 of the 64 possible trinucleotide code words had been defined. Reichard mentioned also the interesting so-called "nonsense" triplets, which were candidates to be the start or the stop codons of polypeptide synthesis. Reichard concluded that the impressive developments in Nirenberg's laboratory over the last few years had made Nirenberg a very strong candidate to a prize. The committee agreed with this conclusion.

In 1966, the recipient of the shared prize in physiology or medicine the previous year, Monod, used his right to propose candidates. He did this in a very informative letter to Gard which presented a number of strong candidates. It first proposed the phage geneticists Luria, Delbrück and Hershey, who came to be recognized by the Nobel Prize in physiology or medicine in 1969. Next he nominated Nirenberg for his already famous work with artificial messengers and more recently, with synthetic triplets. He then expressed a strong feeling that Nirenberg should be associated with two highly qualified candidates. They were Sydney Brenner and Seymour Benzer. The motivation was:

> "Benzer because of this magnificent, profoundly original work which established the linear nature of the genetic material down to the ultimate level of resolution, and that of Brenner because of his major contribution to the work which established the triplet nature of the code and the fact that it was sequential. In addition, both were responsible for the discovery of the non-sense mutants, and Brenner, as you undoubtedly know, used the knowledge thus gained to establish the colinearity of nucleotide sequences and peptide sequences."

He then cited the names of still a number of important contributors to the field, but did not nominate them. One was Khorana who was praised "for his magnificent recent work on the synthesis of specific polypeptides from synthetic polynucleotides."

Monod added later on that Nirenberg's discovery of the effect of poly-U "was probably more a stroke of luck than one of genius of insight" but still his explanation of the finding "signals an exceptionally gifted experimentalist." Brenner received the compliment of being "undoubtedly the most brilliant of the three, both as a theorist and as an experimentalist." In this context it can be added that it was fortunate that Brenner finally received his prize in 2002. Over decades he had made discoveries of Nobel Prize caliber and hence a prize was long overdue. The letter finally mentioned two possible additional candidates. These were Paul Doty and Julius Marmur, who had contributed to the development of techniques to compare the base sequences in different forms of DNA and also in separate DNA and RNA molecules. Such nucleic acid hybridization techniques were mentioned briefly in Chapter 3 and will be returned to in the final chapter. There were three additional separate nominations of Nirenberg, one from the 1958 laureate George W. Beadle, another from J. Burdette, Houston, Texas and finally also one by H. R. V. Arnstein from London. The committee decided to ask its new adjunct member Giuseppe Bertani to do a review.

Bertani was introduced in my previous book[3], but I incorrectly speculated that he might not become involved in the Nobel Prize work. He did in fact become an adjunct member in 1966 and stayed in that position for two more years. His native language was Italian, but his stay for many years in the U.S. had made him fluent in English. I do not know how much Swedish he knew, since we always spoke English when we met. However, he must have known some Swedish since he needed this to follow all the discussions at the committee meetings, which naturally were conducted in Swedish. However, he of course wrote his evaluation in English. This is the first time that the Nobel archives at the Karolinska Institute included a review in this language. With time progressively the committee started to use reviewers outside the committee and also from non-Scandinavian countries. Much later, in the first decade of the present century, it was decided that all reviews should be written in English, also by members who had Swedish as their native tongue. Most of the present members of the committee have worked in an English-speaking country for a shorter or longer time and hence are proficient in this language. The current chemistry committee at the Royal Swedish Academy of Sciences has a somewhat different policy. Swedish-speaking members themselves can choose if they want to write in Swedish or in English. If they are ambivalent the recommendation is to use the Swedish language.

It should be added that since the 1960s English has been the language of science which we all use when publishing new data. And still, when it comes to nuances of formulation it may be a disadvantage not to use your native tongue. I experience this in my many translations of extracts from reviews in Swedish to English. The general impression has been that it works well only to use the English language when writing reviews, but the discussions will always be conducted in Swedish. This change of praxis in the use of Swedish or English of course will be of great help to non-Swedish-speaking historians who may want to use the Nobel archives in the far future. There was one more change introduced when Göran Hansson, at the present time Permanent Secretary at the Royal Swedish Academy of Sciences, formerly professor at the Karolinska Institute, was secretary of the Nobel Committee at the Karolinska Institute in the early 2000s. The question was raised if, in cases when consecutively different reviewers were used, the reviewer selected should have access to all previous reviews, which until then always had been the practice, or not. It was decided that this should be settled from case to case. But coming back on track, let us see what Bertani wrote.

He first noted the existence of earlier reviews, including the two previous reviews by Reichard. He also mentioned that Benzer had been evaluated once before by Gard in 1961. He had concluded that the discoveries were not of sufficient magnitude to motivate a prize, but that the data presented might be reassessed when workers in the field of phage genetics were subjected to further evaluations. Bertani's review of Nirenberg and collaborators was relatively brief and he mostly referred to Reichard's previous reviews. The reviews of Brenner and Benzer were much more comprehensive. However, this work to a major extent concerned matters outside the scope of the present presentation. There may be reasons to come back to them in 2020 in connection with discussions about the 1969 Nobel Prize in physiology or medicine to three bacteriophage geneticists. Suffice to note the already mentioned contributions by Brenner and in part Crick in defining the triplet nature of the code and the existence of additional stop signals, ochre and opal. As presented the same code triplet is used to signal start and to specify the position of methionine. Of some interest in relation to transfer RNA, to be discussed at length in the next chapter, were some elegant experiments by Benzer and collaborators. They purified a population of transfer RNA with specificity for a certain amino acid. Hereafter they chemically modified the attached amino acid X and converted it into another kind of amino acid, Y. When this construct was used in a Nirenberg expression system the protein formed instead of containing amino acid X contained amino acid Y. They were also able to demonstrate that two different populations of transfer RNA carrying the same amino acid but using different triplets of the degenerative code specifying this amino acid, became located in different positions depending on the sequence characteristics of the messenger-RNA. These experimental results validated the specificity and degeneracy of the code. However, we will return to aspects on transfer RNA in the next chapter.

Not surprisingly, Bertani, like Reichard concluded that providing a solution to the code problem using the biochemical approach of Nirenberg and collaborators represented a discovery highly worthy of a Nobel Prize. He emphasized however that the follow-up genetic experiments confirming the various conceptual conclusion were also important. He further cited Monod in mentioning other people of importance to the developments, like Khorana, Jack A. Yanofsky, A. Garen (demonstrated stop codons in parallel with Brenner) and George Streisinger and to these he added Grunberg-Manago and Guido Pontecorvo. It is interesting that he gave Grunberg-Manago more credit than Ochoa for the discovery of the RNA phosphorylating enzyme. All these names just highlight that when a topic in science gets hot, because of an unexpected discovery, there are many contributors to the expansion and deepening of the groundbreaking findings. Bertani concluded that he could not suggest a better combination than Nirenberg, Benzer and Brenner, the one proposed by Monod. However, a little further on he stated that the pioneering work by Benzer and Brenner to a large extent depended on earlier critical work by the fathers of phage genetics. His concluding sentence was "I would therefore have the greatest hesitation in placing the group discussed here ahead of the other group, Delbrück, Hershey and Luria, also under consideration this year." In the end though cracking the genetic code appropriately was awarded before the fathers of phage genetics were recognized for their pioneering work in 1969.

In 1967 the number of nominations for a prize to Nirenberg increased. A. C. Allison's nomination of Nirenberg alone was repeated and there was also another nomination only including his name from James A. Shannon at NIH. He was nominated for the first time together with Khorana by George Broder from Chicago and a similar nomination was submitted by Albert L. Lehninger from Baltimore. Finally there was a nomination of Nirenberg together with both Khorana and Holley. This was submitted by Torvald C. Laurent, a professor of biochemistry at Uppsala University (see pp. 272, 294). It was the first time that the three scientists who eventually shared the prize were proposed together. It is not unlikely that Laurent, the successor of Reichard in the chair of biochemistry at Uppsala University, was familiar with discussions about a prize for the genetic code, both at the Karolinska Institute and at the Royal Swedish Academy of Sciences, at which institution as we shall see Khorana had been proposed for a prize in chemistry already since 1963, and Holley, since 1966.

This year Reichard had returned as adjunct member of the committee and naturally it became his responsibility to shoulder the task of evaluating not only Nirenberg, but also the more recently nominated Holley and Khorana. In addition he was asked to again comment on Matthaei and in addition to include two more nominees in his review, Mahlon B. Hoagland and Paul Zamecnik. Reichard delivered a remarkably comprehensive review covering 42 pages and with three attached figures. It presented the candidates in a kind of historical sequence starting with Hoagland and Zamecnik. This year they had been nominated by the secretary of the committee, Bengt Gustafsson. He had introduced a new principle that highly prize-worthy candidates should be carried over from the previous year by a nomination by the secretary if there should be no external nomination. By this arrangement they were available for a broader discussion, in this case on protein synthesis and the genetic code. In 1965 they had been nominated for the first time and Reichard had already reviewed them then. Hence his 1967 review was only an update. Their major contributions had been made in the late 1950s. They pioneered methods for analysis of protein synthesis in a laboratory system. In the development of this technique they were among the first scientists to identify the critical component later called transfer RNA. However, they were not alone in making this discovery.

As was already mentioned it was made also by two other groups, one of which was headed by Holley. Reichard returned to him later in the report. Already in 1965 Reichard had come to the conclusion that Hoagland's and Zamecnik's contributions might be considered worthy of a Nobel Prize, but that there were other much stronger candidates, like Nirenberg. Still the availability of the laboratory protein synthesis technique they had developed was of decisive importance for Nirenberg's studies. Holley had two nominations jointly with other candidates, as already mentioned. The main thrust of the nominations was that he was a co-discoverer of transfer RNA and the first to provide the complete structural characterization of a biologically active nucleic acid. We will return to the evaluation by Reichard over eight pages of Holley's pioneering work in the next chapter.

Although Nirenberg had been reviewed previously by both Reichard and Bertani and declared worthy of a prize, Reichard supplemented and expanded his review over seven pages. At first he again stressed the importance of the laboratory protein synthesizing technique developed by Zamecnik's group. It included many components like ribosomes, cellular fluid, ATP as an energy generating system and two more kinds of RNA — transfer RNA and messenger RNA. In the system used by Nirenberg and Matthaei they had arranged, as already mentioned, that the activity due to preexisting messenger RNA included in the original material had tapered off so that the activity registered should only derive from the new synthetic messengers added to the system. He then mentioned again the experiment with the poly-U and the synthesis of phenylalanine, the breakthrough finding. Also the critical follow-up experiments with statistically represented synthetic poly-nucleotides, employed by both Nirenberg's and Ochoa's groups and finally Nirenberg's and Leder's pioneering approach of using synthetic tri-nucleotides were recapitulated. The latter approach was exemplified by describing an experiment with three possible combinations of two U and one G; GUU, UGU, UUG. It was found that a binding between radioactively labeled amino acid specific transfer-RNA and ribosomes could be selectively demonstrated in the presence of the following tripeptides: GUU led to a unique binding of valine-tRNA; UGU to a binding of cystein-tRNA and finally UUG to a binding of leucine-tRNA. At the time of writing the review, when also the results of Khorana's work was included, the function of 63 out of the 64 code words had been identified. Only UGA remained to be characterized. It was later found to be another stop codon.

Reichard also described the mechanism of recognition elucidated by Nirenberg and collaborators experiments using the highly purified alanine-specific tRNA. The findings indicated the relatively lesser importance of the third nucleotide and that the major emphasis was on the central nucleotide. The so-called "wobble"-hypothesis initially proposed by Crick was referred to. Finally the generality of the code was discussed and references were given to the fact that it appeared universal, although not categorically so in some details. It did indeed represent a "frozen" advance in evolution, presumably the most critical one ever to develop.

When Matthaei had returned to Germany after his very successful brief period of work in Nirenberg's laboratory he had ambitions to continue to make important contributions to the development of the code. Reichard reviewed this work by Matthaei and his independent group in some detail over four pages. Without going into detail it can be noted that generally he was very critical. No major new findings had been published. The summary was the following (translated from Swedish):

"Matthaei's production after 1962 is according to my opinion not very meritorious. There are a number of incorrect results, which are only in part corrected in later work. It is never made clear what the cause of the errors might have been. Hence Matthaei has not contributed any original data of importance to the solving of the genetic code after he left Nirenberg's laboratory. I do not consider Matthaei worthy of a prize."

It can be added that in a letter by Perutz nominating Khorana and Nirenberg for a prize in chemistry in 1966 it was stated, referring also to discussions regarding Matthaei with Crick and Watson, that "his independent work has not borne out his earlier promise."

These statements ruled out Matthaei's eligibility to share the prize. Although he received four nominations the following year, all of them of course including Nirenberg, but in one case also Khorana and in still another Zamecnik, he was never to be considered as a serious candidate.

The remains of Reichard's long review discussed Khorana's work which we will return to in the next chapter. The extensive review ended with a general discussion over some five pages which there are also reasons to revisit in the following chapter. Suffice to note that Nirenberg had repeatedly been interpreted as being a very strong candidate and self-evidently fit to be represented in a prize for the deciphering of the genetic code.

The archives of 1968 had a different appearance than those of the previous years. The dynamic and well-organized secretary Bengt Gustafsson had created a preprinted form to be used by all the nominators. Most of them abided by this but some persisted in sending letters which later had to be transcribed often as an abbreviated variant into the form. An important quality was precisely the fact that in the form only half a page was available for the presentation of the achievements of the proposed candidate encouraging a strictly focused introduction. This form came to be used in the future. Recently it has become possible to substitute regular mail, by sending in the duly completed form electronically. It took time for the introduction of the latter modernization, since it had to be assured categorically that secrecy could be retained.

The 1968 archives are unique also in another way. There was an explosive development of nominations for one of the candidates responsible for the discovery of the genetic code. It was Nirenberg. He received no fewer than 42 nominations for a prize out of which 33 were for him alone. To be proposed in almost 15% of all nominations is probably an all-time high for a prize candidate in physiology or medicine. It should be added that 23 nominations came from colleagues at the National Institutes of Health. Khorana also had a fair number of nominations whereas proposals for Holley were rather few, which we will return to in the next chapter. The committee decided to let Theorell, who had made a preliminary review of Nirenberg in 1964, do a review. This review also included Zamecnik. The review covered 17 pages but was unique in its disposition. It started out by discussing why there have not been closer interactions between the committees responsible for the prizes in chemistry and in physiology or medicine for an award recognizing the discovery of the genetic code. His general recommendation was that there should be a larger sharing of materials. Hereafter Theorell remarked on the publications by the four candidates during the last year, with the conclusion that not much of importance had been added. This of course is not surprising since the code was essentially completely deciphered in 1966. The major part of the remaining relatively long review was verbatim citations of a large number of selected sections of all previous reviews by both committees. Only in certain selected cases did Theorell give some brief personal remarks. In the context of his own 1964 preliminary review he took an opportunity to comment on Chargaff. He reduced the significance of his findings of an equimolar representation of the nucleotides adenosine and thymidine and also cytosine and guanidine and argued that it was the hydrogen bonding of these pairs of nucleotides shown by Watson and Crick that was the critical discovery.

Theorell's review in fact did not add anything new, but noted en route that although Hoagland's and Zamecnik's contributions might be worthy of a prize, their contributions could not compete with those by Holley, Khorana and Nirenberg. On the last two pages Theorell gave some more extensive comments of his own. He concluded, unsurprisingly, that the time was now mature to recognize Nirenberg, "the prime discoverer of the golden egg, the genetic code," as well as Khorana and Holley. In his summary he wrote (translated from Swedish)

> "Concerning Nirenberg there is agreement between the two committees. However, professor Myrbäck still in 1967 wants to delay a decision since there were certain questions that still remained to be answered. He did not define which they were and I cannot share his opinion that the awarding of a Nobel Prize should be delayed until all problems are solved — if that were the case it would never be possible to award a prize. On the contrary it is a characteristic of every great

discovery that it brings to the fore a number of new challenging problems. The recommendation 'to delay decision for some time' is traditionally used much more by the committee of chemistry. This praxis facilitates the procedure to weed out all the candidates except the one or more who are finally awarded the prize of the year. In the medical committee the large number of candidates declared worthy of a prize (in 1968 they were 52 as mentioned, my remark) leads to considerable difficulties in the final round."

Theorell then summarized that there were three possible alternatives in recognizing the discovery of the genetic code. The first alternative was to give a prize to Nirenberg alone. The second possibility was to let the three candidates share the prize equally. A third possibility mentioned was the option to give Nirenberg half the prize and to let Khorana and Holley share the other half. He did not like the last alternative and wrote in his concluding paragraph (translated from Swedish):

"Personally my greatest sympathy is for giving an undivided prize to Nirenberg. He deserves the foremost honor of having made one of the largest biological discoveries of this century, and the current tendency within the medical committee to split the prize in order to allow inclusion of as many candidates as possible probably is against the original intentions of the donor."

It is now time to get to know the two scientists Khorana and Holley and their important scientific contributions which eventually made them join Nirenberg in receiving the 1968 Nobel Prize "for their interpretation of the genetic code and its function in protein synthesis."

Chapter 7

The Formation of a Trio for the 1968 Prize

THE BOOKS OF LIFE
A LIMITLESS LIBRARY
FOR FUTURE READINGS

The lead character of the previous chapter Nirenberg, could easily have carried a Nobel Prize in chemistry or in physiology or medicine on his own. However, in 1968 he was joined by Robert W. Holley and H. Gobind Khorana for the prize. Let us trace why the Nobel Committee at the Karolinska Institute preferred a prize including all three of them. At the earlier time of writing about the 1968 prize I was communicating with the only surviving member of the committee at the Karolinska Institute at the time, Reichard. He was a very central person in connection with the prize for the genetic code. As has already become apparent he was the main reviewer of the candidates both at the Karolinska Institute and also at the Royal Swedish Academy of Sciences. When the prize was awarded he was selected to be the introductory speaker at the prize ceremony. On May 30, 2018 I received an email from him starting and finishing with a few words in Swedish, but mainly using English. It read:

> "Tack for din (thanks for your) email. I look forward to your new books. Nirenberg was certainly the main candidate among them, as he made the first key experiments but was not generally considered a "great" scientist. Khorana was the most outstanding molecular biologist among them whereas I was surprised about the inclusion of Holley who came in from the sideline. Anyway aside from details, the experiments that led to the deciphering of the genetic code obviously

had to be an early subject of a Nobel Prize. Varma haelsningar (warm greetings) Peter."

Obviously I would have liked to discuss further some of these remarks, in particular the meaning of a "great" scientist, with Reichard. However, sadly 19 days after having sent this email he had died. I assume that Reichard meant the aspects of Nirenberg's character, discussed in the previous chapter, of not being visionary, speculative and imaginative. However, he was full of ideas, a trait he combined with being a hands-on meticulous scientist, setting his goals high and creating a laboratory that provided a great schooling for budding scientists. He did not like to blow his own trumpet and to use the music metaphor from the previous chapter maybe it can be said that he was not a person who liked "to play cadences."

Khorana was a great molecular biologist as we shall see, but in his early career he was standing on the shoulders of Todd and the genetic code would have been solved without his contribution. Still he was a scientist of remarkable qualities. He had set very ambitious goals for himself and he managed to fulfill the most important among them. It was not to crack the genetic code, but to be the first one to synthesize a complete gene, something he did in fact manage to pioneer a few years after his prize. It was on his way towards this goal that he developed techniques to synthesize ever more complex predetermined sequences of DNA. These early variants of synthetic DNA with short defined repeat sequences of nucleotides turned out to be of great value in consolidating the deciphering of the genetic code. However, access to them was not decisive. Inclusion in the group of prize-recipients of Holley, whose work only tangentially involved the genetic code, viz. the existence of a trinucleotide anti-codon represented in a specific transfer RNA, was not an obvious choice, as Reichard commented. It may have represented a means of using the opportunity of identifying the genetic code to also recognize the importance of understanding the structure of this category of RNA. This does not detract from the importance of showing the multiple fascinating functional qualities of the beautiful clover-like structure of this kind of molecule to be described below. Its inclusion in the prize was very enriching. However, as has already been noted there were also other scientists who made major contributions to the discovery and characterization of transfer RNA.

In the tracing of the history of molecular biology it is of course important to make clear that there were Nobel Prizes highlighting the identification of the existence of all the three major forms of RNA discovered in the 1950s to be central actors in protein synthesis; messenger RNA, ribosomal RNA and transfer RNA. The identification of messenger RNA was indirectly recognized already in 1962, when Watson in his Nobel lecture[1] discussed RNA instead of presenting data on DNA. His discourse included the recent observations in his laboratory that gave some support to the existence of messenger RNA in a bacterial system. In 1965 Jacob's and Monod's decisive contributions to the conceptual and simultaneous experimental identification of messenger RNA, and also their naming of it, was highlighted by the prize they received together with Lwoff[2]. Ribosomes which to a large extent are composed of RNA were recognized as morphologically identifiable structures by the prize in 1974 to Claude, de Duve and Palade, briefly alluded to in the final chapter. Much later there were Nobel Prizes in chemistry recognizing insights into the fine crystallographic structure of the transcription as well as the translation machinery as we shall see. For the sake of completeness it should be added that there are still a number of additional important categories of RNA[3].

Todd's Perspective on the Biochemistry of Nucleic Acids

Before the introduction of Gobind Khorana it is appropriate to consider his mentor, the 1957 recipient of the Nobel Prize in chemistry, Lord Alexander Todd. He received the prize "for his work on nucleotides and nucleotide coenzymes" as presented in part earlier[4]. Nucleotides may be best known at present because of their central role as building stones in DNA and RNA, but they also carry a number of other very fundamental functions in cells. One such function is to mediate signaling between different compartments of such entities. It was found that a particular kind of nucleotide, adenosine, had a critical function in this context. In 1971 a Nobel Prize in physiology or medicine was awarded to Earl W. Sutherland "for his discoveries concerning the mechanisms of the action of hormones." In the transport of signals, studied by Sutherland, cyclic adenosine monophosphate (c-AMP) was found to be the critical molecule. Another important

Alexander Todd (1907–1997). [From Ref. 16.]

function is to serve as an energy reservoir by the formation of adenosine triphosphate (ATP). This kind of compound can release energy by the loss of one phosphate group (ATP to ADP, triphosphate to diphosphate), which takes place in mitochondria in eukaryotic cells and in simpler structures in bacteria. The discovery of the chemical background to this central function in cells has also been recognized by a Nobel Prize. In 1997 Paul D. Boyer and John E. Walker received the prize in chemistry "for their elucidation of the enzymatic mechanism underlying the synthesis of adenosine triphosphate (ATP)." Hence nucleotides serve many important functions in nature, but this was not known when Todd became involved in characterizing their chemistry. His unchartered advance into studies of nucleotides was described in his Nobel lecture[5]. Let us follow his discourse.

Todd presented himself as an organic chemist in the tradition of the "great Swedish chemist" Berzelius. Thus he studied living matter with the hope that "it may prove to hold the keys to Life itself." It should be noted that organic chemistry originally was referred to as physiological chemistry and that it was first towards the end of the 1940s that it was relabeled biochemistry. Thus Todd's work during his training period first focused on natural products including colored matters. Chemists at the time liked colored matters since they could be more readily followed than colorless substances when subjected to various separation procedures. Todd also learned early on that synthesis and degradation experiments provided complementary tools in the development of knowledge about a specific organic structure. In the mid-1930s Todd became involved in investigations of the anti-beriberi factor, vitamin B. Vitamins and their different physiological roles have been in the focus of many Nobel Prizes in physiology or medicine during the first part of the previous century, as discussed in my previous book[2]. Todd mentioned the role of certain vitamin compounds as cofactors in enzyme reactions studied by Theorell. He also mentioned the already alluded to role of ATP as a critical energy source discovered about the same time.

The source and nature of nucleotides was originally defined by breaking down — hydrolyzing — different forms of nucleic acids. Four nucleotide bases (residues) could be identified in both DNA (originally thymus nucleic acid) and RNA (originally yeast nucleic acid), but one of them was unique, thymidine (T) in DNA and uracil (U) in RNA. The remaining three nucleotides adenosine (A), cytosine (C) and guanidine (G) were shared. In addition, as their names indicate, the associated sugar component varied, deoxyribose in DNA and ribose in RNA. In the mid 1930s the knowledge of nucleic acid

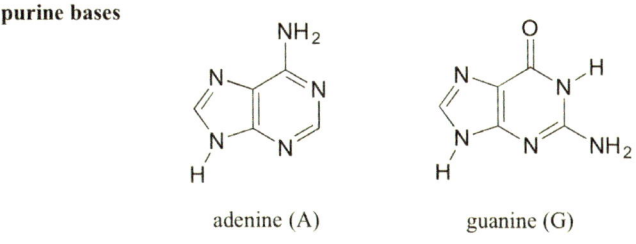

The five different kinds of nucleotides used in the formation of DNA and RNA.

components was in a "primitive state." The introduction of new separation techniques — paper and ion-exchange chromatography, paper electrophoresis and counter-current distribution — changed the picture as discussed in the presentation of Chargaff's contributions in the previous chapter. Todd managed to separate the nucleo*sides* — the base with the attached sugar but no phosphate. He and his colleagues also learnt to add phosphate groups and in the end all the essential nucleo*tides* could be synthesized in the laboratory.

Already in the mid 1930s Caspersson and collaborators, by use of the new technique of ultracentrifugation, had shown that nucleic acids could have very high molecular weights. It became clear that these large molecules were formed by use only of the four identified building stones. Todd's and his collaborators' work gave a first insight into how large molecules could be built, but noted that techniques to further examine them were still in their infancy. He also concluded that the molecules were linear rather than branched. It was then stated with foresight in his lecture "Since the essential difference between individual nucleic acids must reside in the difference *sequence* (my italics) of nucleoside residues in them, methods for sequence determination are clearly of importance for further work." However, it would take a considerable time before this became practically feasible, as described earlier[2].

Todd then made an even broader visionary statement as a provisional conclusion of his Nobel lecture. It reads:

> "Time does not permit me to pursue this topic further and to trace how the chemical information discussed above has been combined with the results of X-ray and other studies to build up current views on the macromolecular structure of the deoxyribonucleic acids. Suffice to say that the double helical structure of the DNA molecule adumbrated first by Watson and Crick on these foundations bids fair to open a new era in molecular biology. For it offers clues to the significance of nucleic acids in the transmission of hereditary characteristics and, taken in conjunction with our greater understanding of the properties and reactions of organic phosphates, it permits an approach to a closer understanding of the role of nucleic acid in cellular processes."

In the second part of his lecture Todd discussed the role of nucleotides as a coenzyme and as a source of energy in general, which need not be discussed in the present context. Suffice to note the first part of the final paragraph. It reads:

> "To-day the nucleotides occupy a prominent place in chemical, biochemical and biological research and new vistas are opening before us which may in a relatively short time lead to a far deeper understanding of the mechanisms of the living cell than seemed possible only a few years ago. And this is surely a matter of profound importance to humanity in its ceaseless struggle against disease."

A Star Biochemist with an Exceptional Background

It is not known when H. Gobind Khorana was born. Retrospectively it was judged to be January 9, 1922. He was the youngest of five children in a family in the small village of Raipur, with some hundred families in the rural Punjab region of what presently is Pakistan. His father was a Hindu tax clerk, a *patwari*, serving the British colonial government. The early schooling was provided by a village teacher under a tree. Khorana practiced what came to be a characteristic microscopic handwriting by positioning himself at the nearby post office, helping illiterate town people to write letters. Encouraged by his father he continued his education and finally received a scholarship that allowed him to study chemistry at the Punjab University. At this university he earned his bachelor's and master's degrees. Thereafter he left for the U.K. on a scholarship from the Indian government, which was taking post-war initiatives to further its independence and to modernize. At Liverpool University he presented his Ph.D. thesis in 1948 under the tutelage of Roger J. S. Beer. Beer took a personal interest in Khorana's future individual and scientific developments. Afterwards Khorana had a short but important post-doctoral period at Eidgenossische Technische Hochschule (ETH), Zürich, Switzerland. There he met Esther Elizabeth Sibler, a life-changing encounter as it turned out. In fact they had coincidentally met the year before in Prague at the first World Festival of Youth and Students. The absence from his home country had made him "feel out of place everywhere and at home nowhere," but Elizabeth, who was later to become his wife, gave him a completely new purpose in life. The two of them soon moved back to the U.K., where he learned how to do molecular biological work at Cambridge University, which was developing to become a Mecca for this kind of science. His schooling under the aegis of Todd provided a critical vector to his development towards becoming one of the leading biochemists studying nucleic acids.

Gobind Khorana (1922–2011), the co-recipient of the 1968 Nobel Prize in physiology or medicine. [With permission from TT, Sweden.]

When it was time to evolve independence he carried on west and moved his science first to the British Columbia Research Council in Vancouver in Canada and then finally to the Institute for Enzyme Research at the University of Wisconsin. Prior to moving to Vancouver he and Esther got married. They were to have three children and Esther remained a close companion in private life and work for over 50 years. Already in Vancouver, Khorana built a research group and developed his pioneering work on proteins and nucleic acids, but it was in Wisconsin that he became involved in deciphering the genetic code, the work that later earned him his Nobel Prize. In 1966 he became an American citizen and four years later he finally moved to MIT to become the Alfred P. Sloan Professor of Biology and Chemistry.

Khorana's contributions to solving the genetic code were based on techniques originally developed when he worked in Todd's laboratory but then considerably further refined in his own laboratory environment. By skilful advancement of the technology he managed to synthesize DNA polynucleotides with a predetermined representation of nucleotide bases. This work using the so-called carbodiimide reaction took off after his move to Vancouver. In the early 1960s he managed to produce a repeat DNA polynucleotide that could be copied into a synthetic messenger, UCUCUC... When this was used in a laboratory protein synthesizing system the polypeptide generated represented a chain alternating between serine-leucine-serine-leucine ... Hence it could be deduced that UCU coded for serine and CUC for leucine. Expanding this approach by testing a full variety of synthetic oligonucleotides it was possible to consolidate the 64 code words most of which already had been deduced from experiments in Nirenberg's laboratory and often confirmed by Ochoa and collaborators in New York. At the Cold Spring Harbor meeting in 1966 it was possible to join data from all groups involved in this work and settle once and for all the code words (see p. 316).

The Review of Khorana for a Prize in Chemistry

Khorana's first nomination for a Nobel Prize was submitted in 1963 to the Royal Swedish Academy of Sciences and recommended that he should receive a prize in chemistry. The nomination was submitted by R. K. Morton, Adelaide, Australia. It proposed a prize for the development of a technique to construct polynucleotides. This newly-developed technique provided opportunities to synthesize deoxynucleotides containing *specific* predetermined nucleotide sequences. The original technique had been developed in collaboration with Todd and published in 1953. In the nomination the further development of the technique by Khorana was described and a list of some 30 follow-up publications was submitted. The committee let Fredga, who had been responsible for all the background reviewing of Todd, leading to his prize in chemistry, do a review. It was brief, covering only three pages.

Fredga remarked that the critical joint publication had been reviewed before in his examination of Todd. He praised the method and noted that in the light of the understanding of the central role of DNA in storing genetic information, it might be of great value. However, the further development of the technique by itself could hardly be considered sufficient to motivate a Nobel

Prize for Khorana. To this could be added that this particular contribution had already been considered in connection with the consecutive reviews of Todd that finally led to his prize. In conclusion, Khorana could not be recommended for a Nobel Prize. Nothing was mentioned about the possible use of the synthetic DNA molecules in studying the genetic code. The next nomination of Khorana for a prize in chemistry was submitted two years later, in 1965, by B. N. Ghosh, University College of Science, Calcutta. The committee did not take any action but referred to the earlier review by Fredga. In 1966 there was a new nomination of Khorana, this time by H. O. Huisman, Amsterdam. He emphasized not only the development of the technique to synthesize DNA but also its possible use in studying protein synthesis and the genetic code. There was also an additional nomination of Khorana by the earlier 1962 Chemistry Nobel Prize recipient Perutz. He recommended a joint prize to Nirenberg and Khorana. Perutz had discussed his proposal for a joint nomination with Crick and Watson, who could not themselves make nominations for prizes in chemistry, since they had received their prize in physiology or medicine. Their general conclusion was "that the Chemistry Prize would be a more appropriate recognition of Nirenberg's and Khorana's work, since the methods used are entirely chemical in nature." This time the committee asked Myrbäck, introduced in the previous chapter, to do a review.

Over ten pages he described Khorana's large numbers of publications. In particular he emphasized his crucial contributions of supplementary relevance to the interpretation of the genetic code. The approach used was the following. First a synthetic DNA with a defined sequence specified by use of preselected consecutive sets of di- or trinucleotides was synthesized. This single-stranded synthetic sequence was then made double-stranded by use of a DNA polymerase. Thereafter an RNA polymerase was used to synthesize an RNA copy of the double-stranded DNA. Finally this synthetic messenger RNA was used in a laboratory protein-synthesizing experiment and the amino acid composition of the peptide produced was determined. Most of the data gave an independent confirmation of the coding specificity of different trinucleotides already determined by Nirenberg and Ochoa. In some cases the results obtained provided critical clarifications in situations of earlier ambiguous results. Although Myrbäck agreed that Khorana's data consolidated the interpretation of the genetic code his conclusion was the following: "Thus it is my opinion that the proposal to award Khorana should not be brought forward and that the committee as concerns the proposal to jointly award Nirenberg and Khorana should delay its decision." The committee agreed.

The Formation of a Trio for the 1968 Prize

In 1967 there were renewed nominations of Khorana for a prize in chemistry, this time by A. P. Nygaard, Bergen, Norway and G. N. Ramachandran, Madras, India, for his studies of polynucleotides and the genetic code. As in the previous year Khorana was also nominated together with Nirenberg by Perutz. The committee asked Myrbäck to make a supplementary review. During the year that had passed Khorana's group had been very active and some 15 new publications and manuscripts were referred to in the report covering more than ten pages. It was emphasized that the application of Khorana's synthetic DNA had allowed a major consolidation of the chart on the 64 codons and their individual specificity. It was further noted that Khorana did not have priority in the discoveries, but that important clarifications had been made in particular cases when earlier conclusions had been ambiguous, "based on indirect and/or incomplete evidences." Experiments with repeat dinucleotides or trinucleotides in the synthetic DNA copied into mRNA were cited to illustrate the definite results obtained. The messenger RNA sequence UGUGUGUG etc. included alternating codons of UGU and GUG and in the resulting polypeptide cysteine alternated with valine. In similar experiments the trinucleotide sequence CUUCUUCUU etc. represented three triplets CUU, UUC and UCU and hence resulted in peptides containing exclusively leucine, phenylalanine or serine, depending upon where the triplet reading had started. At this time Khorana and collaborators were able to conclude that the specificity of 63 out of the 64 possible three-letter codons had been settled by use of their technology.

It had been speculated at the time that some antibiotics like streptomycin and similar compounds might interfere with protein synthesis. Khorana and collaborators were able to demonstrate that this was the case and that in the presence of these kinds of drugs there was a certain pattern in the misreading leading to a defective protein. Finally Myrbäck mentioned one publication with a very different content. Stimulated by Holley's spectacular full characterization of transfer RNA specific for alanine, which we will return to in detail later, Khorana and collaborators had sequenced the transfer RNA specific for another amino acid, phenylalanine. There was great similarity to the principal structure demonstrated by Holley and collaborators and as had been found earlier, 15 out of the 76 mononucleotides showed deviating characteristics from the traditional four kinds of bases. In his 1967 summary Myrbäck declared that nothing new in the form of further advanced technology had been introduced during the year that had passed and that the data, although important and of high quality, in essence did not in principle introduce anything new. Hence a possible awarding of a prize to Khorana should be delayed for future consideration. He then stated that the prime mover when it came to deciphering the genetic code was Nirenberg. The one who came closest to him to be recognized by a prize was Khorana. However, at the time he could not endorse a sharing of the prize between Nirenberg and Khorana. The committee agreed with Myrbäck's conclusion but expressed its intention to follow with diligence continuing developments in research concerning the chemistry of the genetic code. It is of interest to note this hesitancy regarding a prize among chemists as compared to the more positive attitude of members of the committee at the Karolinska Institute to be described later. Could it be that molecular biology with its dependence on biological experimentation still had some way to go before it was accepted as a true subdiscipline of chemistry?

In 1968 725 nominators had been invited to submit a proposal for a prize in chemistry and there were 132 responses. Among these were several nominations for a prize to Khorana. There were nominations from A. R. Battersby, University of Liverpool, J. Halpern, University of Chicago and R. P. Mitra, University of Delhi, India. Two of the nominations included attachments. Battersby wrote "The prospects opened up by these discoveries, and by the availability of ribonucleotides built in known sequence, are immense." In addition, as mentioned, Khorana together with Nirenberg was nominated for the third year in a row by Perutz, a nomination supported separately in nominations by Jacob, Lwoff and Monod. Various kinds of additional supplementary materials were submitted. The philosopher Lwoff wrote (translated from French):

> "The history of biology is distinguished by five great stages: 1) The discovery of the nature of the genetic material; 2) The discovery of the structure of DNA; 3) The discovery of the physiology of genes; 4) The discovery of RNAs involved in the synthesis of proteins; 5) the discovery of the code, that is the language of genetics. It is the latter contribution which is proposed for a Nobel Prize in chemistry."

The committee decided to use an outside reviewer. As mentioned it was Reichard, who knew the field in detail and had already presented his views in the analyses he had made for the Karolinska Institute, the latest one the previous year. His review covered 26 pages and naturally expanded on the review he had previously made for the committee at this institute. He gave comments on Khorana's work over some four pages. After an introduction in which he remarked on the earlier somewhat lukewarm reception of Khorana by the chemistry committee he discussed the impressive recent advance of his work under the headings (translated from Swedish): Organic-chemical synthesis of oligonucleotides; Enzymatic synthesis of defined polynucleotides; and Determination of the code by use of synthetic poly-nucleotides. As a summary Reichard noted that Nirenberg broke new ground in the field by his work with statistical nucleotides and later the linking of transfer RNA and ribosomes by trinucleotides. He then stated that the conclusive definition of the cipher became possible first by Khorana's work on synthetic nucleotides. Clearly both Nirenberg and Khorana were highly qualified for a prize in chemistry.

The Late Nomination of Khorana for a Prize at the Karolinska Institute

Khorana was not nominated for a Nobel Prize in physiology or medicine until 1967. There were two nominations and both included Nirenberg and hence they had already been cited in the previous chapter. One was the spontaneous letter on February 3, 1966, three days after the formal last day for submitting a nomination, to von Euler, secretary of the committee until the previous year. It was from George Broder, Chicago and started "Last evening we had H. Gobind Khorana as a Stieglitz lecturer. He gave a magnificent account of his truly great work on synthesis of specific DNAs and RNAs and the verification of the genetic code. I strongly suspect that he and Marshall Nirenberg have both been nominated for the Nobel Prize in Physiol. & Med. and/or Chem...." This was accepted as a nomination for the prize in 1967. There was also another nomination from the Uppsala professor Laurent, which included all three forthcoming 1968 Nobel Prize recipients, as also already mentioned in the previous chapter. Because of these nominations Khorana was included in the comprehensive review done by Reichard. As always, he gave a very lucid description, in this case of the development of Khorana's work.

Torvald Laurent (1930–2009).

The training in Todd's laboratory was described as essential in initiating

Khorana's deep involvement in biochemical synthesis of DNA nucleotide compounds. However, in the middle of the 1950s he had taken an independent lead in this field of research and as formulated (translated from Swedish): "... it was at this point clear that Khorana, essentially on his own, in a commendable way, had succeeded in solving the synthetic challenges in the field of nucleotide chemistry, which the Cambridge school had not succeeded in implementing in a satisfactory way." But apparently Khorana wanted to set his aims even higher. He wanted to produce chains of nucleotides with a predetermined sequence. In order to achieve this he invented ingenious methods for so-called block polymerization of polynucleotides and eventually he was able to produce an oligonucleotide chain containing 16 nucleotides in a predetermined exact position. Potentially it had now become possible to attempt to crack the genetic code by a new approach. However in this work it would be an advantage to synthesize even longer nucleotides. In order to do this, use of the naturally occurring enzymes like DNA polymerase to produce long chains of DNA, and for the corresponding purpose RNA polymerase to produce long chains of RNA, both needed to be employed. In the end long molecules with a repeat predetermined sequence could be produced. The terms oligo- and polynucleotides were used in this context, but it was not strictly defined when an oligonucleotide — possibly limited to a few hundred nucleotides — should be called a polynucleotide.

The way Khorana managed to provide an independent determination of different codons has already been introduced above. Individual triplets were defined by use of simple repeat sequences like UCUCUCUCUC... or AAGAAGAAGAAGAAG...leading to polypeptide chains characterized as containing serine-lysine-serine-lysine... and alternative chains of only lysine, arginine or glutamine, respectively. In the interpretation of the data it was of course seminal to appreciate that Nature used a triplet code. In yet another additional example of the experimental approach using a synthetic RNA with a repeat tetranucleotide was used. When the iterated repeat UAUC sequence was read a protein with a repeat sequence of four amino acids

UAUCUAUCUAUCUAUC...
tyr-leu-ser-isol-tyr-leu

Synthetic RNA with the repeat sequence UAUC and the resulting amino acid sequence in the polypeptide.

was produced — tyrosine-leucine-serine-isoleucine-tyrosine-leucine etc. Khorana's biochemical approach also allowed a confirmation of the nature of the start codon and the stop codons. Reichard's conclusion was that there was no hesitation at all regarding Khorana's entitlement to a prize. Hence he was much more positive and categorical at this stage than the biochemists on this matter. One may wonder what might have been the reasons for this difference in attitude. It would seem that Khorana's biochemical approach ought to be very attractive also to a traditional chemists, but maybe as already suggested molecular biology was still on its way to establishing itself as a central field within chemistry at large, which traditionally might have had a different major focus. The development of the field was dependent on the use of naturally occurring partially purified enzymes and protein expression systems.

As mentioned in the previous chapter nominations in 1968 were unique in that there was an avalanche of nominations for Nirenberg. Khorana also had a fair number of nominations. There were 12 out of which three were for him alone; six were for a joint prize with Nirenberg and the remaining three for a prize to all three candidates. As already mentioned Theorell was selected to be the reviewer. His final conclusion was that all three candidates deserved to be recognized by a prize, but that he personally preferred to give the prize only to Nirenberg.

The Third Man and a Single Molecule

It is now time to introduce the third member of the triumvirate, Holley, who was eventually recognized by the 1968 Nobel Prize in physiology or medicine. He emerged as a candidate only late in the game. Holley was born in 1922 in Urbana, Illinois. He was one of four sons and his parents were both educators. During his early life the family moved between different parts of the U.S. but eventually returned to Urbana where Holley graduated from High School and later received a B.A. degree in chemistry. Further studies at Cornell University led to a Ph.D., but then the Second World War broke out which led

Robert W. Holley (1922–1993), the co-recipient of the 1968 Nobel Prize in physiology or medicine.

to his becoming involved in studies of penicillin in the laboratory of Vincent du Vigneaud, the forthcoming 1955 Nobel laureate in chemistry, introduced at length in my previous book[2]. Afterwards he trained as Postdoctoral Fellow with Carl M. Stevens at Washington State University, after which he returned to Cornell University as Assistant Professor of organic chemistry and later Associate Professor at the same institution during 1950–57. In his Nobel lecture[7] he emphasized the importance of taking a sabbatical leave. What happened to him was that in 1956 he spent a year with James F. Bonner, a plant molecular biologist at the California Institute of Technology. This turned out to be a decisive experience for the future work that earned him the Nobel Prize. He studied protein synthesis which led to an involvement in experiments focusing on the acceptor molecule for the activated amino acids, the transfer RNA.

As was described above individual amino acids had been found to be activated enzymatically and then to be connected to a form of low molecular weight RNA, by attachment to a terminal adenosine residue of this RNA. As already pointed out, these insights had been gained in experiments by Hoagland, Zamecnik, Berg, Lipmann and others. Studies of the transfer RNA became the central focus of his work when Holley had returned to Ithaca, New York, as Research Chemist at the United States Department of Agriculture, Agricultural Research Service at the U.S. Plant, Soil and Nutrition Laboratory on the Cornell University campus. In 1964 he rejoined the faculty of Cornell University and was chairman of the Department of Biochemistry and Molecular Biology 1965–66.

The goal set in the early 1960s was to isolate and characterize a single species of transfer RNA. He selected alanine transfer RNA, which turned out to be a fortunate choice. Crick believed that it would be difficult to isolate a single particular form of transfer RNA among the many variants projected to exist, since their molecular structure might be too similar. However, it turned out that this category of molecules in an alternating way use different atypical variants of the four kinds of nucleotides, which led to them displaying distinctly different physico-chemical properties. It was this difference in characteristics that allowed the isolation of a specific transfer RNA in a highly purified form. As the work progressed it turned out that it had been possible to prepare material in such a purified form and in such a quantity (one gram) from the yeast cells (a total of 300 pounds) selected to be used in the study that the complete structure of the particular molecule chosen for the analysis could be determined. This was a major achievement, appreciated by the reviewers selected by the Nobel Committees, as we shall see. We will return to the beautiful "clover-leaf" structure revealed to be formed by intramolecular base pairing and the functional implications of the particular structure selected by evolutionary processes. This work, which came to represent Holley's main lifelong contribution to science, was completed in 1964.

Holley as a Candidate for a Prize in Chemistry

The first nomination for a prize in chemistry to Holley was submitted in 1966 by the Swedish biochemist Per-Åke Albertsson, who the year before had become professor of biochemistry at Umeå University. One may ask why he was the first one to nominate Holley. Albertsson was introduced in my second book on Nobel Prizes[6] as a Ph.D. student in Tiselius's department in the late 1950s. I wrote my first scientific publication with him in 1960 describing the concentration of poliovirus by use of the two-phase system he had developed and presented in his Ph.D. thesis. He was not himself a molecular biologist, so it is not unlikely that he received the incentive to nominate Holley from Tiselius, who spearheaded the advance of molecular biology in Sweden as repeatedly pointed out, or from one of his nearby co-workers. The nomination by Albertsson briefly noted that Holley's presentation of the structure of transfer RNA for alanine was the first of its kind ever presented. He mentioned that it was a challenge to obtain a highly purified material for the analysis and that new methods for the sequence analysis had to be developed, although they, in principle, copied the application of techniques originally introduced by Sanger to determine the amino acid sequence of the protein insulin.

The committee at the Royal Swedish Academy of Sciences let Tiselius, chairman of the committee at the time, do a review of Holley. It was brief, covering only three pages. Tiselius remarked that at the time it had been shown that there were separate transfer RNAs for each amino acid and that their limited size might make them amenable to a determination of their unique sequence of nucleotides. He mentioned that two different enzymes capable of splitting RNA, but with different specificities, had been used and that the original breakdown products, 19 and 17 fragments of oligonucleotides, respectively, had been characterized by chromatography using DEAE cellulose in 7 molar uric acid. The summarized results demonstrated that the complete molecule was built up of 77 nucleotides, and hence had a molecular weight of 26,600. The RNA was found to have a very striking capacity for internal base pairing, leading to the formation of a "clover leaf"-like structure

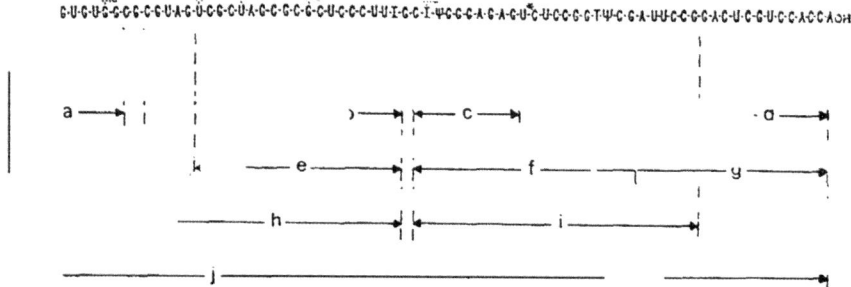

Fragmentation of alanine transfer RNA in different pieces. [From Ref. 7.]

(see p. 303) originally proposed by others. Tiselius noted that it was an interesting structure but recommended that the question of Holley's candidacy for a prize should not be discussed until confirmatory data had been published. This was also the conclusion of the committee.

In 1967 Albertsson repeated his nomination, but he did not provide any additional data. The committee retained its attitude from the previous year, and referred to Tiselius's brief review that year. It wrote: "… to await for some time the development of this highly important investigation in order to possibly obtain a confirmation of the structure by alternative approaches to sequence analysis." In 1968 there was a pronounced increase in nominations of Holley for a prize in chemistry. There were two nominations by his colleagues at Cornell University, Alfred T. Blomquist and Efraim Racker, with attachments and the former one was supported by none other than Vincent du Vigneaud. Finally there were nominations for Holley also by Jean-Pierre Ebel, professor of biochemistry, University of Strasbourg (delayed from the previous year) in a four-page letter in French and of particular interest also by Sanger. The latter nomination was in the form of a remarkably readable handwritten letter. It should be kept in mind that the person Sanger nominated was his own main competitor, since his and his group's major thrust at the time was to develop methods to sequence RNA, as described earlier[2] and briefly mentioned above. The second paragraph of the letter began:

> "The two most important components of living matter are the proteins and the nucleic acids. No sequence had previously been determined in a nucleic acid and Holley's work represents a very important advance since it is the first RNA sequence to be worked out, and demonstrates

the possibility of determining other sequences. Such studies are of particular importance since it is now believed that the complete information for the production of an organism is stored in the form of nucleotide sequences in the gene and copied to the messenger RNA before translation to an amino acid sequence in proteins."

Straightforward prose from a person who was uniquely qualified to make such a statement! Sanger then went on to praise the techniques used and to highlight the important conclusions on the correlation between the structure, the active sites and the mode of actions of transfer RNA. As mentioned the committee called upon Reichard to include Holley in his review of the three candidates. Sanger concluded the letter with a P.S. which reads:

"I have discussed this proposal with Dr Perutz, who has proposed Drs Nirenberg and Khorana for the prize. We both agree that their discovery of the genetic code and Holley's work is of such importance as to deserve a Nobel Prize each and that one should not be divided between three of them. My own opinion is that the work of Nirenberg and Khorana is more appropriate for the prize in Medicine."

Apparently what Perutz and Sanger had hoped for was that Holley should receive the prize in chemistry and that Nirenberg and Khorana should share the prize in physiology or medicine, but this did not come about. As already mentioned in the previous chapter the chemistry committee this year did not give highest priority to the discoverers of the genetic code, since they were aware of the fact that the three prize candidates would be taken care of at the Karolinska Institute

An Exemplary Review of Holley for a Prize in Physiology or Medicine

Holley was nominated for the first time late in the game, in 1967, for a prize in physiology or medicine. As mentioned, he was nominated together with Nirenberg by Lehninger at Johns Hopkins University, Baltimore, and also by Laurent, who made a joint nomination of all three candidates. The committee at the Karolinska Institute included an analysis of all three candidates together with a few other candidates in the extensive review allotted to Reichard. The part of the review concerning Holley covered eight pages and it is exemplary in its lucidity and structure, like the rest of Reichard's broad analysis. Discussing Holley, he first referred to the introductory part of his review in which it was noted that Zamecnik had developed a useful laboratory system to study protein synthesis and that Hoagland had noticed that the first step in the synthesis of a new protein was a chemical activation of the amino acids to be included in the nascent product. Then he referred to a publication in 1957 by Holley which demonstrated that the activated amino acid could not be directly used to form a part of a new protein. An intermediary structure of this activated amino acid associated with a certain kind of RNA needed to be formed. This was deduced from the observation that an energy transfer identified by an isotope-exchange measurement could be blocked by treatment by the enzyme ribonuclease. Similar observations of an association of an activated amino acid and transfer RNA interaction were made also by other researchers, like Hoagland, as emphasized above.

Holley decided to concentrate his efforts on purifying alanine transfer RNA and completely characterizing its chemical structure. This was a very daunting task, but he and his collaborators were lucky in that the presence of a number of so-called atypical nucleotide bases in the molecule resulted in fragments of it behaving in a somewhat unique way. On the other side the identification and chemical characterization of the previously unidentified atypical nucleotides must have been a major challenge. It was known at the start that transfer RNA had a molecular weight of 25–30,000 Dalton and that in one end of the molecule it had a guanylic acid and in the other end, the nucleotide sequence CCA. Various separation techniques were tried, but the one eventually selected was counter current electrophoresis. This technique had been presented by Lyman C. Craig introducing a further development of the partition chromatography technique recognized by the 1952 Nobel Prize in chemistry to Archer J. P. Martin and Richard L. M. Synge. After having tried out the use of different solvents, it was possible for Holley and collaborators to prepare a highly purified alanine transfer RNA and later also additional transfer RNAs, like tyrosine transfer RNA, of a quality allowing a complete chemical analysis. The scientists learnt to be very careful in their experimentation, since it was experienced that ribonucleases, enzymes capable of breaking down the molecules studied, occurred in many environments, including also on the researcher's fingers.

The sequence of the nucleotides was determined by use of a technique largely similar to the one that Sanger had used for determination of the amino acid sequence of the protein insulin[2]. Techniques were developed to prepare different kinds of fragments and reconstruct the intact molecule by identification of overlapping parts of fragments. The transfer RNA was split into pieces by treatment with two different kinds of enzymes with different specificities, one referred to as pancreatic and the other as Takadiastase T1 ribonuclease (p. 299). The various fragments obtained were then separated by chromatography. In the case of larger fragments a supplementary form of enzyme digestion using a snake venom phosphodiesterase was used. This enzyme consecutively removed one nucleotide at a time from one end of the fragment. Eventually, the nucleotide sequence of each unique fragment could be identified. To cut a long story short it was finally possible to identify overlapping parts of the many different fragments and come up with the final structure.

It should be noted that the application of this experimental approach is much more complicated with nucleic acids than with proteins. The former are composed of only four different kinds of nucleotides, as compared to proteins which are built up by 20 different amino acids. But it should also be remembered that the occurrence of atypical nucleotides was of great help in identifying overlap of fragments. However these atypical bases also caused problems of characterization and certain new technologies had to be invented. It was hard work. The final step in identifying the complete structure included one additional experimental approach. It was to use the ribonuclease T1 enzyme under very restricted conditions, four minutes and one hour of exposure, respectively, at freezing point. This led to only one and two cleavages of the intact molecule and it helped to finally define its two halves.

In the complete analysis it was concluded that the molecule was composed of 77 nucleotides, out of which 9 had atypical features. The internal base pairing deduced gave the molecule a "clover-leaf" structure. The deduction of the full structure allowed a number of important general conclusions regarding various functions associated with different parts of the molecule. Firstly, the tail exposing the single-stranded ACCA (marked I) might readily be used to bind the specific amino acid, in this case alanine, to the terminal adenosine. Secondly, the different odd atypical bases were found not to be evenly distributed over the nucleotide chain but they were concentrated to the regions of the molecule where hydrogen bonding preferably should be avoided, viz. the three protruding loops forming the edge of the "leafs" (marked II, III and IV by Reichard). The representation of the atypical bases seemed also to be decisive for the binding of a certain amino acid to a transfer RNA carrying the specific anticodon for this specific amino acid. Thirdly, it could

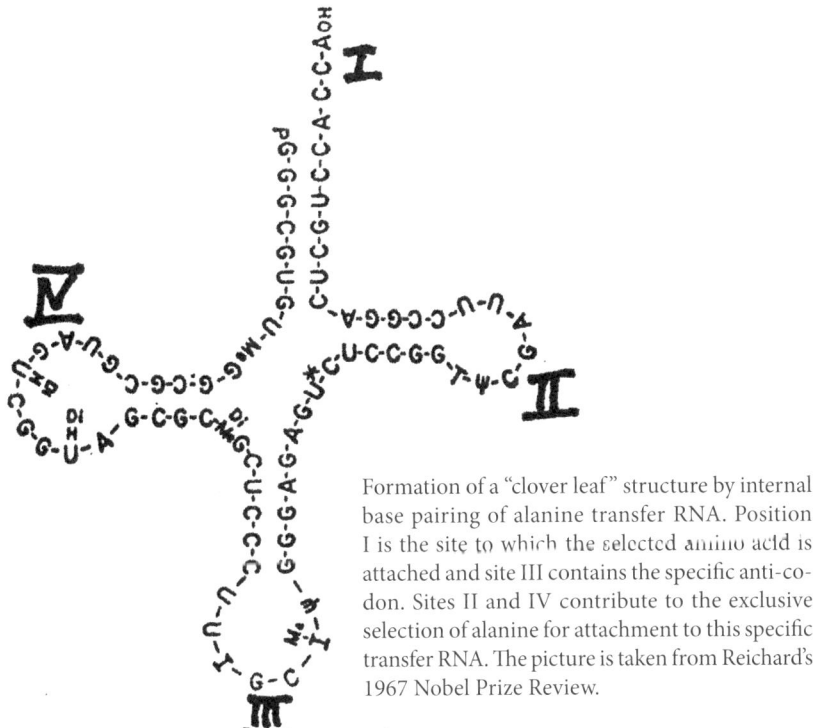

Formation of a "clover leaf" structure by internal base pairing of alanine transfer RNA. Position I is the site to which the selected amino acid is attached and site III contains the specific anti-codon. Sites II and IV contribute to the exclusive selection of alanine for attachment to this specific transfer RNA. The picture is taken from Reichard's 1967 Nobel Prize Review.

be assumed that the latter loops serve special functions. One of them (II) has the sequence G-T-ψ-C-G. Holley's group were able to demonstrate that this feature sequence also occurred in other kinds of transfer RNAs, serving some function shared between all forms of them, like for example binding to a certain ribosome structure. The other two loops have the sequences IGC (III) and CGG (IV), respectively. Since Nirenberg had identified GCX (where X can be either U, C, A or G) as the triplet code for alanine, either one of them could represent the anti-codon by help of which alanine transfer RNA could attach to messenger RNA. At the time it could not be decided definitely where the attachment occurred but in Reichard's review it was considered likely that the critical binding site was IGC, which would be capable of binding in an anti-parallel way with the critical codon G-C-X (site III in figure). This has later turned out to be the correct deduction. The review of Holley ended by stating that "… according to my opinion he to a high degree deserves to be recognized by a Nobel Prize."

Reichard's comprehensive review of the total of six candidates finished

The Formation of a Trio for the 1968 Prize 303

with a section headed "Conclusions and final review" covering a little more than six pages. It is so well written that it is tempting to cite it in its full extent. However, only a condensed version will be presented. The fundamental question of how one gene can direct the synthesis of one protein was highlighted as a preamble. It was a question of a deciphering problem and the genetic code provided an answer to this. In order to approach this experimentally it was necessary to develop a laboratory system for studies of protein synthesis. Many scientists made important contributions to these early developments as repeatedly mentioned, Zamecnik, Hoagland, Lipmann, Berg and others. Their work led to the discovery of the existence of transfer RNA, and the representation of separate versions of such molecules for each kind of amino acid. Reichard then illustrated in a picture, reflecting knowledge at the time, of the "factory," the ribosome, where proteins were produced. He emphasized the crucial importance of triplets, in the form of codons and anti-codons, both in the transcription of DNA into RNA and later also the expression of messenger-RNA in association with ribosomes and transfer RNA. He then wrote the following (translated from Swedish):

> "Much of the general conceptual background to the scheme presented derives from Crick (the adaptor hypothesis) and Jacob and Monod (the messenger RNA concept). The experimental evidence and the deciphering of the genetic code derive from experiments by primarily three researchers: Nirenberg, Khorana and Holley."

Nirenberg was described as having broken the field open, Khorana as having contributed to the conclusive clarification of the cipher. The combination of results obtained by use of Nirenberg's and Khorana's methods provided the final clarification and a conclusive deduction of the complete code. Holley's contribution was of a different kind. He co-discovered transfer RNA and he also clarified the first complete structure of this kind of molecule, the one for alanine. He later also showed the corresponding structure for tyrosine transfer RNA. Reichard also mentioned that later Khorana in addition had deduced the structure of one more form of transfer RNA, namely the one for phenyl alanine. He concluded that Holley's contributions had a greater importance than those of Zamecnik and Hoagland. He also mentioned the importance of collaborators, in particular Matthaei, which was eventually left out as already mentioned. The geneticists Benzer and Brenner were also mentioned but it was concluded that they had to give way to the pioneers in the field of phage genetics, Hershey, Delbrück and Luria. Finally the one-step mutants studied by H. G. Wittman and C. Yanofsky were also referred to but their data were regarded as not reaching the same level of importance as those collectively presented by Nirenberg, Khorana and Holley. In the final recommendation of the latter trio Reichard took notice that the latest addition of seminal data derived from 1966 and hence a prize awarded in 1967 might be considered to satisfy Nobel's original will specifying that a prize should be given to the one "who during the previous year…" has made an important discovery. However, the prize had to wait one more year.

As already mentioned there was one other biochemist who played a major role in developing techniques to sequence RNA. This was Sanger as described in my previous book[1], who appears as a very important nominator mentioned above. This humble giant of biochemistry, who received two Nobel Prizes in chemistry, the first one for his sequencing of the protein insulin and the other (shared) for his development of techniques to determine the nucleotide sequence of DNA, was also involved in the sequencing of RNA. Sanger's RNA work was initially inspired by a new collaborator, Leslie Smith who had arrived from Zamecnik's laboratory at Harvard University where he had gained important experience of working with transfer RNA. Together with George Brownlee and other collaborators Sanger developed a very sensitive so-called "finger-printing" technique using nucleotides radioactively labeled with phosphorous. Because of its efficacy it came to be more widely used, by Sanger's group and others, than Holley's more cumbersome technique, for characterization of additional transfer RNAs, like for example for methionine, for tyrosine and also for phenyl alanine. The nucleotide sequence of the latter form of transfer RNA was determined by Sanger's group, but their data were not published until 1969 after similar data already had been presented by Khorana and collaborators.

The technique developed was also used by Sanger's group for a pioneering characterization of 5S RNA which represents an important part of the complex ribosome structure. At the time when they had managed to unravel its sequence it was the largest RNA — 120 nucleotides long — examined at the time of publication in 1967. In other studies of RNA Sanger together with the Danish scientist Kjeld Marcker discovered that a special form of amino acid, formyl-methionine and not methionine itself served to allow initiation of protein synthesis in prokaryotes. Sanger and his colleagues also characterized parts of the RNA genome of the bacterial virus R17. The genome of this virus contains 3000 nucleotide residues, which was a molecule of a size that could not be sequenced at the time. However, one fragment that could be analyzed coded for the coat protein of the virus particle. Since the amino acid sequence of the coat protein had been determined, it became possible for the first time to verify the validity of the deduced genetic code. I have speculated earlier that possibly Sanger might have been nominated for even a third Nobel Prize, for his RNA work, but so far no documentation to support this has been found. This important work was however recognized by the Hopkins Memorial Medal awarded by the Biochemical Society in the U.K. in 1971. Let us now go back to the final discussions for a prize in 1968.

Decisions on the Prize in 1968

As mentioned above nominations in 1968 were unique in that there was a massive campaign for a prize in physiology or medicine to Nirenberg. Khorana also had a fair number of nominations, but proposals for Holley were few. He only had three nominations and they all included both Nirenberg and Khorana. Two of them were by Swedish scientists, Laurent, who the year before had made a joint nomination of all three candidates and Boman, introduced in Chapter 1. There was one additional nomination by the Nobel laureate Lipmann.

As mentioned in the previous discussions of the way the committees at the Karolinska Institute managed the distillation process towards its final recommendation it was very generous in recognizing contributions in different fields of physiology or medicine as worthy of a prize. In 1966 35 candidates or combinations of them were allocated to this category and in 1967 it had increased to 37. The pressure on the committee increased further in 1968 when as many as 52 candidates or groups of them were competing for the prize. And still it would seem that the final choice this particular year should not have been that difficult. Never before or possibly even later has a single candidate received such a large proportion of the nominations as Nirenberg did. In Chapter 2 a brief summary was given of the different subdisciplines in the field that were highlighted in the reviews made. A particular emphasis was given to the development of insights into human genetics and dysfunctions of importance for diseases in man. The insights into the genetic code obviously had to amplify developments in this field. And still over the years important advances in human genetics of importance for understanding of unique disease processes have not been recognized by Nobel Prizes.

The role of genetic disorders for the development of diseases in man is progressively becoming clarified. The importance of using genome characterization in a *precision medicine* approach to managing many different kinds of diseases, especially including various forms of cancers, was discussed in Chapter 3. Various genetic abnormalities of specific relevance to cancer development were referred to in this chapter. The list of all other kinds of genetic disorders in man associated with different kinds of diseases is long and rapidly growing. Presently there are somewhat more than 2,000 diseases registered as being related to specific changes in the genome. They can be due to the exchange of a single amino acid, as first demonstrated in the case of sickle-cell anemia[6], but they can also be the result of a mutation changing the reading frame as in Tay-Sachs disease. This disease leads to destruction of nerve cells in the brain and the spinal cord. Other neurological disorders like Huntington's disease and early onset Alzheimer's disease are also influenced by gene dysfunctions. Still another example of a congenital disorder is phenylketonuria which occurs in 1/10,000 to 1/15,000 newborn children. By the testing of blood collected from the umbilical cord it is possible to make an early identification of the genetic defect as a source of a serious metabolic disorder. In those children who have this defect, normal brain development can be secured by adjusting the diet to not include the amino acid phenylalanine. This is particularly important during the first years of life when the brain is developing rapidly.

There are also a number of diseases caused by major chromosomal changes; three copies of chromosome 21 in Down's syndrome, already briefly mentioned in Chapter 2, and a gain and loss of the X chromosome in Klinefelter's and Turner's syndromes, respectively. Finally there are also a number of genetic diseases due to defects in the separate small DNA associated with mitochondria. In the fertilized egg this organelle exclusively derives from the mother. When a mitochondrial genetic defect is identified it is therefore possible to remedy the situation by an exchange in the laboratory of the cytoplasm of the fertilized egg before implantation. This technique has become allowed for use in the U.K. Because of the dramatic improvements in possibilities to sequence and characterize the human genome it can be predicted that our understanding of genetic diseases in man will increase dramatically in the coming years. It will be interesting to see how the possibilities to correlate genotype and phenotype will develop. It is somewhat surprising that the important and growing field of recognizing association between various forms of genetic disorders and certain diseases as yet has not been recognized by a Nobel Prize in physiology or medicine. Let us now return to the final discussions by the committee in 1968.

In its final position document, including the already mentioned 52 prizeworthy candidates the committee in September 26 summarized that it would come back to its final decision. On October 16 it announced that it recommended dividing the prize equally between Holley, Korana and Nirenberg. One may wonder why the committee needed this extra time. It would seem self-evident that the cracking of the genetic code needed to be recognized as soon as possible after its discovery. Since only two years had passed after it had been finally described the Karolinska Institute was relatively close to the one year proposed by Nobel in his will. The Institute could retrospectively be proud of including among its prizes both the discovery of the structure of DNA and also the revelation of the genetic code. It might be added that whereas there was an impressive number of nominations for the identification of the genetic code, there was no corresponding "campaign" for recognizing the double helix structure of DNA[6]. Later on the chemistry committee at the Royal Swedish Academy of Sciences came to be more aggressive and took control of most of the prizes recognizing the fast developing field of molecular biology.

Time for the 1968 Nobel Festivities

When Nirenberg travelled to Stockholm he was accompanied by his wife, his sister Joan and his uncle Arthur. Sadly his parents did not live to join in this happy event. His mother, Minerva, had never learned about Nirenberg's success as a scientist, since she had already died of cancer in 1960. His father Harry towards the end of his life managed to stabilize his finances by shifting from farming to the development of warehouses. He was also able to share in the first success in Nirenberg's scientific endeavors. However, he found the loss of his wife so heavy that he committed suicide in 1961. Khorana and Holley were joined by their wives and also some of their children.

The Opening Address at the prize ceremony was given by the Chairman of the Board of the Nobel Foundation Ulf. S. von Euler, himself a Nobel Prize recipient two years later. He first gave praise to two earlier recipients of the Nobel Prize in physiology or medicine, Henry H. Dale and Corneille Heymans, who had been his mentors. He expressed a great admiration for the contributions by Dale, a legendary neurophysiologist described at length in my previous book on Nobel Prizes[2]. In reference to the Nobel Prize to Heymans he mentioned that the prize had been awarded in 1939 and hence because of

Nirenberg with his wife Perola at the time of the prize ceremony. [With permission from TT, Sweden.]

the war situation had to be presented to him in his home country of Belgium. Then he discussed if Nobel Prizes had a stimulating effect on the science concerned. The answer given was somewhat vague although it was emphasized "that a scientist does not work for the prize. The urge to do research is more deeply founded." This answer might even have been more categorical. At the time of a major discovery there is no thought in the mind of the scientist of any reward, only the great joy of being the first to gain insight into a major new understanding. Von Euler then appropriately discussed the huge increase over time of scientists involved in research in different branches of natural sciences and also emphasized the enlargement of the fields of research encompassed by different disciplines. He said (translated from Swedish):

> "Thus physics has come to include the field of Astrophysics in a broader sense, and the subjects Physiology or Medicine are now taken to comprise Genetics, Biochemistry and Molecular Biology. Psychology and Behavioural Sciences are also parts of Physiology but have not so far been eligible for an award of this kind.
>
> These alterations have taken place gradually as an adaptation to the change in structure in the fields of science. Methodological advances have often played a decisive role for this type of development. Team work has become more frequent but in spite of this it is still often the achievements of single scientists which represent the

The Formation of a Trio for the 1968 Prize 309

breakthroughs and are responsible for the advancement of science. It seems that Alfred Nobel had this in mind in the first place when he instituted his prizes."

The prize in physics this year was awarded to Luis W. Alvarez from Berkeley, CA and the prize motivation was exceptionally long; "for his decisive contributions to elementary particle physics, in particular the discovery of a large number of resonance states, made possible through his development of the technique of using hydrogen bubble chamber and data analysis." The important use of this technique for the development of high-energy physics was presented by Professor Sten von Friesen. The chemistry prize as mentioned was awarded to Lars Onsager, of Norwegian origin, but active at Yale University, New Haven, Conn. He was introduced by Stig Claesson, professor of physical chemistry, Uppsala University. Following the prize for the discoverers of the genetic code the Nobel laureate in literature was recognized. The committee of the Swedish Academy had over the years aimed at finding representatives from different language groups, a problem of a general nature when it comes to evaluating world literature. Special translations could be commissioned by the Academy. In the particular year of 1968 the prize recognized an author of Japanese origin, Yasunari Kawabata. A review of the archives reveals that there had been particular tense discussions in the Swedish Academy this year. A leading candidate like Samuel Beckett did finally receive a prize the following year, but other literary giants like André Malraux from France and not least W. H. Auden, from Great Britain, would never receive the Nobel Prize for literature. Kawabata, dressed in a beautiful male kimono, received the prize from the hands of His Majesty the King. I have mentioned this author in one of my previous books[2], not only because Japan happens to be one of my favorite countries but also because once in Kyoto, my wife and I stayed in the room in a Ryokan hotel where Kawabata did most of his writing. It was a solemn experience.

It can finally be added that 1968 was the last year when the ceremony appropriately reflected Alfred Nobel's will. The ceremony ended as it should with the prize in literature. Although it is not possible to add to a will, the following year a new external prize had been added. It was the prize in economic sciences in memory of Alfred Nobel, established by the National Bank of Sweden on account of its 300 year jubilee. It was managed by the Royal Swedish Academy of Sciences. Since then this prize has been attached to the ceremony and in all aspects borrowed the plumage of the proper Nobel

Prizes. It is referred to as "The Sveriges Riksbank (Bank of Sweden) Prize in Economic Science in Memory of Alfred Nobel" which is impossible to refer to and hence to the public it becomes the Nobel Prize in Economics, but there is no such thing! Regrettably it came into existence under time-compressed circumstances. Whereas the true Nobel Prizes are defined as to their content by Nobel's will, the statutes for the prize in economy simply states that it should be awarded "in the spirit of Alfred Nobel," whatever that means. It should be added that the National Bank of Sweden has no influence on the expenses required to be met to fulfill the prize obligations it has taken on. The value of the Nobel Prizes is decided each year by the board of the Nobel Foundation, on which the bank has no representation. Teasingly one may ask how many other such agreements the National Bank of Sweden may have. After this emotional detour let me return to the thematically exceptional Nobel Prize in physiology or medicine of 1968.

Reichard introduced the prize recipients at the ceremony. He referred to the original history in the late 19th century when the Swiss physician Friedrich Miescher discovered a substance he called nuclein, later renamed nucleic acid, rich in phosphorous but lacking sulphur, often found in proteins. He also mentioned the Czech monk Gregor Mendel, who in the middle of the same century, pioneered the identifications of units of heredity, rediscovered some 50 years later and soon thereafter named "genes." He alluded to the difficulties along the road until it was finally understood that the two discoveries were connected, that DNA represented the central storage of the genetic information. The rapid development of the field of molecular biology was emphasized with the 1968 prize being the sixth in a row of prizes during a time period extending over only ten years. Departing from a reminder of the Rosetta Stone he then briefly referred to the contributions by the three prize recipients. Towards the end he said (translated from Swedish):

> "The interpretation of the genetic code and the elucidation of its functions are the highlights of the last 20 years' explosive evolution of molecular biology which has led to an understanding of the details of the mechanism of inheritance. So far the work can be described as basic research. However, through this work we can now begin to understand the causes of many diseases in which heredity plays an important role."

After this it was time for the three laureates to receive their prizes from the

Holley (a) Khorana (b) and Nirenberg (c) receive their prize from the hands of His Majesty the King.

hands of His Majesty the King.

At the banquet Nirenberg, appropriately representing the three laureates, gave a short speech. He emphasized that to have been made a Nobel laureate was overwhelming. He said "It was as though we all were involved in an experiment that, *quite unlike most of our experiments*, (my italics), had worked." As a true scientist he further said "The sustained satisfaction of course, comes from the work itself, from generating the ideas and selecting those that match Nature's temple, from excitement and adventure of exploring."

Two days later it was time for the Nobel lectures. They were presented in sequence following the alphabetic order — Holley[7], Khorana[8] and Nirenberg[9]. Nirenberg spoke last but there are good reasons to mention his presentation first. He had opened the field and it was his early breakthrough results that had stimulated the inquisitive and skilled biochemist Khorana to get involved. Appropriately the title of Nirenberg's presentation was simply *The Genetic Code*. He emphasized the importance of his two years of learning to study protein synthesis in the laboratory with the support and advice from many senior colleagues. He also gathered impressions from many laboratories outside NIH before he embarked on his mission of using synthetic nucleotides in his work. He learned to make sure that the original endogenous protein synthesis was phased out before he challenged his system with the selected synthetic nucleic acid material. His control was RNA from TMV. It was under these conditions that he, together with Matthaei, made the original discovery that single-stranded poly-U led to a synthesis of a protein only containing phenylalanine. This finding broke open the first of the two phases of experiments that eventually after six years allowed the description of the complete genetic code, the code meaning of the 64 triplets of nucleotides. In the first follow-up phase of examination of synthetic poly-nucleotides containing only random nucleotide synthetic polymer including two or even three nucleotides were studied. They allowed a deduction, sometimes tentative because of the statistical representation of nucleotides, of many more codons (see p. 261). He referred to the first 1963 summary at a Cold Spring Harbor meeting of both his own group and of Ochoa's group. Representative code words deduced from the experiments with the single, dimer or trimer nucleotides already then allowed a presentation of the codons for 19 out of the 20 amino acids. At the time there was no code word proposed for glycine. He also presented the early evidence for the triplet nature of the code described by Crick, Brenner and collaborators[10].

He then focused on the second phase of the studies by himself and his many collaborators. It was the approach used by him and Leder which was based on studies of the binding between transfer RNAs and ribosomes in the presence of different combinations of polynucleotides. Eventually it had been found, as mentioned, that, conveniently, it sufficed to use only tri-nucleotides of predefined sequence as potential connecting reagents. He referred to the many collaborators contributing to the advances made by the group. The progress of the work allowed a complete presentation of the significance of the 64 code words. He emphasized the importance of Khorana's findings supporting the final outline of the genetic code. The remains of the lecture were devoted to a discussion of punctuation, a term used to present the connected processes of transcription and translation, physically separated in eukaryotes. The discovery of start and stop codons was also presented. Finally Nirenberg discussed redundancy, the fact that many amino acids were specified by the first two nucleotides only, the universality (and age) of code used and finally reliability and rate of transcription. The presentations of different matters were mostly made in a dry factual form and little room was left for speculations and conjectures on the impact of knowing the genetic code for the future reading of the books of life and possibly even for the writing of new copies of them. The last paragraph started

> "The genetic code is now essentially deciphered. I have been fortunate in having the collaboration of many enthusiastic associates during the course of our studies. To do justice to the years of effort and the important contributions made by associates and numerous colleagues throughout the world is virtually impossible in the available time."

The impact on the deciphering of the genetic code was to be increasingly amplified during the coming years. Some of these major impacts are summarized in the last chapter.

Khorana started his Nobel lecture by referring to the corresponding lecture given six years earlier by Crick[10] which focused exclusively on the genetic code. He also gave the classical historical references to Avery, Hershey and Chase, as well as to Caspersson and Brachet. Todd was highly instrumental in Khorana's early involvement in the experiments to prepare chains containing predetermined sequences of nucleotides, but in the Acknowledgments he brought another mentor to the fore. He wrote "I wish to make a personal acknowledgement to one more scientist. Fortunately, I was accepted by Professor V. Prelog of the ETH, Zürich, as a postdoctoral student. The association with this great scientist and human being influenced immeasurably my thought and philosophy towards science, work and effort." It can be mentioned that in 1975 Prelog received one half of the Nobel Prize in chemistry "for his research into the stereochemistry of organic molecules and reactions." However, it was of course critical to future developments that Khorana later worked in Todd's laboratory in the early 1950s when he and other collaborators in 1952 elucidated the internucleotidic linkage in nucleic acids, only a year before the unraveling of the double-helix structure of DNA by Watson and Crick. It was also at this time that the hypothesis of a possible relation between the linear structure of nucleic acids and of the sequence of amino acids in proteins was born.

Khorana appreciated the significance of this conjectural proposal and he reflected on a possible approach to test this critical hypothesis by means of biochemistry. In his lecture Khorana defined the biochemical outsets, viz. the possibility to form specific linkages between individual nucleotides, in four steps. As to development in time he concluded: "… synthesis of short chains of deoxyribopolynucleotides with predetermined and fully controlled sequences became possible in the early sixties." Stimulated by the 1960 observation by Matthaei and Nirenberg he decided to use synthetic polynucleotides of DNA with a defined repeat nucleotide sequence to contribute to completely defining the code words in the genetic language. The work was divided into three stages; synthesis of short chains of DNA containing predetermined repeat nucleotide sequences; conversion of these single stranded molecules to longer chains of double-stranded DNA; and finally the use of the latter molecules to synthesize single-stranded RNA representing the repeat sequences. The latter molecules were then employed in the protein-synthesis technique adopted in the laboratory. Eventually all the possible sixty-four ribonucleotide triplets had been synthesized and tested. The genetic code was confirmed and consolidated by this systematic approach.

Towards the end of the lecture Khorana also presented more recent work developed after the genetic code had been solved. This included the studies by him and his group of the full structure of phenylalanine transfer RNA. After Holley and colleagues in their pioneering work for the first time had shown the structure of a transfer RNA, a number of groups attacked the same problem using other kinds of transfer RNAs as objects of study. This is not an uncommon phenomenon in science and is often referred to as a "band-wagon" effect. It may be referred to as "horizontal" research in which well-established methods are used to solve a well-defined problem. In this kind of research valuable

	THE GENETIC CODE				
1 ST LETTER	2 ND LETTER				3 RD LETTER
	U	C	A	G	
U	PHE	SER	TYR	CYS	U
	PHE	SER	TYR	CYS	C
	LEU	SER	C.T.	C.T.	A
	LEU	SER	C.T.	TRY	G
C	LEU	PRO	HIS	ARG	U
	LEU	PRO	HIS	ARG	C
	LEU	PRO	GLN	ARG	A
	LEU	PRO	GLN	ARG	G
A	ILEU	THR	ASN	SER	U
	ILEU	THR	ASN	SER	C
	ILEU	THR	LYS	ARG	A
	MET (C.I)	THR	LYS	ARG	G
G	VAL	ALA	ASP	GLY	U
	VAL	ALA	ASP	GLY	C
	VAL	ALA	GLU	GLY	A
	VAL (C.I)	ALA	GLU	GLY	G

The genetic code. [From Ref. 8.]

information can be accumulated but no new doors are opened. Khorana, like Holley in his presentation, mentioned that at the time of lecturing 11 fully characterized transfer molecules of various kinds of specificities already had been presented, including the structure of phenylalanine transfer RNA determined by himself and his colleagues.

In the concluding part of the lecture it was mentioned that several laboratories had managed to crystallize transfer RNAs and that hence the three-dimensional structure would be forthcoming. Khorana also discussed the initiation of protein synthesis, highlighting that the identification of the single code word for methionine, AUG, was critical in this context. A special form of transfer RNA, referred to as initiator transfer RNA, carries this amino acid. This was first identified by Marcker and Sanger, in the form of formylmethionine, particular to bacteria, as already mentioned. The start methionine is cleaved off in most proteins and when this amino acid is to be included in other positions in a protein, a kind of transfer RNA with different characteristics than the one signaling initiation is used. Khorana also mentioned data suggesting that GUG, which codes for valine also occasionally was used for initiation of protein synthesis, suggesting a certain degeneracy of the first letter. He finally also referred to ongoing work in his own and in other laboratories on the relative role of the 30S and 50S unites of the ribosomes, a field that was to develop as more data on the chemical structure of these components and their three-dimensional folded structure was to accumulate. Appropriately Khorana concluded his lecture in the following way:

> "Nevertheless, the problem of the genetic code, at least in the restricted one-dimensional sense (the linear correlation of the nucleotide sequence of polynucleotides with that of the amino acid sequence of polypeptides), would appear to have been solved. It may be hoped that this knowledge would serve as a basis for further work in molecular biology and developmental biology."

This hope indeed was not in vain. Probably no other Nobel Prize in physiology or medicine has had such an impact. Reading and writing the books of life endlessly continues to provide insights into ever-expanding fields of fundamental knowledge.

Holley gave a comprehensive historical background to the attempt by himself and his group to characterize the molecular structure of an isolated transfer RNA, which only indirectly had a bearing on the genetic code. By use of the separation technique selected they managed to obtain reasonably pure preparations of the transfer RNAs for alanine, tyrosine and valine. As already described they elected to use alanine transfer RNA in their continued work. He then commented on the different cleavage experiments and the figure presented at his Nobel lecture (see p. 299) illustrated the use of overlapping fragments to mark the position of each one of the close to 80 nucleotides representing the molecule. Once the sequence of the molecule had been determined it was time to identify its different functional parts. They were already referred to above in the evaluation by Reichard of the work. Naturally there was a particular focus on the amino acid-attachment site and on the anti-codon site, serving to connect to the specific triplet codon in messenger RNA. Holley finally noted the suggestive clover leaf structure, first proposed by other scientists. He cited references presenting data on sequences of altogether 11 transfer RNAs illustrating the avalanche effect of the breakthrough findings by him and his group. And finally he mentioned the already cited importance of a sabbatical leave, which in his case had led to the foray into transfer RNA research. Towards the end of his lecture he stated the following:

"In these times of highly competitive research, few scientists have the satisfaction of carrying through a research program that takes 9 years. Without minimizing the pleasure of receiving awards and prizes, I think it is true that the greatest satisfaction for a scientist comes from carrying a major piece of research to a successful conclusion."

Life After Cracking the Genetic Code

All three recipients of the 1968 Nobel Prize in physiology or medicine were relatively young at the time of the award. Holley and Khorana were 46 and Nirenberg only 41 years of age. The average age of Nobel Prize recipients in this category is well above 50[4] and seems to increase with time. The scientific challenge all three of them had taken was very intense. The goal to identify the specific importance of each one of the 64 theoretically possible triplets was exceptionally well defined. It was a totally absorbing effort and it took of the order of six years to reach the goal. In the same vein identifying the structure of the first transfer RNA to be fully characterized was a uniquely pioneering achievement. Towards the end of my previous book on Nobel Prizes[2] the fable of the Greek lyric poet Archilochus illustrating that "the fox knows many things, but the hedgehog knows one big thing" was cited. It was concluded that scientists making major breakthroughs were highly focused or even obsessed by the challenge they had taken on — they were "hedgehogs." The three laureates discussed in the previous and the present chapter are highly representative of such a monolithic behavior, although Khorana had a more long-ranging ultimate goal as will be further discussed. However, it might be conjectured that after the totally involving efforts leading to the cracking of the genetic code their continued contributions to science might be less spectacular. Still all of the three laureates were such ingrained scientists that they remained obsessed by their trade and stayed in the laboratory. They did not become general spokespersons for the kind of science they had successfully pursued. That was taken care of by others, like Crick. None of them turned into science administrators.

Probably Khorana was the most successful among the three of them[11] in the aftermath of the prize. He stayed well within the field of molecular biology. He wanted to synthesize a whole gene, as mentioned, and he managed to pioneer this achievement. In 1972 he and his colleagues, who at the time had moved to MIT, Mass., filled an entire issue of the leading journal of the field, *Journal of Molecular Biology*, with 15 related articles totaling 313 pages! They had managed to synthesize a functional transfer RNA gene. In a way it was en route to managing this project, initiated in 1960 that Khorana, kicked-off by the Nirenberg-Matthaei breakthrough finding, made his seminal contributions to the cracking of the code. In the mid-1970s he changed field completely. He decided to determine the fine structure of membrane proteins. After careful screening, he chose to work on bacterial rhodopsins, light-driven membrane channels. The molecular background to activation of this protein by retinine, defined by Wald's work, was introduced in Chapter 5. The complete amino acid sequence was determined, the first one of an integral membrane protein. Thus he and his colleagues contributed another "first." In the analysis of the function of the protein he took advantage of a method for site-directed mutagenesis discovered by a former post-doctoral student of his, Michael Smith. The findings by Khorana and his collaborators in the characterization of the chemistry of the membrane protein had a major impact on the development of insights into the field of so-called G protein-coupled receptors. These different advances were later recognized by Nobel Prizes in chemistry and in physiology or medicine. In 1993 Smith was recognized with half a prize in chemistry "for his fundamental contributions to the establishment of oligonucleotide-based mutagenesis and its development for protein studies" and in 2012 the prize in physiology or medicine was awarded to Robert J. Lefkowitz and Brian K. Kobilka "for studies of G-protein-coupled receptors."

Khorana has been described as exuding "an almost childlike enthusiasm and energy, but with a laser-like focus" and his experimental approach was described as characterized by "boldness and audacity"[11]. His life's work was impressive as was his role as a mentor. It is said that more than two hundred graduate students and postdoctoral fellows were trained by him. His most seminal publications with valuable summarizing comments have been compiled in a book[12]. Khorana's remarks reflect a man of exceptional erudition and refined sensitivity. He concluded his remarks with an extract from the poem *Four Quartets* by T. S. Elliot. It reads:

We must be still and still moving
Into another intensity.
For a further union, a deeper communion.
Through the dark cold and the empty desolation.
The wave cry, the wind cry, the vast waters.
Of the petral and the porpoise.
In my end is my beginning.

Khorana died in 2011 at the age of 89 years.

Nirenberg also had a long and active life as a researcher after the prize. However, he completely changed his field of research. He wanted to study the brain in cell cultures or as he said to Gallo in the parking lot at NIH "to study neurons in a test tube." Development of techniques to grow cells in cultures using synthetic or semi-synthetic media is an art that has developed over many decades. Since propagation of cells in the laboratory is a prerequisite for studies of intracellular parasites, like viruses, the development of the discipline of virology has been dependent on availability of cells in cultures for propagation of this kind of infectious agent. It was first in the 1950s that effective and convenient techniques for growth of cells were developed, partly due to the availability of the recently introduced antibiotics, like penicillin and streptomycin. Adding them to the liquid medium used to nourish the cells prevented the growth of potentially contaminating microorganisms. Many new viruses were identified in this decade and effective and important vaccines were developed and produced on a large scale in such cultures. The kind of cells generally used were of epithelial or fibroblast origin.

The possibility of retaining various kinds of specialized properties displayed by the more than one hundred forms of cells in various differentiated tissues represented in our body was another matter. It should be noted that in the laboratory, under so-called *in vitro* conditions, there is an absence of various kinds of naturally-occurring stimulating and inhibitory substances which surround cells when they represent a part of the intact body, *in vivo*. Still for example heart muscle cells cultivated in the laboratory can be demonstrated to continue to contract rhythmically for days and weeks and nerve cells can be shown to exhibit electrical activity for a very long time. Nirenberg wanted to find out to what extent fundamental properties of nerve cells could be maintained and examined in culture.

Thus Nirenberg retained high ambitions for his research. Like other great scientists we have encountered earlier, Crick and Edelman[6], he decided to explore the human brain. All the big and challenging questions were there. What is consciousness? How does memory work? Challenging questions; very hard to get a grip on, even by the best of human brains. Nirenberg was eager to apply the most recent technique and to go back to fundamentals. Hence he decided to grow nerve cells and examine their interactions in the laboratory; used the fruit fly system to study the embryonic development of the brain; attempted to block the expression of specific messenger-RNAs by use of so-called interfering RNA, which had been designed to have a nucleotide sequence contrary to the one of a selected messenger etc. This phenomenon will be returned to briefly in the final chapter. In a number of these projects he teamed up with Phillip G. Nelson, who was to become one leader in the Behavior Biology Branch in the National Institute of Child Health and Human Development at NIH. Their first encounter was accidental.

In 1966 Francis Schmitt arranged a Neuroscience Research Program meeting in Boulder, Colorado. At this meeting Nirenberg, Nelson and the prominent neurobiologist Erik Kandel, who were to share the Nobel Prize in physiology or medicine in the year 2000, had a meeting. Nirenberg who already at this time had started to initiate a reorientation of his laboratory towards work in the field of neurobiology asked Kandel how he could build up electrophysiological competence in his laboratory. Kandel, serving like a Jewish matchmaker, a *shidduch*, pointed to Nelson. This became the beginning of a collaboration over many years. A neuroblastoma cell line, C-1300, was selected for the work. The introduction of an electrode into the cells demonstrated that they still — they had been propagated since 1940 — retained a capacity to exhibit electrical excitability. This finding triggered the idea to dissect the genetic program underlying neurobiological activity.

In the continued studies it was discovered, perhaps not surprisingly, that a single kind of cell exhibiting electrophysiological activity in the laboratory could not represent the complexity of the intact nervous system. As in experimental studies of fusion of non-transformed and transformed cells (see Chapter 3) attempts were made to study the effect of fusing electrophysiologically active cells and regular fibroblast cell lines. Again the complexity of the system examined precluded any major advance of new knowledge. One general limitation of the neuroblastoma cell line used was that individual cells showed major variations in behavior. They had varying chromosome characteristics — they were *aneuploid*. The role of opiate tolerance, dependence and withdrawal, a theme that has taken on an increased relevance with time, not least in the U.S., was also subjected to cell culture studies. Again the simplified system turned out to have inherent limitations. Many other examples of initiatives taken by Nirenberg to approach central problems of neurobiology could be given. Perhaps, however, it is best to cite the comments on Nirenberg's personality traits given in one memoir[13].

One trait was ideas. His critical collaborator Leder noticed that there was a new idea every two or three minutes and others noted that "Ideas were his stock in trade and he took them very seriously, and had both many ideas and good ideas." Nirenberg had the habit to use notebooks in which every evening he put down his major ideas of the day. There are some 49 volumes of these "diaries of ideas" at the National Library of Medicine at NIH. It has been pointed out that his "gentle scholarliness was coupled with an uncompromising high scientific standard and a positively voracious capacity to process information (his briefcase always carried data from one or another postdoc to be taken home and evaluated)." He also had a capacity to stock and process large amounts of information, but he never managed to identify something in the realms of a "neural code."

Nirenberg's pleasant personality influenced many in his surroundings. When Craig Venter moved to the NIH in 1984 he became very enthusiastic about the conditions for pioneering research. On the floor below him was Nirenberg's laboratory and in his biography[14] Venter wrote "He and his NIH peers could teach us new things and help spark our own novel ideas." When NIH established a Human Genome committee, headed by Venter, Nirenberg was one very active member of the group.

During the 1990s Nirenberg's family life was sadly overshadowed by the fact that his wife Perola developed Alzheimer's disease. This kind of progressively developing dementia represents one of the major challenges of modern medicine. It occurs at a high frequency and the relative number of cases will increase as the average life span of humans is progressively prolonged. I have a particular sympathy for Nirenberg's attempts to manage problems at home as Perola's disease worsened. At the time of writing my own wife Margareta has been taken care of for more than one year at an institution for dementia sufferers because of her Alzheimer's disease and eventually she died in November 2018. We fought for a number of years to manage the progressive infringement on the integrity of her personality, but finally had to give up. In Nirenberg's case he apparently managed to care for his wife at home, which however required major adjustments in his lifestyle and professional involvements including no travelling. After ten years of her family fighting the disease Perola died in 2001, leaving a grief-stricken Marshall. However, the last ten years of his life were brightened by a new relationship with Myrna Weissman, a professor of epidemiology and psychiatry at Columbia University, New York. In 2005 she became Nirenberg's second wife. Nirenberg who had no biological children of his own suddenly had four stepchildren and a number of step grandchildren, a completely new and very enriching dimension of life which he did indeed enjoy.

On Veterans' Day November 11, 2009 Myrna had arranged a large reunion party at the Cosmos Club in Washington inviting some 350 people who had contributed to Nirenberg's life and career. This club occupies a precious building at Embassy Row close to Dupont Circle in the city. It is a very archaic environment and the Club excels in having as many Nobel Laureates as possible as their members. However, when Nirenberg had been offered membership he declined because the club did not admit women. Eventually this was changed, but it took a long time because of major resistance among certain of its members. Thanks to Barbara Cullington, a highly reputed science journalist and friend, who had become one of the slowly growing number of female members, I have been invited to stay overnight at the club a number of times. It is a high quality experience, in part like travelling back in time. The reunion feast was a great success and Nirenberg handed out 27 replicas in solid brass of his Nobel Prize gold medal to the collaborators who had had the largest influence on his career as a scientist. The atmosphere was very cordial but what the guests did not know was that Nirenberg had a terminal rare form of intestinal cancer. He died of this disease only two months later.

Finally, one more testimony to the richness of Nirenberg's personality may be given. Hamilton Smith already cited above, worked together with Nirenberg in a prize selection committee in Spain during the early part of this century. He described how "Marshall was one of the most reserved, unassuming, and conscientious scientists I have known, and we became great friends." Hence it was natural that when Smith, a leading scientist at the J. Craig Venter Institute in La Jolla, California, in late 2009 applied for an adjunct professorship in medicine at University of California at San Diego, he asked Nirenberg to write a letter of recommendation. When Nirenberg had died Smith was contacted by his second wife Myrna who told him not to worry about the letter. It had been written the day before Nirenberg had died.

After the successful all absorbing effort by Holley and his group to present the complete structure of a transfer RNA including the important identification of a number of nucleotides with an aberrant chemical structure Holley shifted his field of research. It remained for others to make comparative studies of various kinds of transfer RNA and to determine the crystallographic structure of this kind of molecule. The latter task appeared at first to be difficult, but eventually in 1975 Aaron Klug, Cambridge managed to produce a good model of phenylalanine transfer RNA from yeast. Similar results were obtained in competitive research by Alex Rich, MIT, Mass. When Klug became the sole recipient of the Nobel Prize in chemistry in 1982, he lectured on the structural analysis of virus particles and of chromatin[15], but he did not mention his transfer RNA work. Holley moved to La Jolla, California, in the late 1960s to become a Resident Fellow at the Salk Institute, an institution introduced in my first book on Nobel Prizes[4]. His research group was interested in the importance of serum factors for the growth of cells in cultures. Holley was instrumental in creating the Salk Institute Cancer Center founded in 1970 based on an NIH Cancer Core Grant. He himself headed this center for five years as an American Cancer Society Professor of Molecular Biology. He was also an Adjunct Professor at the University of California at San Diego.

From 1972 his laboratory was named The Molecular Biology Laboratory and the work pursued was focused on growth control of mammalian cells. He and his colleagues continued to examine the role of nutrients like serum factors and also small molecular weight substances on the growth of cells in the laboratory. He was also interested in the epithelial growth factor mentioned in Chapter 3, since it had a bearing on the development of certain kinds of cancer. The importance of these growth factors was recognized by the 1986 Nobel Prize in physiology or medicine as discussed in that chapter. Holley's laboratory encompassed a small scale operation. It was mostly managed by the technicians working with him and there was only a single graduate student over the years. Holley published two to three papers annually mostly in the journal managed by the National Academy of Sciences, of which he of course was a member. This work was phased out in the late 1980s, because of Holley's

Later in life Holley developed his talent to create miniatyre sculptures, primarily of female dancers. The statues were photographed by the author and the one to the left can be found in the library of the Salk Institute and the one to the right in the home of Richard and Nicky Lerner in La Jolla.

declining health. He, a non-smoker, sadly died of lung cancer in 1993 at the age of 71.

During the latter decades of his life Holley took the opportunity to develop his talent for making small sculptures, preferably of ballet dancers. I have seen one of his sculptures in the library at the Salk Institute and two more in the home of Richard and Nicky Lerner in La Jolla. Although made of iron they give an impression of lightness, defying gravity. In summary it may be said that Holley did not make a large number of major contributions in life, but those he did he did well.

It is now time to reflect on the immense consequences of our capacity to read and write the books of life on our insights into its existence and history. Our curiosity encourages us to attempt understanding how it all started and how it may have developed through the remarkable process of evolution. As mentioned various evolving forms of life that have prevailed at a given time of the journey of Earth have been challenged by major alterations of conditions for their existence but life has always prevailed in some form. Very late in this process a species emerged — us — who can reflect on and scientifically analyze the molecular basis of life processes and look far back in time. As will become apparent from the last chapter many advances in science have become possible because of the development of new technologies. Important insights have been gained but there will always remain new discoveries to be made. Paradoxically the basis for the pursuit of scientists is the appreciated existence of as yet undefined knowledge. The precision and refinement of this pursuit is the mass of irrefutable acts accumulated by earlier generation of scientists, a remarkably successful story, which is highlighted by the world-unique Nobel archives. The famous expression that scientists could see further because they were standing on the shoulders of giants was used already in 1675 by Isaac Newton. However, it is even older and was initially formulated already in the 12th century by Bernard of Chartres and read in Latin "nanos gigantum humeris insidentes."

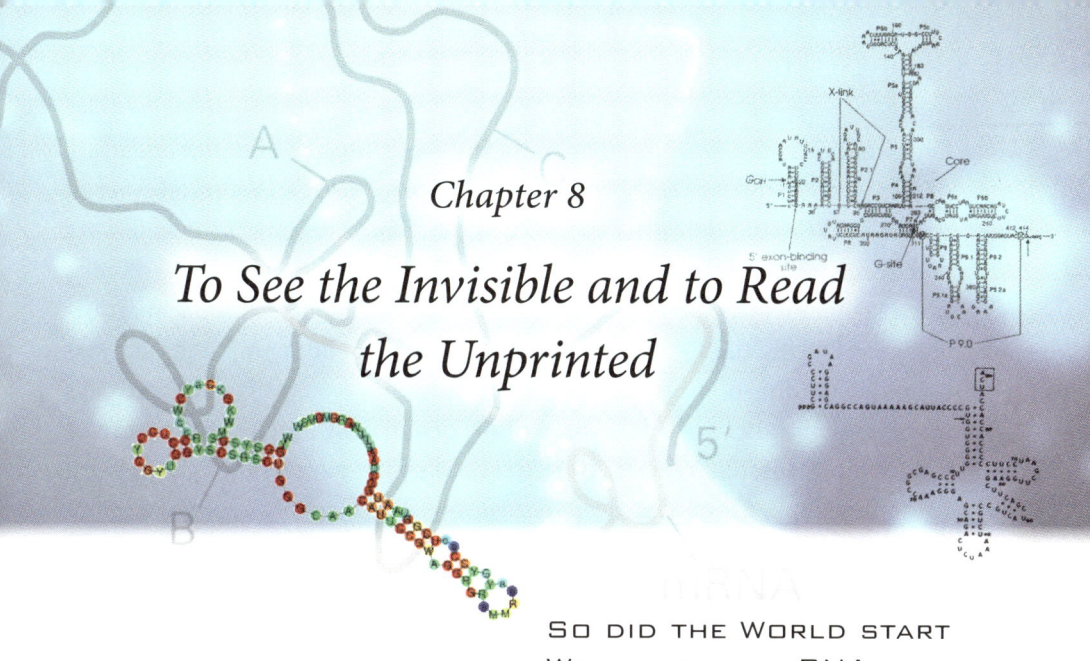

Chapter 8

To See the Invisible and to Read the Unprinted

SO DID THE WORLD START
WITH A HUMBLE RNA
NOT WITH DNA

Humans are a curious species. Already early in civilization it was noticed that water could change the angle of light. When glass was first produced some 2,000 years ago it was observed that this material too could bend light. The first lenses were produced and were given their name because of similarity in form to a bean, Latin *lentil*. It took some thousand years before the first eye glasses were produced but in between the Basra-born Arab polymath Hasan Ibn al-Haytham, Latinized to Alhazen had described the fundamentals of optics. As we have learnt in Chapters 4 and 5, what we see is light that has bounced off the objects filling our surroundings. Alhazen was a prominent representative of what is referred to as the Islamic Golden Age. He should also be mentioned because he may have been the first proponent of the scientific method in which a hypothesis is proven by experiments or by confirmatory mathematical procedures. This concept was reintroduced with force by the later Renaissance researchers launching the modern era of science-based societal developments. Galileo Galilee is considered as the pioneer in this field. He used his primitive binoculars to identify four moons circulating around Jupiter and this led to the revolutionary change from an Earth-centric to a heliocentric view of our solar system. Whereas Galileo looked outwards another scientist of the 17th century looked inwards.

It was Antonie van Leeuwenhoek, the father of the first light microscope. Because he had been appointed as chamberlain to the sheriffs of Delft,

Leeuwenhoek had become financially independent and could concentrate on his hobby, which was to grind perfect lenses. He became a forerunner of what Lewis Carroll described in his classic *Alice's Adventures in Wonderland* and *Through the Looking-Glass, and What Alice Found There*. Leeuwenhoek's microscope identified a new world encompassing single cells and microorganisms. During the 19th century developments in staining methods allowed an understanding of the general architecture of the eukaryotic, nucleated cells and eventually also towards the end of the century the much smaller form of prokaryotic cells, the bacteria. The latter derive their name from a Greek word for staff or cane, a term also used in its diminutive form for a special form of spore-forming bacteria, the bacilli. The foremost pioneer in identification of bacteria capable of causing disease was Robert Koch, who received the 1905 Nobel Prize "for his investigations and discoveries in relation to tuberculosis."

The limit of resolution of the light microscope was calculated by the German physicist Ernst Abbe and expressed in a formula. It stated that the maximal resolution was half the wavelength of the light used. Abbe was so proud of this achievement that the formula was engrained on his tombstone. However, even what is cut in stone can sometimes be challenged later. In 2014 the Nobel Prize in chemistry recognized Stefan W. Hell from Germany and two scientists from the U.S., Eric Betzig and William E. Moerner. To the surprise of the scientific community they managed to develop technologies which allowed a higher resolution than what was originally thought to be possible. The introductory speaker at the prize ceremony, Måns Ehrenberg, expressed their achievement as follows (translated from Swedish) "Hard as steel seemed Abbe's law and stern were its prescriptions, but the Norns had a different view, and fate paved the way not for just one, but for two ways to surpass Abbe's magical limit"[1]. The techniques developed used fluorescent molecules and were referred to as Photo Activated Localization Microscopy (PALM) and Stimulated Emission Depletion (STED). This advance serves to remind that dogmas can sometimes be broken by use of heterodox thinking. However let us now return to the conventional light microscope and the critical identification of bacteria as a cause of infectious diseases. Once this important advance had been made it became apparent that there remained a number of such diseases that clearly were of an infectious nature, although no bacteria could be identified. Like the "Scarlet Pimpernel" they remain elusive for some time and they were referred to as "viruses" from a Latin name for toxin.

As exemplified by Rous's studies, described in Chapter 1, in the beginning viruses were defined by use of fine filters that did not allow the passage of bacteria. Hence they were initially referred to as "ultrafiltrable agents," as repeatedly mentioned. The authority on viruses during the 1920–30s Rivers, whom we met in Chapter 1, defined their nature by three negative criteria: they were *not* retained by filters withholding bacteria, they could *not* be observed by a conventional light microscope and they did *not* have a capacity to replicate on artificial substrates that allowed the growth of bacteria. Hence their nature remained enigmatic and there was a need to visualize them. This became possible by the introduction of the electron microscope during the 1930s. This important invention was recognized by the Nobel Prize in physics in 1986 awarded to Ernst Ruska jointly with two other scientists, who had developed the so-called scanning tunneling microscope. One may ask why, as in the case of Rous, it took more than 50 years to recognize Ruska's work.

A Seemingly Analysis of the Wrong Nominee

The first nomination for a prize recognizing the use of the electron microscope was addressed to the Karolinska Institute in 1942 and submitted in French by Luigi Villa. The letter lacked date and no geographical origin was given. Villa (1896–1992) was an Italian physician, active in introducing modern physicochemical techniques into medicine in Italy. During the early 1940s he was associated with the ancient University of Pavia, where he had done his early studies. At this institution he began his efforts to modernize medicine based on the recent traditions established by the neurobiologist Camillo Golgi, the 1906 recipient (shared) of the Nobel Prize in physiology or medicine. Villa then moved to the young University of Milan, founded in 1924. Later the school of medicine at this university progressively developed to become the most influential school in this discipline in Italy. Villa made major contributions to this development. Towards the end of the war he became director of the Granelli pavilion which was added to the Medical School. He was also an associate member of the Lombardo Institute of Science in Milan, founded by Napoleon in 1797, which might have given him an opportunity to nominate a Nobel Prize candidate in physiology or medicine. Villa was very active in furthering the modernization of Italian medicine taking advantage of cutting-edge biochemical technologies. Thus he appreciated the new tools for ultrastructural analysis as well as the progressive introduction of molecular biological techniques. He was president of the Italian Society of Internal Medicine from 1958 to 1973. His opening lectures at the annual meetings of the society represent important documents of their time.

Finally the Luigi Villa Centre, also referred to as the Centre of Molecular Pathology Applied to Clinics was founded in 1969 by a formal cooperation between the University of Milan and two extramural clinical institutions in Milan. The introduction of electron microscopy in Italy at the time of the Second World War was in a building-up phase. A Siemens instrument had been made available already during the later part of the war, but it was not in operation until 1946. Villa probably had got to know about the new technology and most likely was familiar with Helmut Ruska's studies of viruses. Hence he submitted a nomination of Helmut and not Ernst Ruska together with Bodo von Borries. Translated from French to English the nomination read as follows:

> "Expressing my gratitude to the honorable Nobel Committee for Physiology or Medicine, which has sent me an invitation to propose a candidate for the Nobel Prize in physiology and medicine in 1942, I have the honor of proposing the scientists who have discovered the use of electronic rays for ultrastructural studies and have made the electron microscope, the exceptional importance of which is known to the whole world.
>
> The authors of the discovery are H. Ruska and B. v. Borries.
>
> At this moment I am incapable of supporting my proposal with writings and documents in accordance with article 8 of the statutes of the Nobel Foundation. I only want to point out that notes about the electronic microscope are widespread in world literature and do not need any documentation.
>
> With my sincerest respects,
> Luigi Villa"

The nomination was somewhat vague and emphasized the possibilities of studying sub(light)microscopic structures, presumably of organic origin. However, no examples of such applications were given. The visualization of virus particles was not mentioned, but besides a few pictures of bacteria they represented the only material of medical importance examined by the new instrument and described in the literature at the time. Henschen was selected to make a preliminary review. It is impressively detailed, but although it claims to review H. Ruska it discusses his brother with the initials E. A. F., the father of the electron microscope later recognized by the Nobel Prize in physics. The review described the events during the end of the 1920s and early 1930s and the contribution of many scientists to various steps in the early development of the instrument. This led Henschen to the conclusion that Ruska (E.) and von Borries played an important role in the development of the first electron microscope, but that there were also many other critical players. He correctly noted that the achievement of developing the electron microscope was an invention and not a discovery and he drew the logical conclusion that the advances could not be recognized by a Nobel Prize in physiology or medicine.

Virus Particles Visualized for the First Time

Ernst Ruska gave his own version of the original development of the electron microscope and its first application to studies of organic material in his Nobel lecture[2] He began his involvement in the development of this instrument in the late 1920s. It was carried out when he was a student at the Technical University in Berlin and his supervisor was Max Knoll. He focused his development work on creating two coupled electronic lenses. This led to the design of the first primitive electron microscope. This instrument showed a markedly superior resolution compared to a light microscope. His Ph.D. thesis was completed in 1933 and then he continued his work on electron optics, first at Fernseh Ltd in Berlin-Zehlendorf and four years later at Siemens-Reiniger-Werke AG in the same city. At Siemens he fathered the first commercially available electron microscope, at an early stage often referred to as the Ruska microscope.

Max Knoll (1897–1969) and Ernst Ruska (1906–1988), the fathers of the electron microscope. [From Ref. 24.]

In the early planning of objects to examine by the new instrument it was decided to include organic material. Such material does not by itself provide enough contrast to the electron rays. It has to be fixed, for example by formalin treatment, and stained with some heavy metal salt. It was mentioned in the lecture that already by use of the early prototype model two budding scientists, Heinz Otto Müller (a student of electro-technical engineering) and Friedrich Krause (a medical student), studied the wing structure of the housefly, diatoms (microscopic calcium-containing green algae) and even bacteria. Sadly these two gifted students did not get an opportunity to fully develop their studies, since they were both killed in the Second World War.

Toward the end of the 1930s E. Ruska managed to raise resources which allowed him to convince the Siemens Company to set up a laboratory for visiting scientists to study biological material. From the beginning it was led by the already mentioned Ernst's brother Helmut, who was a young medical doctor at the time. He had a clinical tutor Professor Siebeck, who used his influence to launch the project and ensure that Helmut could start a research career. The goal was to use the new instrument for medical and biological applications. The choice of objects for the initial studies was limited. Complex material like intact cells remained for studies after the war when techniques to fix and stain them with electron dense material had been developed and in addition very thin slices of the fixed material could be prepared by use of a particular cutting instrument, the ultramicrotome. The decision was wisely taken to examine different kinds of materials expected to contain viruses. Siebeck mentioned a number of important viruses to study and he also included agents possibly important for virus-induced tumors — "like Rous's sarcoma, chicken leukemia, human lymphogranulomatosis." He also mentioned "the nature of phage (bacterial viruses, my remark)" as a potentially promising area of research.

The first publication demonstrating the structure of viruses appeared in 1938 and had both brothers and von Borries, who by the way had just become their brother-in-law, as authors. It demonstrated the size and structure of two relatively large and brick-shaped, related viruses, ectromelia, a poxvirus from mice, and vaccinia, the virus used to immunize against smallpox. This was the first time that virus particles had been convincingly identified! A year later there was an important publication describing for the first time the elongated structure of TMV. As already mentioned this virus was the first candidate ultrafiltrable infections agent identified and it came to play a major role in the historical development of knowledge about virus structure and biochemistry. It was the object of Wendell Stanley's studies that earned him a shared Nobel Prize in chemistry in 1946[3]. The importance of the recognition of the structure of TMV by electron microscopy was emphasized in the review of five virologists by Svedberg described in Chapter 1. In the pioneering publication on the structure of TMV the brother Ernst was *not* a co-worker, but both brothers contributed to an article on "The significance of electron microscopy for virus research" (translated from German) published in *Archiv für die Gesamte Virusforschung* (incidentally the first journal established to focus on virus research) half a year later. Another pioneering publication by Helmut alone appeared in 1940 describing for the first time the tadpole structure of bacteriophages. Two years later the relatively larger and rounded structure of varicella/zoster virus was published, but this information may not have been available at the time of Villa's nomination in January of 1942. In 1960 a major improvement in means of visualizing virus particles was introduced. This was the previously discussed[3] negative contrast technique examining virus particles suspended in an electron dense solution instead of attaching the stain to them.

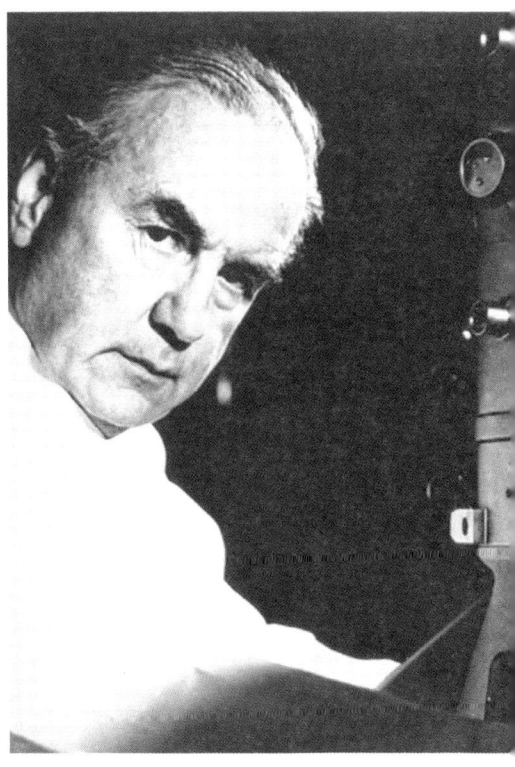

Helmut Ruska (1908–1973), a pioneer in demonstrating virus particles by the electron microscope.

Ruska's Nobel lecture[2] gave a fascinating description of how, after many years of development work, including many collaborators, not least his mentor Knoll and his fellow graduate student von Borries, he managed to produce a commercially available microscope at Siemens in 1938. Amazingly Siemens managed to continue producing a series of electron microscopes in the midst of the war in Berlin, a total of 40 before the war ended. In the autumn of 1943 instrument number 26 was delivered to Arne Tiselius in Uppsala! In part it was paid for by the Swedish Government with some support from the Gustaf V's 80-years fund. This fund was established in 1938 and had the goal of supporting research in two major disabling diseases affecting the public at large, poliomyelitis and rheumatoid arthritis. However before the instrument was delivered both Tiselius and his close collaborator Sven Gard had visited the center for ultrastructural studies established in Berlin and even performed experiments with the instruments available locally. At the time when they were pursuing their studies in Berlin Helmut was serving as a medical officer of the Wehrmacht for nine months somewhere near Moscow.

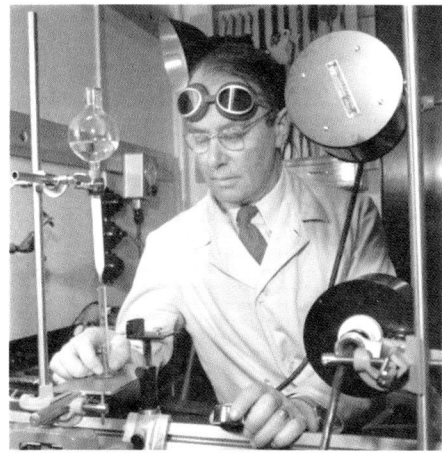 

Arne Tiselius (1902–1971) and Sven Gard (1905–1998).

Gard was my predecessor as professor at the Karolinska Institute. We have met him repeatedly in all my books on Nobel Prizes[3-5]. Tiselius and Gard were interested in the structure of the medically very important virus causing polio at the time. The subject of Gard's Ph.D. thesis was purification of the virus from the brains of infected mice. They published on the observation by use of the microscope of elongated particles in the preparations. Later these structures were found to represent contaminating material of cellular or possibly microbial origin. The proper spherical shape of the small virus was not identified until the late 1940s.

Helmut Ruska's description of the size and morphology of virus particles represented an important milestone in the development of the young discipline of virology. It made a major difference that virus particles could be observed and that it was discovered that they show major variations in size and shape. It is not unlikely that it was this kind of discovery that Villa wanted to highlight by his 1942 nomination of H. Ruska and *not* the development of the electron microscope as an instrument that Henschen assumed was his task to review, and which he shouldered in a qualified way. However, we will never know this for sure. The problem was of course that Villa's nomination was incomplete and allowed the misinterpretation of initials. One may also ask why he proposed to include von Borries in the nomination since he was a co-author only on two of the five publications, also including Ernst Ruska, describing different virus structures. So in retrospect, was Helmut Ruska's contribution of the quality that it could have been recognized by a Nobel Prize in physiology or medicine? It was after all the first visualization of a complex biological structure of potential relevance to medicine. In spite of this the answer to this question is no.

Although the identification of the particle nature of viruses was important it was not a discovery of the order of magnitude that should motivate a serious consideration for a prize. Insights into correlations of structure and function might have added to the strength of the nomination. Still there is no doubt that Helmut pioneered a completely new field of research. It should be added that he continued his involvement in ultrastructural analysis of viruses and microorganisms throughout the war and after the defeat of Germany. He presented his thesis in 1943 and was appointed lecturer in medicine at Berlin University the same year. Already at this time he made the proposal that viruses should be classified based on morphological criteria. This was very insightful but it was not until 1962 that such a virus classification system was introduced.

In his follow-up work he focused in particular on bacterial viruses, the bacteriophages. In 1948 Helmut was appointed professor at Berlin University (renamed Humboldt University in 1949). In 1952 he applied for work in the U.S. He used the U.S. order No. 63 which stated that individual scientists of German origin with exceptional skill in science and technology should be used for development of American science. One example of an individual included in this category was the aerospace engineer Wernher von Braun. It should be added that the person considered should not be incriminated in war crimes. The Ruska brothers were not tainted by any affiliations with the Nazi movement. Helmut was well received in the U.S. He led a research group for six years at the New York Department of Health in Albany and had many contacts with other laboratories involved in the use of the electron microscope. Afterwards he returned to Germany to become Director of Biophysics at the University of Düsseldorf and also became involved in university administration. He died in 1973.

Construction of an Electron Microscope

The life of the brothers Ruska and their importance for the development of ultrastructural studies has been described in detail[6,7] and the whole history of the development of the electron microscope also has been presented in a recent comprehensive review[8]. The latter review brings up some interesting comments on Léo Szilárd, the larger-than-life physicist who among many involvements also contributed a lot to Monod's and Jacob's work on gene regulation. Because of this, Szilárd, a highly original individual, was briefly introduced in my previous book on Nobel Prizes[5]. It turns out that he was one of the first to conceptualize the potential of possibly developing what later became the electron microscope. In 1928 when Hans Busch had introduced the concepts of geometric electron optics, but without reflecting on applications, Szilárd's dynamic mind started to spin. A friend of his, Dennis Gabor, another uniquely talented physicist originating in Budapest, Nobel Prize recipient in physics in 1971 "for his invention and development of the holographic method," gave the following historical reflections in 1974 apropos the origin of the electron microscope:

> "There were people with more imagination than Hans Busch, in particular my friend Léo Szilárd, probably one of the brightest inventive minds that ever existed. It was from him that I first heard the words 'electron microscope'. I think I remember a conversation which I had with him in 1928, in the Café Wien, in Berlin, almost verbatim. Szilárd: 'Busch has shown that one can make electron lenses, de Broglie has shown that they have sub-Ångström wavelengths. Why don't you make an electron microscope, one could see atoms with it (…)"

Szilárd did in fact apply for a patent in 1931 for the construction of an electron microscope, including some highly imaginative ideas, but at that time Siemens had just cornered the market by submitting applications for two critical patents in the field. And apropos Gabor and his invention there was a hologram depiction of Alfred Nobel in the offices of the Nobel Committee at the Karolinska Institute in the 1970s and 1980s. It gave an impressively realistic three-dimensional presentation of the famous donor, but if you looked at it at an angle Nobel appeared like Woody Woodpecker!

In 1965 it seems that a coordinated effort was made to reawaken E. Ruska's and possibly also his collaborators' candidacy for a prize in physiology or medicine. There were no fewer than seven nominations of him, all from German scientists. Two of them were for E. Ruska alone, one of them was for him and his early collaborators and in the remaining proposals, other strong candidates were mentioned in parallel. Thus in one nomination the proposal included J. H. Matthaei and Marshall Nirenberg, discussed at length in Chapter 6. In another nomination parallel proposals were given for the German physiologist Richard Wagner and also for Gerhard Schramm. The latter scientist was one of the discoverers of the existence of infectious RNA in TMV. This important finding came very close to becoming recognized by a Nobel Prize as discussed in my previous books[3,4]. The nomination of E. Ruska which included the already mentioned collaborators also proposed B. von Borries (surprising since he had died in 1956) and Ernst's brother Helmut, but in addition Manfred von Ardenne. At the time of developments of the electron microscope by Siemens in the late 1930s there were two other competing groups in the area. One of them was a private laboratory led by von Ardenne and he did in fact pioneer the development of the first scanning transmission electron microscope. The committee decided to let the professor of medical physics Arne Engström do a preliminary review.

Engström started by referring to the nomination in 1942 of von Borries and H. Ruska and the preliminary review by Henschen. He traced the source of Henschen's rather deep insights into the field of electron microscopy. It was a document from the University of Heidelberg to be used for cultural exchange (Nazi propaganda) published the same year. Engström cited two critical paragraphs from Henschen's review. One concerned whether this so-called "electron optical apparatus" could be considered at all for a Nobel Prize and the other raised the question of priorities to the "discovery/invention." Engström himself used another source to evaluate priorities. It was the appendix "Industria" to *Frankfurther Zeitung* of February 1943. In summary it described that the trajectory of the electron beam was first calculated by Hans Busch in 1927 and further developed by two other engineers. However, they did not draw any practical conclusions from their work. E. Ruska's contributions were presented in the following way (translated from German):

> "Time was now ripe to build the electron optical design that in its most beautiful application was used in the development of the electron microscope. The 'magnetic electron optics', which was mainly developed by E. Ruska in the High Voltage laboratory of the Technical High School of Berlin (equivalent to university, my remark), later in collaboration with B. von Borries at Siemen-Werke leading to a development of the (first) magnetic Siemens Electron Microscope; the 'electron optics', the care of which was established at the Research Institute of the Company in its primary line of priorities, having its origin in contributions by H. Mah, led to the development of the electron microscope."

Engström then emphasized that the performance and use of the electron microscope to a large extent had superseded the expectations humbly expressed by Henschen. He wrote (translated from Swedish):

"From a general point of view the construction of the electron microscope represents one of the most important instrumental additions to the fields of biological and medical research and has to a major extent contributed to their powerful expansion during the latter decades. Almost in all disciplines of biomedical research electron microscopic observations has had a decisive influence."

He exemplified this by reference to virus structures, to cellular structures and to molecular biology. It was emphasized that the development of this very important instrument should be recognized by a Nobel Prize. He then hesitated and emphasized that a lot remained to be understood to fully clarify priorities for the development of the electron microscope and that this extended examination did not belong within the realms of the committee for physiology or medicine. No further review was made and the discovery/invention was concluded again not to be a responsibility of the latter committee.

One may wonder if Gard, who was a member of the committee at the time, made some comments referring to his experiences in wartime Berlin of the developments of the electron microscope. In addition he most likely should have been informed about the fact that von Borries had died already in 1956. It is surprising that neither the nominator nor the reviewer had noticed this fact. In 1968 Ernst Ruska was again proposed by three German nominators and one of the nominations also included his brother Helmut and mentioned the importance of the electron microscope for the study of viruses. Not until 2037 will it be possible to get a full overview of the nominations for a prize in physics or in physiology or medicine for the development of the electron microscope, an invention at last recognized by the joint Nobel Prize in physics including Ernst Ruska in 1986.

Interest in constructing an electron microscope during the 1940s could also be recorded within Sweden. One of the pioneers in this field was Fritiof S. Sjöstrand. He got to know about the developments in the field in 1938 and established collaboration with the Swedish 1925 Nobel Prize in physics (the prize for 1924) recipient K. Manne G. Siegbahn. The microscope they jointly attempted to construct was not ready to be employed in Sjöstrand's Ph.D. presentation, which instead therefore concerned fluorescent microscope studies of auto-fluorescence of cells fixed under certain conditions and was presented in 1944. The year before Sjöstrand made an important contribution

Manne Siegbahn (1886–1978), the recipient of the Nobel Prize in physics in 1924 and Fritiof Sjöstrand (1912–2011).

to the development of the technique of cutting thin sections of tissues, which could also be used for electron microscopy. His further training continued at MIT in 1947–48. During 1949–1960 he was an associate professor of anatomy (prosector) and hence he was a teacher in this discipline when I started my medical studies in 1956. Since he did not become professor and chairman at the Institute he eventually settled in 1959 at the University of California at Los Angeles where he built an influential group for electron microscopy studies. Two years before that he had founded the *Journal of Ultrastructure Research*. He was to become a good friend of Helmut Ruska. It is interesting that pioneering electron microscopy research in Sweden was initiated by Sjöstrand in collaboration with the Nobel Prize recipient Siegbahn and by Gard in collaboration with the forthcoming (1948) Nobel Prize recipient Tiselius. Apparently Sjöstrand's contributions were very fundamental, since in 1968 he was even nominated for a Nobel Prize in physiology or medicine by a Dr R. K. Mishra from the All India Institute of Medical Sciences, New Delhi. This nomination also included the Venezuelan scientist H. Fernándes-Moran, at the time professor of biophysics at the University of Chicago. The latter scientist in fact had his original training in Sweden working with the professor of neurosurgery Herbert Olivecrona and in particular the already mentioned Swedish inventor of an electron microscope Siegbahn.

Possibilities to Examine the Structure of Cells

The later developments of ultrastructural studies of cells are illustrated by the 1974 Nobel Prize in physiology or medicine to a set of scientists active at the Rockefeller Institute, Albert Claude, Christian de Duve and George E. Palade. They developed techniques to separate various populations of cellular components with different functions and could identify their organized presence in different compartments of cells. The prize motivation was "for their discoveries concerning the structural and functional organization of the cell." In fact the Nobel lecture[9] given by Claude, of Belgian origin, revealed an interesting connection to studies of Rous virus that we got to know in Chapter 1. It was formulated in the following way:

> "That Friday, the 13th of September 1929, when I sailed from Antwerp on the fast liner 'Arabic' for an eleven-day voyage to the United States, I knew exactly what I was going to do. I had mailed beforehand to Dr Simon Flexner, Director of the Rockefeller Institute, my own research program, handwritten, in poor English, and it had been accepted. My proposition had been to isolate and determine by chemical and biochemical means the constitution of the Rous chicken Tumor I 'Agent', at the time still controversial in its nature and not yet recognized as a bonafide virus. This task occupied me for about five years (apparently without any critical findings, my remark). Two short years later the microsomes, basophilic components of the cell ground substance, had settled in my test tubes, still a structure-less jelly, but now captive in our hands."

This is a very characteristic presentation of how science often develops in the hands of an imaginative scientist. The premature, failed attempt to determine the structure and chemical composition of Rous virus, clarified only some 30 years later, led to pioneering discoveries about one of the fundamental components of cells, the microsomes, as they were originally referred to. Interestingly Claude's work was performed in James Murphy's laboratory at the Rockefeller Institute. In contrast to Rous, Murphy was not enthusiastic about the possible role of viruses in cancer as referred to in Chapter 1. He represented the opinion that chemical or environmental insults were the critical factors. Claude's work depended on the newly available centrifuges at the time, the ultracentrifuges.

To study the elusive virus in the cancer cells he prepared four major fractions of the cell material by centrifugation at different speeds. It soon became apparent that in order to interpret the results found in studies of cancer cells it was important to first examine normal cells. It was in these studies of his "control" cells that Claude made pioneering discoveries. Of critical importance in these forays into analysis of the composition of normal cells were possibilities to examine the intact cells and fractions of them, separated by ultracentrifugation, by an electron microscope offered by the Interchemical Corporation in New York. For the first time it became possible to map out the architecture of cells and fulfill a criterion later emphasized by Richard Feynman, the 1965 Nobel laureate in physics, namely "only what I can see I can understand."

In 1946, about a year after Claude's demonstration of the first picture of the ultrastructure of components of a cell, Palade was invited to join in the work at the Rockefeller Institute. In the presentation of his scientific career[10] Palade described how he and Keith Porter in the late 1940s focused on the use of electron microcopy. He noted "Porter and I worked out enough improvements in microtomy and tissue fixation to obtain preparations which, at least for a while, appeared satisfactory and gratifying. A period of intense activity and great excitement followed since the new layer of biological structure revealed by the electron microscope proved to be unexpectedly rich and surprisingly uniform for practically all eukaryotic cells. Singly, or in collaboration with others, I did my share in exploring the newly open territory and, in the process, I defined the fine structure of mitochondria, and described the small particle component of the cytoplasm (later called ribosomes, my remark)…" The latter structure was what Claude originally referred to as microsomes. It was much later to be identified as the organelle in which messenger-RNA was translated into proteins.

In 1962 de Duve was appointed professor at the Rockefeller Institute, joining Palade in the pioneering studies of the ultrastructure of the cell. De Duve discovered another organelle in the cell, the lysosome, a membrane-bound vesicle with a responsibility to digest waste material in cells. Since then our insights into the architecture of cells has developed extensively and for example the identification of different transport mechanisms in cells have been recognized by Nobel Prizes in physiology or medicine. There will be reasons to return to this when the complete archives for the 1974 Nobel Prize have become available. In addition there are recent findings demonstrating that compartmentalizing in cells need not be dependent on the use of membrane structures. So-called phase transitions are common in non-living matter. It has now been discovered that liquid-liquid phase separation forming condensed liquid-like droplets may appear and represent non-membrane enclosed structure of physiological importance in cells. This may markedly change our view of the topography of cells. However it is now high time to return to the fundamental development of the electron microscope by the group led by Ernst Ruska.

The Lukewarm Reception of Ernst Ruska by Physics Committees

The first nominations for a Nobel Prize in Physics to Ernst Ruska for his contribution to the development of the electron microscope were submitted in 1955. The physics committee did not call for a special review. In fact it should be added that the policy of using preliminary or complete reviews varies considerably between committees for the different prizes in the natural sciences. committees responsible for the prize in physics over time have used the instrument of preliminary and extensive reviews much less frequently than the other two committees. Thus in this year the committee simply noted in its overall review that there were too many potential candidates for a prize highlighting the development of the electron microscope to allow separating out Ruska as a single strong candidate. Similar statements were made over the years as more nominations for Ruska were submitted; three proposals in 1957; one each in the years 1960, 1961, 1965 and 1966. The committees remained lukewarm and wrote for example in 1965 (translated from Swedish):

> "In connection with a suggestion to recognize the inventor of the electron microscope, Ruska has been mentioned a number of times since 1955. However, already at that time it was noted that the supporting material of particular relevance to the invention of this important experimental tool include a list of names and that Ruska does not (sic!, my comment) take such a position among them that a reward with a Nobel Prize should be considered appropriate."

Although there was no nomination of Ernst Ruska in 1967 there was one for Knoll. The nomination was by G. Möllerstadt from Tübingen in Germany. However, it did not concern the transmission electron microscope, but instead the development of the first scanning microscope. Knoll published two articles in 1935 and 1939, not in association with Ruska, on this new microscope principle. In the literature this variant of the electron microscope is credited as a 1937 invention by the already mentioned Manfred von Ardenne. The committee treated this as another nomination of a prize for the development of the transmission electron microscopy and withheld its previous lukewarm attitude. A reference was made to the existence of other potential candidates for a prize in this field and also to the fact that the principle of using rays of electrons for high resolution microscopy had been published by others already in the mid 1920s. As in 1967 there was no renewed nomination of Ruska in 1968. Future archive material studies will reveal when his candidacy was reactivated and upgraded. Apparently committees responsible for the prize in later years changed their perspective on Ruska's contributions since twenty years later he finally did receive his prize.

In summary it can be noted that it helps to be long-lived. Ruska was 80 years old when he received his shared prize and he was recognized for work done 50 years earlier! He died only two years later. A comparison can be made with Rous, who was also recognized for a discovery made more than 50 years earlier, when in 1966 he received his prize at the age of 87, as discussed in Chapter 2.

The Three-Dimensional Structure of Complex and Aggregated Macromolecules

As presented in one of my preceding book on Nobel Prizes[4] the development of crystallographic methods allowed an examination of the detailed molecular structure of even highly complex organic molecules. This technique has been refined with time and eventually it became possible to examine very complex molecular aggregates, like the machineries for transcription and for translation. The leading crystallographers involved in this work has been recognized by Nobel Prizes in chemistry. In 2006 Roger Kornberg, the son of the 1959 Nobel Prize

Roger Kornberg, the Nobel Prize recipient in Chemistry in 2006. [From *Les Prix Nobel en 2006*.]

recipient Arthur Kornberg, received a prize "for his studies of the molecular bases of eukaryotic transcription." The molecular aggregates he examined by crystallography comprised six components of the transcription complex in the baker's yeast cell system; the central polymerase II and five additional transcription factors. In a way it is remarkable that he was selected to be a single recipient of the prize. There were many discoveries that led to a detailed understanding of the transcription mechanisms, as for example Robert G. Roeder's discovery of different kinds of RNA polymerases as well as other findings of importance in this context.

A few years after the prize for Kornberg, the unraveling of the three-dimensional structure of the highly complex translation machinery represented by the ribosome was also recognized by a Nobel Prize in chemistry. It was in 2009 that the prize recognized Venkatraman Ramakrishnan, Thomas A. Steitz and Ada E. Yonath "for studies of the structure and function of the ribosome." The ribosome is a very complex structure containing roughly two-thirds RNA and one-third proteins and there are certain differences in the composition and relative representation of nucleic acid and protein components in prokaryotic and eukaryotic ribosomal structures. Thus, for example, in bacteria the smaller 30S subunit is composed of 21 proteins and RNA in the amount of somewhat more than 1,500 nucleotides, whereas the larger 50S subunit is built up of some 34 proteins and RNA to the amount of almost 3,000 nucleotides. In the eukaryote the corresponding values are for the 40S subunit 33 proteins and RNA in the amount of 1,900 nucleotides and

Venkatraman Ramakrishnan, Thomas A. Steitz and Ada E. Yonath, the Nobel Prize recipients in chemistry in 2009. [From *Les Prix Nobel en 2009*.]

the larger 60S subunit 49 proteins and a larger piece of RNA including 4,700 nucleotides and in addition a smaller piece with an additional 160 nucleotides. As already emphasized it has been found that it is the RNA components that have a central position in the structure and carry the main responsibility for the metabolic functions by which the polypeptide chain is built up using the specific interaction between the messenger RNA and the amino acids carried to their specific site of inclusion in the protein by transfer RNA. The ribosome is capable of carrying out a translation of messenger RNA into new proteins even if the major parts of the proteins associated with the ribosome structure have been removed[11].

At the time of writing in October 2017 another prize for the development of electron microscopy was awarded. It was a prize in chemistry to Jacques Dubochet, Joachim Frank and Richard Henderson "for developing cryoelectron microscopy for the high resolution structure determination of biomolecules in solution." They had introduced major technical improvements of different kinds resulting in possibilities of achieving a resolution down to the atomic level, something that could only be dreamt of earlier. It was very appropriate to recognize this with a Nobel Prize in chemistry, since in the will this is specified to recognize "a discovery or an improvement." One of the impressive technological developments was the freezing of small samples in such a way that ice crystals do not develop, but instead so-called vitrous glass, like in a window, is formed and surrounds the molecules. The first publication of application of this technique to the study of viruses appeared in 1984 with Marc Adrian as the first author and Dubochet as the second. Since Adrian died in 2013 it was Dubochet who was privileged to catch the limelight.

Other technical advances, also used for studies of viruses, have exploited the possibility of using advanced computer programs to orient randomly tilted members of a population of macromolecules so that a homogenous picture can be deduced. By this technique, referred to as *tomography*, like in computer-assisted tomography used in clinical medicine, virus particles could be reconstructed three-dimensionally. A similar approach had been taken earlier by Aaron Klug, the 1982 Nobel Prize recipient in chemistry, as described previously[4]. He had developed "crystallographic electron microscopy" for studies of nucleic acid-protein complexes, including viruses.

Altogether the different technical advances recognized by the chemistry prize in 2017 allow the identification of individual atoms at a resolution of a few Ångströms. This unit length, 10^{-10} of a meter, was introduced by a Swedish physicist and astronomer Anders Ångström (1814–1874) in his work with spectroscopic analysis. The brothers Ruska presumably would have been very surprised to learn that the instrument Ernst and collaborators had developed eventually would allow studies of organic molecules in their smallest atomic details, whereas originally they could only see a rough outline of whole virus particles. These developments exemplify the fact that the remarkable advance in science to a considerable extent is dependent on technological innovations.

Anders J. Ångström (1814–1874).

The Evolution of the Genetic Language Used Since the Dawn of Life

We will now make a dramatic change in perspective. Moving from the world of visible structures we will focus on how organic molecules can serve to store and transmit information. This will take us back to the very origin of life when simple carbon-dependent molecules managed to establish a primitive form of self-replicating systems. It was the small failures of these simple self-replicating systems that allowed a progression into increasingly more complex structures. Embryonic life forms started to compete for their existence with other similar forms. The winner in this competition was the one that was most successful in evolving a genetic language, to store information to be transmitted from one generation of self-replicating molecules to the next. Our capacity to explore the books of life has opened immense realms of insight into the tinkering of evolution and there is much more to come. In the following we will take stock of where this remarkable growth in knowledge has taken us.

In 1966 the Nobel laureate of 1958 George Beadle and his wife Muriel wrote: "The deciphering of the DNA code has revealed our possession of a language much older than the hieroglyphs, a language as old as life itself, a language that is the most living language of all — even if its letters are invisible and its words are buried deep in the cells of our bodies." This poetic formulation emphasizes the remarkable insight that the understanding of the language of life represents and all the opportunities it offers to both read and write the books of life. It can justifiably be stated that no other discovery has provided such opportunities to develop the life sciences. The advances in the field of molecular genetics are overwhelming and new fundamental insights into the dance of genes offer an unlimited and rapidly growing understanding of all forms of life. A lot has been learned and still there appears to be an almost unlimited journey ahead of us.

There are many ways in which the genetic code can be depicted. Very often the nucleotide base triplets of the code, the *codons*, are presented within a box with three sides defining the first, the second and the third RNA base constituting them (see pp. 261, 316). This arrangement was first recommended by Crick. However, for simplicity the additional picture below lists the twenty

	AGA						GGA			
	AGG						GGC		AUA	
GCA	CGA						GGG		AUA	
GCC	CGC						GGG	CAC	AUC	
GCG	CGG	GAC	AAC	UGC	GAA	CAA	GGG	CAC	AUC	
GCU	CGU	GAU	AAU	UGU	GAG	CAG	GGU	CAU	AUU	
Ala	Arg	Asp	Asn	Cys	Glu	Gln	Gly	His	Ile	
A	R	D	N	C	E	Q	G	H	I	
UUA					AGC					
UUG					AGU					
CUA				CCA	UCA	ACA			GUA	
CUC				CCC	UCC	ACC			GUC	UAA
CUG	AAA		UUC	CCG	UCG	ACG		UAC	GUG	UAG
CUU	AAG	AUG	UUU	CCU	UCU	ACU	UGG	UAU	GUU	UGA
Leu	Lys	Met	Phe	Pro	Ser	Thr	Trp	Tyr	Val	stop
L	K	M	F	P	S	T	W	Y	V	

The specification of amino acids by the genetic code.

amino acids and all the triplets that determine their position. As can be seen there are eight amino acids, which in principle are determined only by the first two nucleotides of the codon. In the third position either one of the four bases can be used. The fact that the third position seems to have a relatively lesser importance presumably reflects the very early evolutionary origin of the code. It is referred to as the "wobble" phenomenon, a term introduced by Crick. In these cases a change of the third base will not cause a mutational change in the protein produced. The assignment of codons to different amino acids is not random. Chemically similar amino acids use codons that are related. As a consequence this means that an error in the first letter would lead to incorporation of an incorrect but similar amino acid into the polypeptide chain, potentially lessening the damage. This association emphasizes a somewhat larger relative importance of the nucleotide in the second than in the first position in a codon. One may ask why a certain codon has been selected for a particular amino acid. Since evolution by definition is non-deterministic, no good answer can be given to this question. It is similarly meaningless to ask if the relative robustness of the code is optimized by the fortuitous evolutionary events. It has been calculated that the number of potential codes is enormous because of the large number of variables involved. Hence clearly there are a huge number of variants that would be expected to be even more robust than the one that eventually was selected. Still it apparently serves its purpose well since it has allowed a continuous development of new forms of life over more than three billion years under dramatically changing environmental conditions.

There are three codons at the farthest right of the picture which do not code for any amino acid. They are *stop* codons, marking the end of the polypeptide synthesis. The first one was named by the original discoverers, Richard Epstein and Charles Steinberg. These scientists had a friend by the name of Harris Bernstein, whose family name is German for *amber*. This was the name given to the first stop codon discovered. When the second one was identified by Brenner he called it *ochre*, in the spirit of color naming. The third one discovered after this was called *opal*. Interestingly there is no separate codon to signal start. In most cases it is the same as the single codon used to define the amino acid methionine, AUG, as already mentioned. In most cases the methionine is cleaved off in the final product. It does not suffice to have methionine as an isolated signal to initiate synthesis of a specific protein. The initiation of expression of a certain gene is a complex matter depending on a number of signals.

It is beyond the scope of this presentation to discuss the way in which the production of a specific messenger-RNA from a stretch of DNA, referred to as an open reading frame (ORF), is initiated and controlled. To manage this involves many controlling signals elicited from sites in nearby DNA but also at some distance. Thus a gene includes segments of DNA at different locations in the genome. We will return to this question later in a discussion of how a gene can be defined. Clearly it is critical when and where in time and space a particular reading frame is made available for transcription. The pattern of expression varies extensively in the developing organism. In the mature animal it also varies between cells representing different specialized tissues. However, all cells retain the original complete and unchanged genome. The only exception is certain cells of the lymphatic system. An essentially unlimited capacity to interact with foreign antigens is generated by randomized reshuffling of selected fragments of genes in cells serving different immunologically specific functions in the lymphatic system[4].

Since there are two strands of DNA and a single stretch of this molecule can be read in three different frames, a certain stretch of double-helical DNA potentially can give rise to six different proteins. The emergence of such potential products need not be mutually exclusive. In particular in viruses and in bacteria one sometimes finds that overlapping reading frames are being used. The drawback of such an arrangement is that a single mutagenic change may result in two different potentially defective products. However, the world of viral reproduction appears to be relatively promiscuous and therefore may live with this limitation. The error rate of their nucleic acid replication is higher than in a cell and the final product is genetically heterogeneous. Hence the virus product has been referred to by Eigen (p. 216) to represent a *quasi-species*.

In principle the same genetic language is used in all forms of life since it evolved in the very early phase of development of self-replicating systems using a separate storage of information in selected molecules. It is of course tempting to speculate about how the standard genetic code has evolved. This needs to be reflected on in the perspective of theories on how life may have originated on Earth. The postulated primary role of RNA in this context will be presented below and it is only after appreciating this that the relatively secondary role of introduction of proteins has been deduced. It is the development of the interplay between these two original central actors in the origin of life that has accidentally come to define the code and also progressively allowed the balance in allocation of different responsibilities for functions between them.

Originally Nirenberg was unwilling to believe the generality of the code he and his collaborators had discovered. However, as data accumulated only a few minor variations were identified. All in all the relations found provide solid evidence that all forms of life on Earth have the same ancestry. The first minor variation discovered was found in mitochondrial DNA. It was mentioned previously[5] that Sanger in his early application of the DNA sequencing technique he invented found a unique situation in mitochondria. In this vestigial trace of an old bacterial world UGA did not represent a stop codon, but instead coded for tryptophan. Since then some 23 nonstandard

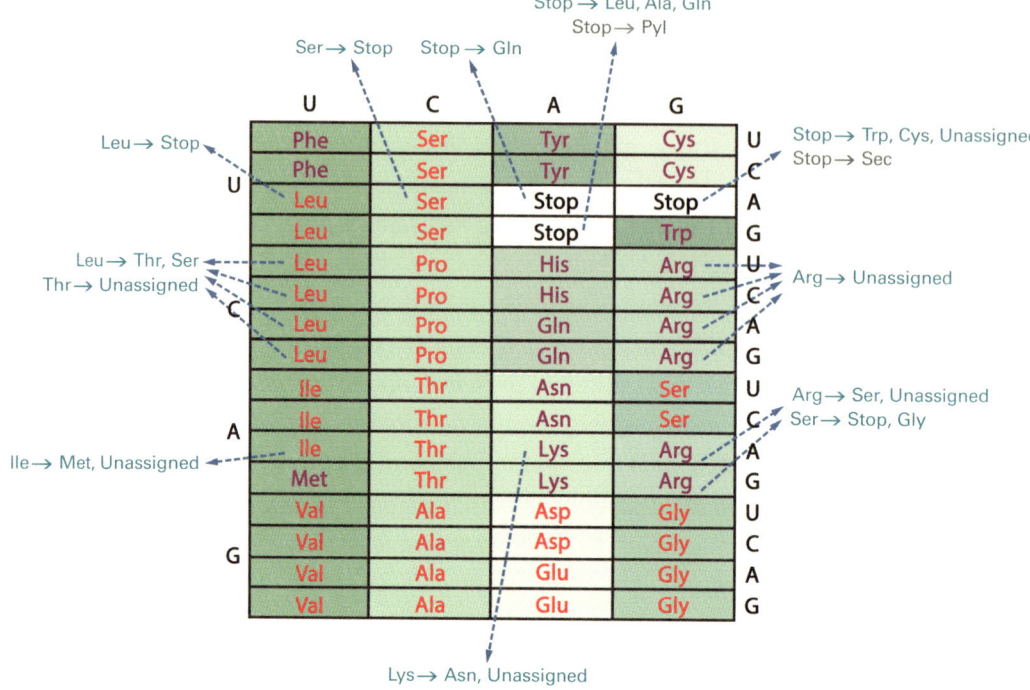

Deviations of code word use. [From Ref. 12.]

codes have been identified[12, 13]. As shown in the figure there are in summary: 8 cases of reassigned or acquired stop codons; 8 cases of codon loss and ten reassignments of codons from one amino acid to another. A situation similar to that in mitochondria has been found in the small intracellular microorganism mycoplasma. The latter observation has relevance for studies of the minimal cell for which DNA can be synthesized in the laboratory. The value of examining a synthetic microbial genome, as uniquely pursued at the J. Craig Venter Institute, has already been introduced[5]. Taking into consideration the unique codon used in the selected mycoplasma organism it has now become possible to synthesize the approximately 70 proteins for which as yet no functions have been identified. The access to these proteins will be of considerable help in finally defining their specific operative responsibility.

It also should be mentioned that two examples of the use of non-canonical amino acids have been found in nature. These are the amino acids selenocysteine and pyrrolysine and there are different mechanisms for accommodating their incorporation. To this can be added that it has been possible to experimentally alter the code in bacteria to accommodate additional atypical amino acids[14] beyond the 20 generally used. It remains to be proven if this approach may open up new biotechnological advances towards solving particular problems.

As presented above Holley and Khorana in their Nobel lectures mentioned that at the time 12 different transfer RNAs had been chemically characterized, 10 more in addition to the alanine transfer RNA originally determined by Holley and collaborators and the phenyl alanine transfer RNA soon thereafter determined by Khorana's group. This immediately raised the question about how many forms of transfer RNA may exist and the associated question of how they are formed. In eukaryotes they are synthesized in the nucleus by an RNA polymerase. Their original size is larger than the final product and we will return to the significance of the final trimming that leads to the functional molecules that are present in the cytoplasm. One special characteristic of transfer RNA is that about one in ten nucleotides are chemically modified. In total as many as 50 different types of nucleotide modifications have been described and they all have a functional importance in the context in which they appear. The uniqueness of individual populations of transfer RNA was already introduced above.

Particularly challenging appears to be the evolution of transfer RNA with alternating, but correlated specificities at the two ends of the molecule. In fact it must be the overall shape/charge of a certain transfer RNA which decides which amino acid it will specifically associate with. Once this has been structurally defined the critical enzyme, the aminoacyl-tRNA synthetase, comes into action and attaches the selected amino acid to the specific site of the molecule, which in its other end carries the critical anti-codon. One can speculate on the early evolution of interactions between the RNA with independent replicative capacity and various amino acids eventually finding a relation between information storage and expression of this in increasingly more complex forms both of RNA and not least of diversified protein structures.

It can also be envisioned that the genetic code might have evolved to first only involve a limited number of amino acids the positioning of which might have been determined only by primitive codons in the form of duplex nucleotides. The already mentioned wobble phenomenon may be interpreted to indicate this. Since there are 61 codons specifying different amino acids one would expect that there might be at least as many forms of transfer RNAs. However, the wobble phenomenon involving the third nucleotide allows a reduction of the number of required specific molecules in some systems. Hence it has been found that in certain bacteria the organism can manage with as few as 31 different transfer RNAs, but in us humans representing a vertebrate animal as many as almost five hundred genes for transfer RNA have been identified in our genome. Still the number of anticodons represented is limited to 48.

Before leaving this fascinating field it should be added that it was Crick who in a seminal 1968 publication[15] shone a spotlight on what has come to be referred to as the frozen accident in the evolution on the genetic code. Once the code had been defined in general terms during early evolution there was no way back. Any mutational change would have had too serious deleterious effects. All reflections on the evolution prior to the emergence of the first kind of cells are highly speculative, as already referred to.

Insights Into an Unknown World of RNA

We will now return to the question of the relative role of nucleic acids and proteins in the complex machinery of transcription and translation. An impressive amount of new information has been gathered over time because of the access to crystallographic techniques allowing the identification of the three-dimensional representation of different molecules in space and of their detailed interactions briefly alluded to above. Of course from the start the foremost goal was to be able to experimentally determine the sequence of nucleotides in a strand of DNA, to read the book of life. As described earlier[5] techniques to do this were presented in 1977 by Allan Maxam and Walter Gilbert and also by Frederick Sanger and collaborators. However, already before that, surprising findings about principles of gene expression had been made using the developing techniques of molecular biology. The techniques used had different origins. One of them was so-called hybridization experiments. An important scientist in the development of this technique was Sol Spiegelman, a narcissistic scientist introduced in my previous book on Nobel Prizes[5]. He was nominated together with two other scientists in separate disciplines for a Nobel Prize in physiology or medicine in 1968 by the Soviet virologist Victor Zhdanov. Before we move on to discuss this nomination a few comments on Russian (Soviet Union) contributions to the advancing fields of biology and medicine might be in place.

As far as natural science is concerned Russian (Soviet Union) scientists who have received a Nobel Prize have been recognized almost exclusively in the discipline of physics. This probably reflects the combined facts that at least for some of the population basic and higher education can be of good quality and that in politically prioritized areas it has been possible to make major contributions. It should be added that in the tradition of Great Russian literature, several prizes in this discipline have also been awarded. However, in chemistry there is not a single prize recipient of Russian origin. The Royal Swedish Academy of Sciences in 1906 was very close to awarding the prize in chemistry to Dmitri Mendeleev, the father of the periodic table, but regrettably it missed this opportunity[16]. In physiology or medicine the great scientist Ivan P. Pavlov is the single true representative of Russian biomedical science. Ilya Mechnikov, the 1908 prize recipient together with Paul Ehrlich was of Russian origin, but spent the second part of his professional career at the Pasteur Institute in Paris. It seems that the field of medicine has not been prioritized by Russia (Soviet Union) during the 20th century and the absence of influential geneticists or physiologists in part can be explained by the terrible period of Lysenkoism, referred to in my previous book on Nobel Prizes[5]. Still in my own field of virology I have got to know a number of qualified and colorful colleagues of Russian (Soviet Union) origin. Zhdanov was one of them.

Viktor Zhdanov (1914–1987). [From Ref. 24.]

He was of Ukrainian origin and wrote his thesis on the epidemiology of hepatitis A virus infections. Between 1950–61 he was Deputy Minister of Health in the USSR and during this time he took an important initiative in the WHO to launch a program for eradication of smallpox. As already described[4] this was successfully concluded in 1977. Zhdanov was also involved in the early efforts to establish a system for classification of viruses. In addition he was one of the four founding fathers, together with representatives from the U.S., the U.K. and Finland, when the First International Congress of Virology was arranged in Helsinki, Finland in 1968. Zhdanov was a politically trusted scientist and could travel. He had opportunities to visit the U.S. He was also permitted by the police to let foreigners visit his home. On one occasion I had the privilege of being one of his guests. It was a memorable evening. He showed me his unique stamp collection hinting at the fact that it would be easy to carry this valuable asset along in case of forced changes of living conditions. He also proudly announced that he had danced with the U.S. Vice President Hubert Humphrey's wife Muriel. When Zhdanov died in 1987 his life partner, Alice Bukrinskaya, also a virologist, invited a small group of us for a memorial conference. It was held in the magical city of Zagorsk, one of the gems in the so-called Golden Ring, surrounding Moscow.

Another high-profile character among Russian virologists was Mikhail Chumakov. He made pioneering contributions to studies of vector-borne virus infections. On one occasion he became infected in the laboratory and lost the major part of his hearing and became paralyzed in one arm. His integrity as an individual on repeated occasions caused him difficulties in his relations with the political authorities. In 1952 he was dismissed as head of the Ivanovskii Institute of Virology in Moscow because he refused to fire Jewish associates in connection with the so-called Doctor's plot. But Chumakov came back and it is said that on later occasions when delegations came to correct him and even threatened to replace him, he simply took out his hearing aid and put it in his pocket!

There was a need for professionals in the country who could organize vaccine trials and implement the use of such products in the 1960s. This was a field where Chumakov came to make important contributions. He played a major role when Albert Sabin wanted to test his live polio vaccine preparations on a large scale. It was done in the Soviet Union and in other countries behind the Iron Curtain at the time and the results obtained allowed the vaccine to become licensed for use in the U.S. Somehow Chumakov managed to continue making important contributions in the restricted Soviet Union environment. He had many children and most of them emigrated, mainly to the U.S. The discipline of virology in general has had difficult times in Russia (Soviet Union) like biological and medical sciences in general. Hopefully future developments will allow the country, using the human resources it has, to again make important contributions in the biomedical sciences, possibly even leading to Nobel Prizes in chemistry and in physiology or medicine. Time now to return to Spiegelman and the review that the committee asked Reichard to do for this work.

Spiegelman was a very dynamic molecular biologist. A major step forward was made in his career when he together with Ben Hall presented a technique that allowed a comparison of the nucleotide sequence of two single-stranded DNA molecules or between one single-stranded DNA molecule

Sol Spiegelman (1914–1983).

and an RNA molecule. This technique became very important, for example, in clarifying the role of DNA and RNA in protein synthesis. It had previously been shown that if double-stranded DNA was heated at close to 100°C and then rapidly cooled, the two strands could be separated. Hereafter it was possible to show that the single-stranded form of DNA derived, for example, from a bacterial virus could associate with RNA extracted from cells infected with the same virus. In cases when the two strands of nucleic acids contained complementary sequences of nucleotides they could form hybrids. Experiments of this kind provided the means for one of the first confirmatory demonstrations of the existence of specific messenger RNA. The technique could also be used for the demonstration of the presence of several sites in cellular DNA coding for the two kinds of ribosomal RNAs and for transfer RNAs with varying amino acid specificity.

Importantly it could also be documented that both strands in a DNA double-helix potentially could be used for formation of messenger RNA with different specificities. As was described in Chapter 2 the RNA polymerase used for replication of RNA viruses, be they single- or double-stranded, is of viral origin. This was proven by use of the hybridization technique. Reichard praised the importance of Spiegelman's contributions in developing the technique. He argued that his collective contributions, not a specific discovery, made him worthy of a prize in physiology or medicine. This was also the conclusion by the committee, again perhaps displaying a relatively generous attitude in its use of the concept "worthy of a prize." However, maybe the fact that Reichard could be relatively persistent in his argumentation played a role in the decision taken by the committee. As we shall see the hybridization technique allowed wider interpretations by examining hybrids formed between DNA and RNA by the electron microscope. Surprising findings were made.

It was believed ever since the Nobel Prize for Beadle and Tatum in 1958 that there should be a strict correlation between one gene represented by a stretch of DNA and one protein (enzyme). However, it turned out that nature could be more imaginative than this. One of the early kinds of DNA sequenced by Sanger and collaborators in attempts to develop methods to characterize genomes was large sections of the DNA of the phage phiX 174. To their surprise it was found that there were overlapping genes. There was a stretch of DNA that coded for two proteins, formed by use of different reading frames. Later it was found that a similar effect could also be obtained by stuttering of the polymerase leading to a change in reading frame, again a phenomenon first seen in viruses. Such an overlap of coding sequences of course makes the system more vulnerable. A single mutation may have serious effects on more than one protein, but apparently the virus replication systems could live with this. There is still one more particular way in which the message of a specific RNA can be modified. This is *RNA editing*, which simply means that single- or a few nucleotides are modified in a specific messenger RNA after it has been produced. Later it was found that there was another, quantitatively much more important way, by which a single stretch of DNA could lead to the production of more than one protein.

As mentioned it took time to separately identify and characterize messenger RNA in eukaryotes. One puzzling finding by James Darnell and

James E. Darnell and Lennart Philipson (1929–2011).
[Photo by Malin Philipson.]

collaborators was that the molecules produced by transcription of DNA into RNA appeared to be much larger than the final operative molecules. It was then found that a large fraction of the newly synthesized RNA carried a long stretch of some 200 copies of exclusively adenosine residues at one end and also had a particular molecular configuration in the other end, a so-called cap. This observation allowed the development of new techniques for purification of this category of RNA molecules. In the continued pioneering studies by Darnell, the foremost virologist in Sweden, Lennart Philipson made some important contributions. Being a visiting scientist with Darnell at Rockefeller University he introduced a simplified system to study the genetic expression of DNA. Instead of examining the large mass of cellular DNA representing some

20,000 genes, fewer than 40 genes specific for adenovirus DNA, expressed in cells infected with this virus, were studied selectively. Further examination of the processing of the precursors of messenger RNA into the functional molecule offered some surprises.

In these developments still one more unexpected finding, the discovery of the restriction enzymes, needs to be mentioned. This kind of enzyme provides bacteria with a defense mechanism, which allow them to protect themselves against invading viruses. Some of these enzymes have a high degree of specificity and can cleave DNA only at particular nucleotide sequences in DNA. Not infrequently the site has palindrome characteristics, which

Werner Arber, Daniel Nathans (1928–1999) and Hamilton O. Smith, recipients of the 1978 Nobel Prize in physiology or medicine. [From *Les Prix Nobel en 1978*.]

means that a nucleotide sequence appears the same regardless of which end it is read from. The identification of restriction enzymes was recognized by a Nobel Prize in physiology or medicine in 1978 for Werner Arber, Daniel Nathans and the already mentioned Smith "for their discoveries concerning restriction enzymes and their application to problems of molecular genetics," referred to already in my first book on Nobel Prizes[3]. The use of this kind of enzyme allowed molecular geneticists to cut and paste in DNA. In later years even more advanced tools to redesign DNA have been discovered. This is the CRISP/Cas9 technology introduced already in my previous book[5]. The latter discovery has not as yet been recognized by a Nobel Prize.

Early comparative studies of the relations between the original adenovirus

Richard J. Roberts and Phillip A. Sharp, the recipients of the Nobel Prize in physiology or medicine in 1993.

DNA and the messenger RNAs it directed the production of were made using various kinds of molecular scissors. Some critical contributions to these developments were made by one of Philipson's top students, Ulf Pettersson. These early experiments served as seeds for high impact experiments to come. The original tentative mapping was made by use of the genome scissors, the restriction enzymes, but then still further tools, the briefly mentioned techniques of hybridization and also electron microscopy widened the opportunities for experimenting. A highly unexpected finding was made both at the Cold Spring Harbor Laboratory and at MIT. This discovery was recognized by the Nobel Prize in physiology or medicine in 1993. The prize motivation was very short "for their discovery of split genes" and the recipients were Richard J. Roberts and Phillip A. Sharp.

The original observations were made in studies using the already introduced adenovirus-infected cells as a model system. The unexpected finding was that internal pieces of the original RNA formed by transcription were cut out in order to form a trimmed mature version of messenger RNAs. These revolutionary insights developed in an explosive way during the spring of 1977 and culminated at the Cold Spring Harbor symposium in early June the same year. The title of the symposium was "Chromatin," but because of the recent developments the focus was shifted. All discussions came to revolve around the unexpected discovery of processing of RNA, sometimes in multiple ways, leading to shortened forms of the final product. When the original virus-DNA was allowed to combine with the trimmed form of mature messenger RNA it could be shown by the electron microscope that there were loops of apparently single-stranded DNA that stuck out from the combined hybrid molecule. They represented the cut-out portions of the original RNA transcript copied from the virus DNA. Rarely has the temperature of discovery been so high and the claiming of priority been so intense. Sharp introduced the traditional sailor's term *splicing* to name the phenomenon and Gilbert proposed that the pieces of messenger-RNA cut out should be called *introns* and the remains re-associated to form the edited messenger RNA, *exons*. The existence of this phenomenon meant that a protein gene in DNA could be expressed in more than one way by alternative splicing, representing various kinds of processing. The phenomenon of splicing turned out not only to apply to the virus model

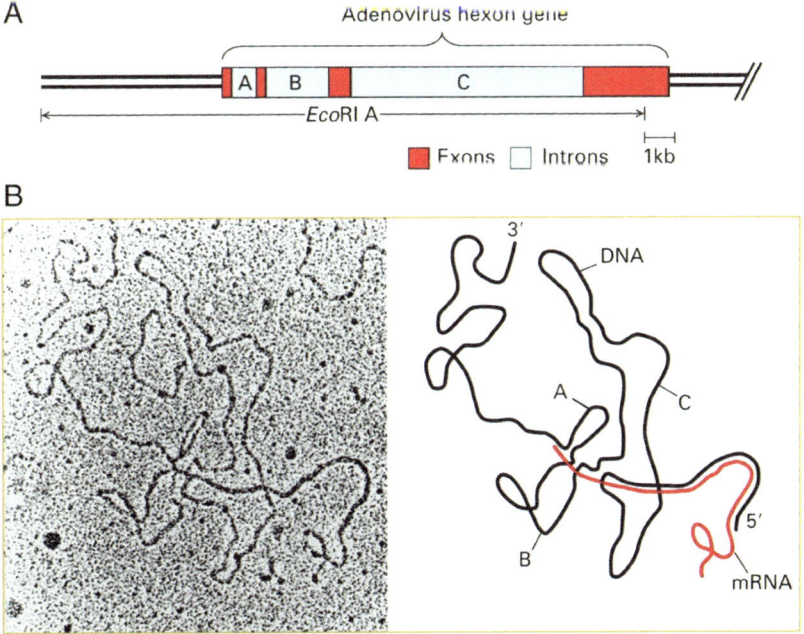

Electron micrograph of a hybrid between adenovirus DNA and one of its messenger RNAs for production of one surface components, the hexon. Only the red parts of the originally transcribed RNA are combined into a functional Messenger.

system used for its detection but also to be a regular cellular phenomenon.

Surprisingly it has been found that introns markedly dominate over exons in the open reading frames in the eukaryotes. Later a special structure in cells which provided the machinery for the trimming of the messenger

RNA was identified. It was referred to as the *spliceosome*. The structure of this messenger RNA processing molecular machinery is in fact even more complicated than that of the ribosome. It is dynamically formed in association with the premature form of messenger RNA located in the nucleus of the cell. Cryoelectron microscopy has provided some insights into the structure of this molecular complex. Components contributing to the assembly of this structure are five different kinds of *s*mall *n*uclear RNAs (snRNA) which are referred to as U1, U2, U4, U5 and U6, because they are all rich in uridine, and then in addition about 80 different proteins. Much remains to be learned about the fine structure and modes of operation of these complexes, which in practice act as a nuclear RNA cutting and ligation machinery. The phenomenon of splicing allows one open reading frame of DNA to result in the production of more than one protein, potentially even a large number. Hence the fact that the human genome has about 20,000 open protein reading frames does not at all mean that the potential of the whole genome to produce individual specific proteins is limited to this number. As more was learned about RNA processing it turned out that there was an even larger surprise in store.

Prokaryote organisms do not contain nuclei and as a consequence they do not have spliceosomes. However, in certain cases they do demonstrate a primitive form of splicing. Highly surprising it turned out that RNA molecules *on their own* could modify their structure. They could self-splice. This completely unanticipated finding was recognized, four years prior to the prize for eukaryote splicing, by the 1989 Nobel Prize in chemistry for Sidney Altman and Thomas R. Cech. The prize motivation was short but powerful — "for their discovery of catalytic properties of RNA." From my subjective vantage point I consider the discovery recognized by this prize, one of the most important ever.

Thomas R. Cech, one of the Nobel Prize recipients in Chemistry in 1989.

The reason for the time delay in recognizing the discovery of splicing by a prize is open to speculation. It should be noted that although the existence of ribozymes was discovered after splicing it was recognized by the prize in chemistry earlier than the prize in physiology or medicine to Roberts and Sharp, as mentioned. The most likely

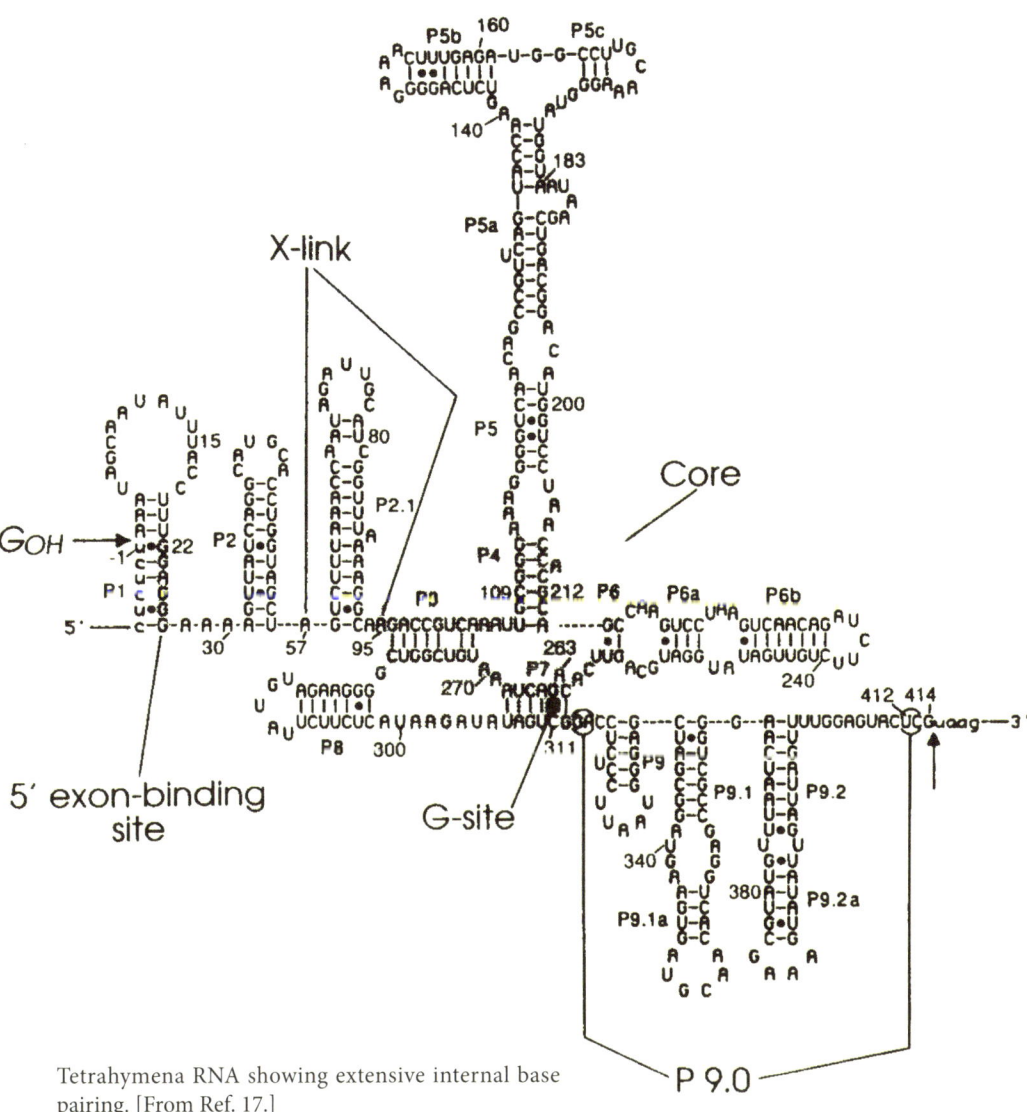

Tetrahymena RNA showing extensive internal base pairing. [From Ref. 17.]

explanation is that a number of years were needed before the Nobel Committee at the Karolinska Institute could decide on who contributed what to the discovery and hence on the proper prize recipients. In January 2044 it will be possible for science historians to see how the reviewers used by the committee decided to untangle the complex priority claims to the discovery.

Cech used the ciliated protozoan *Tetrahymena thermophila* as his model organism. He compared the original and the shortened mature products of ribosomal RNA. A truly astounding discovery was made. When the isotope-labeled intact RNA was isolated in a very pure form to be used as material for enzymatic digestion it was found that a piece in the middle of the molecule spontaneously

To See the Invisible and to Read the Unprinted 361

had been removed by a double cleavage and a re-ligation of the free ends at the gap in the molecule. This was the first evidence that RNA on its own could serve as an enzyme. RNA with this capacity was later referred to as a *ribozyme* and the removed part of the molecule logically was called an *intron*. The chemical reaction was referred to as transesterification. It is a critical capacity of single-stranded RNA to form extensive and complex internal base pairing that endows it with the capacity to display highly specialized biological activities, as for example various kinds of enzyme functions. The proposed folding of RNA of tetrahymena origin was illustrated in Cech's Nobel lecture (see previous page).

Sidney Altman, co-recipient of the 1989 Nobel Prize in Chemistry.

Altman independently made a similar observation. He used purified transfer RNA in the experiment that allowed his unexpected discovery. He gave a good description of the actual discovery in his Nobel lecture[18]. In late 1969 Altman arrived at the MRC Laboratory of Molecular Biology in Cambridge to join in crystallographic studies of phenylalanine transfer RNA. However, he soon learned from Brenner and Crick that the structure had just been solved. So, what to do? He got involved in studies of acridine-induced mutants of tyrosine transfer RNA, a system established in the laboratory. Using a rapid phenol extraction technique he managed to isolate two forms of the transfer RNA, one the mature form, and one, as it turned out, the larger precursor form. This work was further developed when he returned to the Department of Chemistry at Yale University in 1971. After some ten years work at that institution, he made the critical discovery that eventually led to the Nobel Prize. It had already been proposed from Darnell's laboratory that transfer RNAs might be produced from larger precursor molecules as mentioned. Altman was now able to experimentally examine a special variant of this phenomenon. He identified an enzyme named ribonuclease P in *E. coli* which in the presence of high concentrations of Mg^{2+} could bring about the cleavage. It could be added that other work had also highlighted the importance of a high concentration of mono-valent salt as an alternative means to provide the necessary so-called shielding effect. He could determine the site of this cleavage. It turned out to be the second example of the existence of a self-splicing phenomenon, but the

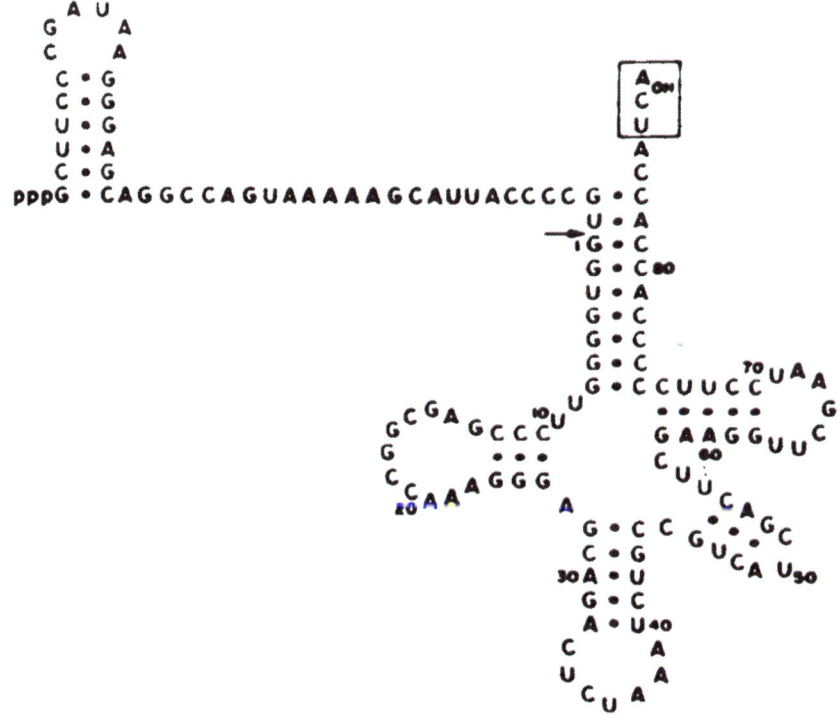

The precursor of tyrosine transfer RNA. The arrow points to the site at which the molecule is cleaved by a ribozyme to give the mature form of the transfer RNA. [From Ref. 18.]

biochemistry was different than the self-splicing identified in ribosomal RNA.

The phenomena of RNA processing have been found to have a major importance for genome expression. Since the original identification of ribozyme activities many new enzymes of this kind have been identified. A general prerequisite to allow the expression of such activities is a pronounced capacity to form structures by internal base pairing (see pp. 361, 363 and 370). As in the case of the molecule transfer RNA many other RNA molecules exemplify the fact that relatively rigid three-dimensional structures can be formed. They can potentially be characterized by crystallography, or why not by the recently introduced cryo-electron microscope technique recognized by the 2017 Nobel Prize in chemistry (see p. 345). Ribozymes are rare in cells and found only in selected contexts, unsurprisingly also in an RNA virus, hepatitis delta virus, to be further discussed below. The insight into the existence of different kinds of ribozymes has opened a completely new world of concepts. This is referred to as *The RNA world*.

RNA and the Origin of Life on Earth

The original speculations about a possible role for RNA in qualified biological activities because of its capacity for internal base pairing allowing formation of complex secondary structures were made by Carl Woese, Francis Crick and Leslie Orgel, all introduced in my previous book on Nobel Prizes[5] and in the case of Crick also in my second book[4]. Crick and Orgel enjoyed each other's company. For a number of years they met once or even twice a month for dinner at Piatti, an Italian restaurant close to La Jolla shores. It has been in existence for decades and is often visited by scientists in the area. The localities employed by Piatti were previously used by another restaurant Gustaf Anders, a name which had a special ring for Swedish visitors. I have had the pleasure of meeting both Crick and Orgel on a number of occasions. Gustaf Arrhenius, grandson of the 1903 Nobel laureate in chemistry, and a respected marine chemist who was interested in the origin of life, living in La Jolla often hosted these encounters. The beautiful home he and his wife Jenny had created in the isolated hills of the 1950s overlooking La Jolla provided rich and unique encounters with leading scientists and cultural personalities of global fame. On November 28, 2018 I received an email from Gustaf on account of my wife's recent death. He wrote some kind words complimenting her self-sacrificing

Leslie Orgel (1921–2007) and Francis Crick (1916–2004) sharing a lunch at the Piatti restaurant in La Jolla Shores. [Photograph by Gerald Joyce.]

and warm personality. He finished the email (translated from Swedish) "I am now myself prepared to follow her in the mystical event that death entails." A few weeks later he himself had died. It could be added that he had a rich and long life. He was born in 1922.

We will now deepen our perspective and see what can be deduced from our knowledge about vestigial forms of RNA represented at the very beginning of life. The key question is how organic molecules could develop to store and transmit information. This will take us back to the very origin of life when simple carbon-dependent molecules managed to establish a primitive form of self-replicating system. It was the small failures of these simple self-replicating systems that allowed an evolution of progressively more complex structures. Embryonic life forms started to compete for their existence with other similar forms. The winner in this competition was the one that was most successful in evolving a genetic language, to store information to be transmitted from one generation of self-replicating molecules to the next.

The origin of life represents one of the three major challenges to the human intellect in our attempt to grasp the meaning of our existence. Another major challenge is to comprehend the origin of the Universe. In this case a lot has been learned but many major questions remain unanswered. The fact that the Universe is expanding from an original Big Bang some 13.8 billion years ago is an important starting point. The comprehension of the development of all the elements organized in the beautiful periodic table also represents fundamental knowledge. We can identify all the critical building blocks for the emergence of life. However, all the elements described in the periodic table only account for close to 5% of all the material constituting the Universe. The rest is "dark energy" and "dark mass," representing about 68% and 27%, respectively. Their nature still to a major extent remains enigmatic.

The third critical area of knowledge which also continues to remain enigmatic in most of its qualities is human self-consciousness. This was discussed in my previous book on Nobel Prizes[5]. Although it is very difficult to describe its nature in a meaningful way we can note that it puts us in a unique position among animals. We are the only species that can in a meaningful way ask the different existential questions and by use of science search for the answers. As to the three challenges we have been most successful in gaining insights into the origin of life. The reason for this is the fact that with an ever-increasing efficacy, today we can read and write the books of life. The origin of this successful journey of knowledge is our insight into the language of genes, the discovery of which was discussed at length in the two preceding chapters. Our capacity to explore the books of life has opened immense realms of insight into the tinkering of evolution and there is much more to come. In the following we will take stock of where this remarkable growth in knowledge has taken us. In particular we will reflect on the challenging question of the origin of life.

It is apparent that these early molecular evolutionary events should have taken place in some aqueous environment and further there must have been some mechanism for increasing the relative concentrations of the potentially interacting molecules. A possible role of lipid micelles, a likely precursor to the forthcoming membrane structures surrounding cells or some charged surface allowing a preferential adsorption of selected groups of molecules are the likely candidates to providing conditions to increase the relative concentration of critical reagents. There was a need to counteract what by an alliterative German phrase has been referred to as "die Verdamte Verdünnung" — the damned dilution.

Identification of the surprising double functions of RNA, being a carrier of information but potentially also serving as an enzyme, led to a rapid involvement of many scientists interested in the emergence of life. A recurring question in discussions of the origin of life has been what came first, the proteins or the nucleic acids. Accepting the double functions of RNA provided a solution to this "chicken-and-egg" problem. The dual functions of RNA have opened a wide field of research which will be only briefly commented on in the present context. Suffice to note that for example Gerald Joyce and collaborators have designed self-replicating RNA molecules. It is to be noted, however, that it was in the context of studies of the fundamental mechanisms of protein synthesis that it became apparent that RNA was the central actor. Thus the emergence of life may have started with evolutionary phenomena involving only RNA molecules and that thereafter there was a secondary mobilization of the building stones of proteins. Since protein was not the first

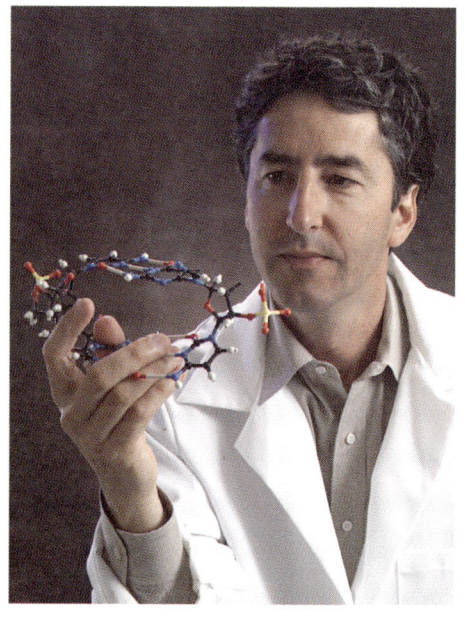

Gerald (Jerry) Joyce. [Private photo.]

category of molecules in the emergence of life but the second, maybe a more proper name would have been not *pro*tein, from Greek the first, but *deuteron*, the second. But it is probably too late to change the terminology now. The term protein was first used in 1838 by the Dutch chemist Gerhard J. Mulder, presumably on a proposal from Berzelius. When I went to school egg white was used as a general term for proteins.

As outlined above the dual functions of RNA to serve as both an information-carrying molecule and also to carry different enzymatic functions allowed the emergence of life to get started. Progressively the other main molecular actors, the proteins and DNA, also became involved. It is logical to postulate that at some stage the major part of enzymatic functions were taken over by proteins. Enzyme functions as we know them today are mainly carried by this kind of molecules. They have a much wider range of chemical reactivity than the enzymatic activities carried by RNA, the ribozymes. This is in part due to the fact that they are composed of twenty different kinds of building stones as compared to only four in RNA, but in particular because the chemical reactivity of the many diverse forms of amino acids is much larger than that of nucleotides. It can be projected that with time more and more complex and functionally diversified protein molecules could evolve to take over and expand the enzyme replicating activities of the progressively developing primitive forms of life.

There are many speculations about how the complexity of protein structures might have developed progressively during the very early molecular evolution. Clearly a relatively small number of amino acids, probably overrepresented in the original biogenic soup, were involved at the beginning. Some of these spontaneously appearing amino acids may have had particular structural properties that allowed a direct interaction with certain RNA nucleotides or combination of a few of them. Besides this stereo-chemical theory for the possible early evolution of the genetic code there are other speculations regarding particular coevolving metabolic processes and also on means for minimizing the errors of insertions — primitive mutations — which must have represented a major challenge to the embryonic genetic code.[12]

There is also the additional highly speculative phase of the later evolutionary exchange of RNA molecules as the dominant information-carrying molecules for the more stable double-helix (originally probably single-helix?) DNA. As already mentioned in Chapter 2 the genetic material of viruses represents essentially any imaginative form of nucleic acid. It can be DNA or RNA, single or double-stranded, a plus-strand or a minus-strand and intact or in fragments. By some process the information content of the ancestral nucleic acid became presented in a messenger-RNA form for the intracellular protein synthesis by transfer RNA and ribosomes. Hence it is possible that the original single-stranded RNA evolved into a functional double-stranded RNA for more stable information storage. This structure then could have evolved step-wise into a double-helix hybrid of RNA and DNA, eventually leading to the emergence of double-stranded DNA. An alternative early step might have been a direct association of single-stranded RNA with a single strand of DNA to form a hybrid molecule. It needs to be added further that the exchange of RNA for DNA requires two critical chemical changes. One is the addition of a methyl group to uracil to form thymidine and the other is to change the associated sugar from ribose to deoxyribose. This is brought about by the enzyme ribonucleotide reductase, which was the object of Reichard's studies throughout his long scientific life.

However there must have been many steps along this speculative road of early molecular evolutionary development of life. There is not a complete agreement on whether the extensive involvement of proteins evolved before or after the introduction of DNA. Opinions vary among scientists as to whether ribozymes would be capable of transmuting RNA into DNA. It is of great interest in this context that a ribozyme with reverse transcriptase activity was recently demonstrated by Joyce and collaborators[19]. Cech[20] has proposed that proteins progressively took over functions displayed by ribozymes, but not in all cases. There are four central mechanisms in which there are reminders of the fact that the origin of life started with the RNA world.

The first one involves the actors in protein synthesis. As we have seen ribozymes still today have a central role in the processing of information carried by the messenger RNA by aid of transfer and ribosomal RNA and also in the case of self-splicing of messenger RNA. A ribozyme is involved in the ribosomal function that leads to the formation of the polypeptide chain. The second fundamental mechanism depending upon a ribozyme is the spliceosome-dependent messenger RNA processing phenomenon in eukaryotic cells allowing the generation of several proteins from a single open reading frame of DNA. It has furthermore been found that synthesis of DNA cannot start on its own. It is dependent on the presence of a short primer of RNA. Finally there is a particular problem with the ends of a replicating molecule of DNA as already discussed in Chapter 3. The way the machinery of replication producing the two anti-parallel stands of the double-stranded molecule works is that it results in a lack of DNA in one of the strands at the end. If this is not compensated for the DNA of our chromosomes would become progressively shorter following repeated cell divisions. Again evolution has selected a mechanism to compensate for this shortening and this is the use of an enzyme, referred to as telomerase, which is dependent on RNA. Hence this enzyme is a form of reverse transcriptase, similar to the kind of enzyme discovered in the many different forms of retroviruses. The discovery of telomerase was recognized by a Nobel Prize in physiology or medicine in 2009 as already described in Chapter 3.

Discussions about the developing early RNA world need to include viruses. Viruses use DNA or RNA as their genetic material, but, strikingly, they never include both kinds of nucleic acids. As was described in Chapter 3, the world of viruses offers insights into the existence of all possible forms of copying between single- and double-stranded DNA or RNA, including reverse transcription of RNA into DNA, as also already mentioned, and not least RNA copied from RNA to produce the genome of both single-stranded and double-stranded RNA viruses. The already repeatedly cited Baltimore, in an ingenious way has used the particular nature of nucleic acids in viruses as a means of classifying them in distinct groups using the principally different kinds of metabolic steps leading to messenger RNA as the distinguishing denominator. In a way these highly diverse forms of viral genomes can be seen to illustrate the rich tool box that probably became available in the early molecular evolutions in the RNA world. It is highly likely that the diversity of RNA viruses has a meaning as successfully surviving archeological remains of the RNA world. In fact it has been found that the genome of an RNA virus can carry a ribozyme activity, as mentioned, but this is a very special case. Such an enzyme activity has been identified as being associated with an infectious agent referred to as hepatitis D (delta) virus. This is a defective virus that can only replicate in a host which simultaneously is infected by hepatitis B virus. The concomitant replication of the two agents leads to a more severe form of the disease.

The genome of the hepatitis D virus is exceptional. It is a circular RNA containing about 1,680 nucleotides. It has a high GC content and it is 70% self-complementary. It can direct the synthesis of only one protein, delta antigen, and in order to transport the virus nucleic acid to infect new cells it borrows the envelope of its helper virus, the hepatitis B agent. The ribozyme of the hepatitis D agents has unique morphological characteristics with extensive base pairing. It is referred to as a "hammer-head" enzyme because of the morphological similarity to the head of hammerhead sharks. The same kind of enzyme has been

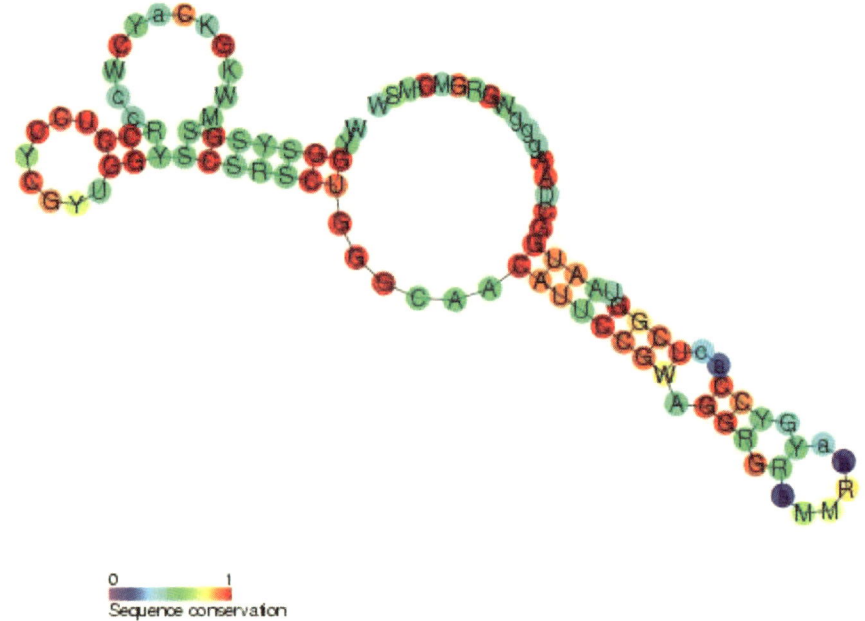

Hepatitis delta virus ribozyme.

identified in an even more remarkable kind of infectious agent. They are called *viroids* and were discovered by Theodor O. Diener as the cause of a severe disease in potatoes. Since then other forms of viroids have been detected in different kinds of plants. In contrast to viruses, viroids completely lack a capacity to code for proteins. Hence they are referred to as a separate group of biological entities, named *subviral agents*. They act as an independent RNA molecule which includes 246-301 nucleotides. Their way of replication is exceptional. As emphasized RNA in our cells are formed by a synthesis using DNA-dependent RNA polymerase. It is only RNA viruses which can code for RNA-dependent RNA polymerases. However, in some way viroid RNA has learnt to exploit the

Theodor O. Diener, the discoverer of subviral agents.

cellular DNA-dependent RNA polymerase to accept it as a matrix and assist in its replication. Since the viroids contain a ribozyme of the hammer-head kind it has been speculated that they might represent another "living fossil" of the primordial RNA world[21]. However, this has been contested since viroids and viroid-like structures so far only have been found in higher plants and in certain vertebrates — the hepatitis D virus mentioned above. Of course it has been speculated that the hepatitis D agent may have its origin in plant viroids. As long as similar RNA entities have not been found in the evolutionary ancestors, like unicellular eukaryotes and not least the prokaryotes, the bacteria and the archaea, the verdict on this "fossil theory" remains out.

Of course viruses as we know them are cellular parasites, but what is discussed here is their pre-cellular existence as possibly independently replicating nucleic acids, a process potentially slowly involving some primitive proteins. It should be emphasized that the multitude of different RNA viruses most likely is a reminder of the time when RNA was the only information-carrying genetic material. It was only later when the first primitive forms of cells had developed that the originally naked RNA started to interact with proteins in a more mature and advanced way and was packaged into virus particles that allowed a safe transport of their genetic material between cells. This allowed a *lateral* transmission of genetic material between the first primitive forms of cells, which probably played a major role by supplementing the *vertical* gene transfer from parent to offspring cells for the early evolutionary developments in the nascent primitive and highly simplified world of life.

An example of lateral transfer of cellular genes was already described in Chapter 2 in the review of tumor viruses and oncogenes. It was also introduced in my previous book on Nobel Prizes[5] discussing the phenomenon of latency of bacteriophage infections. There may be reasons to come back to the central role of viruses and virus-like structure in evolution, to elaborate a *virocentric* view of life, in a possible forthcoming fifth book on Nobel Prizes discussing the 1969 Nobel Prize in physiology or medicine to Max Delbrück, Alfred D. Hershey and Salvador D. Luria "for their discoveries concerning the replication mechanism and the genetic structure of viruses." Viruses which mostly have become known as our foes, taking another perspective in reality should be considered as our friends. If it was not for them, advanced species like us humans would never have evolved on Earth! But let us now return to molecular biological fundamentals.

The developments of the insights into the double physiological roles of RNA, its capacity to both carry information and to display enzymatic activity, has an influence on how we allocate functional responsibilities to the central information-carrying molecule DNA. Hence the terminology *coding DNA* that was introduced after identification of the existence of somewhat less than 20,000 open reading frames for the synthesis of proteins in the human genome and similar findings in other systems needs to be reconsidered. Of course the term as introduced meant *protein*-coding. But since RNA can also serve the functions of representing different kinds of structural components and also to serve as variable kinds of enzymes, although in both cases to a much more limited extent than proteins, there is in principle no difference in the functional capacity of proteins and RNA. This insight has consequences for the precise definition of the term *gene*. The insight into the splicing phenomenon also adds to the complexity. The fact that alternative splicing of RNA allows the involvement of a specific stretch of DNA in the potential coding for several different kinds of functional proteins raises the question if it carries more than one gene function. One would expect that with the impressively deep insights into the molecular expression of biologically active molecules the definition of the concept gene would have become increasingly precise. However in certain situations the opposite seems to have become the case.

Progressively we have learnt that there are well defined genes for the production of the different specific proteins that a certain organism can synthesize. However, it should be noted that as already indicated it is not only the open reading frame controlling the synthesis of a specific polypeptide that represents the whole gene. There are also a number of factors controlling the activation of a specific gene located in other parts of the genome that are associated with the gene concept. The existence of transcription factors was mentioned already in Chapter 3. But in addition as noted we have to account for the stretches of DNA that lead to structural and catalytic RNA as equally accountable as genes. To this come a number of different forms of RNA which have not been discussed in the present context. One more Nobel Prize in physiology or medicine should be mentioned. It is the prize in 2006 to Andrew Z. Fire and Craig C. Mello. This prize was given "for their discovery of RNA interference — gene silencing by double-stranded RNA." They had made the surprising finding using the *C. elegans* system introduced by Brenner that certain forms of double-stranded RNA had a capacity to silence specific genes. The fact that this finding potentially meant the access to a new approach to specifically silence selected genes made it an interesting prize in physiology or medicine. Discussion of this multitude of often different categories of RNA, sometimes very small encompassing only some 15–20 nucleotides is well

Andrew Z. Fire and Craig C. Mello, the recipients of the 2006 Nobel Prize in physiology or medicine [From *Les Prix Nobel en 2006.*]

beyond the scope of this book, but excellent presentation of their nature and functions, as far as it is known, can be found in many books on molecular biology, such as for example the one by Darnell, *RNA. Life's indispensible molecule*[22]. Their specific or more general regulatory functions in many cases still remain to be evaluated.

So, in the light of the increasing insights into the complexity of functions of RNA, what is the proper definition of a gene? Berg and Singer have attempted a definition of the gene concept in a book from the early 1990s[23]. They wrote "A eukaryotic gene is a combination of DNA segments that together constitute an expressible unit. Expression leads to the formation of one or more specific functional gene products that may be either RNA molecules or polypeptides. Each gene includes one or more DNA segments that regulate the transcription of the gene and thus its expression." This is probably as close as we can get at the present time, but a lot still remains to be explained, as for example the role of the many different categories of less well characterized small RNAs. However, to reiterate, a simplified way of identifying a crucial informational significance of a certain stretch of DNA is of course to state that whenever a structural change in a certain part of the genome — a mutation — has an effect on a particular or a group of biological activities it serves a function, one way or another, in the expression of a gene.

Paul Berg, co-recipient of the 1980 Nobel Prize in Chemistry and Maxine Singer. [From Ref. 24.]

To the Greatest Benefit of Mankind

Nobel's will, written in Swedish in 1895, is remarkable in its brevity and scope. The aim in using the financial returns of the trust to be established was to further the advance of humankind in two ways. One way was to encourage

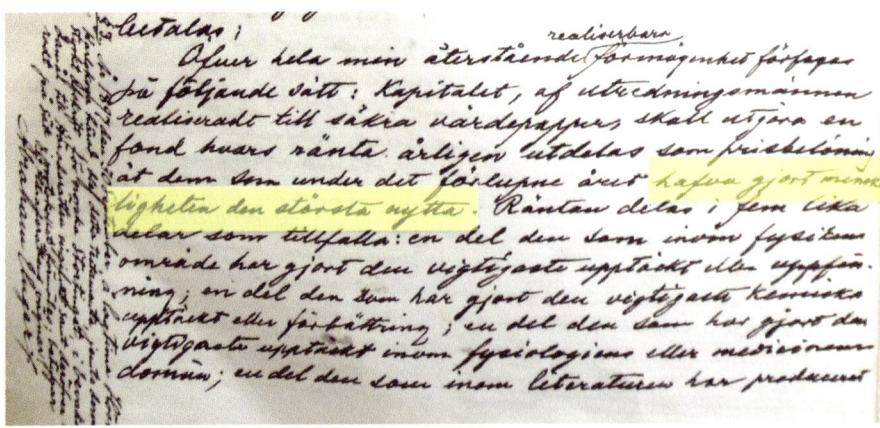

Extract from Nobel's will written in Swedish. The part of the text which translates into "to the benefit of mankind" is colored in yellow.

the growth of knowledge gained by research in the natural sciences defined by discoveries, inventions and improvements. The other way was to stimulate the development of the humanities by encouraging literature providing insights into the depth of human nature and to reward efforts to further peaceful interaction between peoples. The original idea was to identify a person in his or her prime at about 35 years of age and to provide a salary for some 20 years to allow the prize recipient to continue to serve as a beacon in initiating activities to further the development of human civilizations. It is striking that the will written at a time when nationalistic movements were particularly pronounced stated that "no consideration whatever shall be given to the nationality of the candidates, but that the most worthy shall receive the prize, whether he be Scandinavian or not."

It may be reflected that the goals set were too idealistic, but at least the way the progress of developments of the natural sciences over the almost 120 years during which the prizes have been awarded, tells a remarkable story of success. The discoveries in the search for new knowledge recognized by the prizes in the natural sciences can be seen as road marks in describing the history of disciplines in modern human societies. The advances recognized by the prizes include both very important practical achievements, fulfilling the motto "for useful knowledge" that Benjamin Franklin introduced when the American Philosophical Society was founded in 1743, and also fundamental knowledge of relevance for reflections on existential questions.

About ten years after Nobel had signed his will two central concepts were introduced into the natural sciences. These were the terms "atom," from Greek átomos, with the original meaning indivisible, and "gene," to represent the unit of heredity deduced from Bateson's term "genetics," which, as already mentioned, has its origin in a Greek word meaning "to give birth." By a remarkable journey we have learned about the innermost structure of matter with various incomprehensible minute concentrates of energy which by their movement in what seems to be an almost limitless space provide an explanation even for the nature of the hardest of matters. And when it comes to the genes we have learnt to read and write their language appreciating an overwhelming diversity of functions and life forms. We have also traveled back in time gaining insights into the remarkable and unpredictable successive events when evolution has been at play. It is a story of an almost unlikely survival of waves of different imaginative forms of microorganisms and animals under conditions challenging the mere existence of even the hardest representatives of life. Evolution is a game of chance and necessity and progressively it has been learnt that survival is not only a matter of competitive advantages. Not infrequently it is also a question of finding ways for cooperation facilitating a mutual survival of life forms.

Humankind, living on a speck of the incomprehensibly huge universe, has taken on to itself to explain the origin of its place in the world of stars. It has managed to identify the formation and organization of all the different kinds of elements representing the stardust into the harmonious periodic table. It is from this stardust that we are formed and to which we will return as we finish our travel. The cracking of the genetic code as described in the later chapters of this book is an extraordinary and remarkable achievement. We can now even speculate about how life started on Earth, some 3.8 billion years ago by a self-replicating RNA system and episodically follow its often troublesome and precarious journey managed by the unpretentious and persistently tinkering evolution. The remarkable intuitive inventiveness of the process of evolution releases awe and enthusiasm of the human spectator, who, of course, is a fully integrated part of the system. The life forms that have emerged and survived display ingenious and capricious webs of life on Earth. They take on all shapes and sizes, but they also present beauty. The latter admittedly is to be recognized by the eyes of the beholder, but nonetheless it serves a critical role in the non-trivial choice of a mating partner. It still remains to use the technology to read the books of life to make a full inventory of all forms of life in which it is present on Earth. Much is left to be identified and admired. Half of all organic material is represented by the invisible microorganisms and viruses and presently we have only a very limited insight into their presence and diversity.

At the time of writing the Nobel Prizes in 2018 were announced. The prize in chemistry was referred to as a (r)evolution. As already presented in Chapter 3 it recognized Frances Arnold, George Smith and Gregory Winter. They had managed to take control of evolution in the laboratory. The scientists had succeeded in producing more efficient enzymes and antibodies by use of evolution controlled by man. They may become of use to the future production of biofuels and pharmaceuticals and in medicine the antibodies developed have already served for the treatment of autoimmune diseases and also for attempts to manage advanced forms of cancer and there is much more to come. These discoveries represent excellent examples of how we can learn from Nature and of the value of being capable of reading and writing the books of life. The possibility of developing enzymes with improved catalytic and other biological activities was already proposed in 1984 by Eigen, whom we met in Chapter 5. Inspired by this proposition many scientists have become involved in demonstrating its potential value by various experimental approaches and hence one can speculate that there should have been many contestants for the prize. This field of research as well as many others illustrates the large challenges of present-day committees to filter out the most worthy recipients for the prize among a large group of highly qualified nominees.

During the unpredictable course of evolution the true surprise occurred when an unprecedented species appeared which had a brain system allowing not only consciousness, but critically also *self*-consciousness. The possibility to look backwards and forward in time in the intuitive and conscious planning for actions has markedly improved possibilities for survival even under dramatically changing conditions. Added to this there was a critical social dimension. Humans could form highly productive collectives and their constructive interactions were amplified and cemented by the development of language. One of the most precious assets that came to be shared was knowledge. The progressively improved conditions for reproduction and survival of the original groups of hunters and gatherers and later the communities of early settled individuals advanced the evolution of culture. There was a relentless increase of intergenerational transfer of knowledge of supplementary importance to survival besides what was controlled by the genes and the traditional inheritance. Empirical knowledge was carried from generation to generation and eventually we learned to manage plants and animals in a way that allowed coherent settled civilizations some 12–10,000 years ago. A single farmer could produce food sufficient not only for himself and his family but also for many of his fellow humans. The idea that man or human collectives could own a part of nature was established. It became possible to stake out your land and the artifices of collective ownership on a larger scale led to the formation of nations. A prize paid for the increased density of humans in the aggregated settlements was the emergence of acute infections in an epidemic form.

Way back in time the efficient memory and the verbal proficiency of the storyteller made him or her a very important individual in the group. In the flickering light of the evening fire or under a star-saturated sky at night fantastic stories explaining the origin of man and the surrounding world with all its surprises were spun. A seed for the establishment of religions, first of polytheistic and later of monotheistic nature, was planted. The first attempts were made to explain natural phenomena like the starry sky, like rain and thunder, like earthquakes and tsunamis. Distilling natural religions to monotheistic creeds meant the establishment of an important power instrument. Wars between larger gatherings of people representing different monotheistic religions have led to tragic episodes of bloodsheds on a huge scale continuing into the present time.

The need for registration and management of larger amounts of information eventually led to the emergence of a written language and from then on the culture of documenting and transmitting cultural, non-genetic qualities between generations has accelerated, today epitomized by electronic media. Accessing knowledge can be highly seductive. Of course it has a critical importance on the way we formulate and manage our daily life, but equally importantly it provides a great substrate for our interaction with other humans in our sphere of existence. Hence there is a deep humanistic dimension of knowledge accumulated also by research into the basic natural sciences. Besides genes linking parents and offspring there was a progressively expanded wealth of knowledge transferred between generations, as mentioned. This increasingly important cultural inheritance sometimes is referred to as *memes*. The way we manage our responsibilities as members of what today is a global society and the way we individually search for answers to existential questions is contingent upon the richness and the kind of knowledge we have managed to accumulate.

By use of our knowledge of the genetic language we have learned about the original development of modern man in Africa and his repeated migrations out of this continent originally for the first time some 80,000–60,000 years ago. However, also earlier kinds of hominids managed to make such an exit, but without finally establishing themselves on a global scale. Our dark-skinned ancestors were impressively successful adventurers in the long term, but they also often had to pay a price. The later development of different features of our brothers and sisters after the exodus has been determined by genetic selection due to change of environmental conditions, the emerging infectious diseases they were exposed to, the kind of food that was available for their survival and finally their exposure to challenging climatological conditions sometimes amplified by natural disasters like major volcanic eruptions. Not infrequently only a fraction of the tribe managed to survive, occasionally referred to as a "bottle-neck" phenomenon. It is only by deep insights into our history that we may control our tendency to develop inherently destructive attitudes of "us and them."

During the previous century and progressively into the present there has been a dramatic change in the way we are able to manage diseases and secure a high quality of life for human beings. All in all the quality of life has been gradually improved in an impressive way in modern times and the length of life has increased from generation to generation. The Nobel Prizes in physiology or medicine can serve as road marks to outline these impressive achievements. The way we can prevent and treat infectious diseases; the way we can secure survival of both mother and child at birth; the way we can manage cardiovascular diseases; the way we can control many hormonal diseases and as discussed at length in the first three chapters of this book, the progressively improved means of managing the often life-threatening forms of cancer diseases etc. has completely changed conditions for the survival of humans on Earth. It is in particular since the Second World War that we have witnessed an unprecedented development. We have become the dominant species and as a consequence we are now in control of evolution on Earth. This is a major responsibility, and I have repeatedly returned to the important role of our stewardship in my writing about Nobel Prizes.

Originally there was a single unit of knowledge. Philosophy means love of knowledge. Hereafter knowledge was divided into different disciplines, physics, chemistry, biology etc. in part to facilitate research in and teaching of the designed subject. Hence Nobel proposed separate prizes in physics, chemistry, physiology or medicine etc. It should be noted that he emphasized in parallel physiology and medicine, reflecting the insight that only by understanding the conditions of normal functions can we interpret disease processes. It has become apparent with time that the borders between different disciplines become increasingly fuzzy. This can be recorded historically in the increase of nominations for a certain discovery within more than one discipline. As a consequence the committees at the Royal Swedish Academy of Sciences and at the Karolinska Institute have annual contacts to discuss their responsibility for certain shared individual candidates. The volume of work of the committees has expanded markedly with time. This trend will continue, since the number of scientists in our world is progressively increasing and the volume of data predictably will continue to grow. For the foreseeable future there will remain a lot of "dark matter" to explore also in the field of life sciences. For example the forms of gene representation and expression in the globally represented microorganisms and viruses so far has only been subjected to a limited exploration. And what about possible unpredictable forms of life on anyone of the more than thousand so far recognized exoplanets discovered only outside our solar system in not too distant galaxies. Current predictions include the possible existence of more than 11 billion potentially habitable Earth-size planets in the whole Milky Way!

It is tantalizing to reflect on the possibility to look around the corner to see what the next major discovery might be! But the great charm of science is that this will remain hidden until another large step unexpectedly has been taken. And when that discovery is made many fellow scientists will say "yes of course that must be the explanation." Sometimes it is suggested that possibly many of the major discoveries have now been made and that the future may have less in store. This is definitely wrong. The enormous strength of the archive materials used to write the present kind of books on Nobel Prizes of more than 50 years ago is that they give a true presentation of the state of knowledge in the particular year examined and hence provide a firm reference to how inerasable new insights aggregate year by year. It can be projected that the written documents of the work of the present committees 50 years from now will be of equal value for historical analyses to those used for the present books. Knowledge will continue to grow and this will be of value for applications that improve the development of human civilizations, but equally important, it will help us to create societies of dignity, with respect for the value of each and every individual. To this should be added the cornucopia of joy that the growing knowledge potentially provides to all of us. Knowledge is a remarkable immaterial social glue in rapidly evolving human societies and it shows a number of attractive features. It is light to carry, it is a great asset in our interaction with other fellow human beings and it does not encourage greed.

But there is one more aspect to be considered. That is that knowledge can be not only used but also misused. The saying that knowledge is power — *scientas potentas est* — is attributed to Francis Bacon. The influential psychiatrist of the early 20th century Ernst Kretschmer wrote a book entitled *Geniale Menschen* (Brilliant People). In this book he compared the professional involvement of scientists with that of the artist or the prophets. He referred to the fact that these categories of humans "… are often restrained by a *daimonium*, by intense passions and to a surprising extent intuition." To Socrates daimonium was his inner voice that restrained him from doing the wrong things. But what is right and what is wrong must be interpreted in a cultural context. It has changed over time and will continue to change. In this context it can be added that the first Nobel conference arranged by the Nobel Foundation in 1969 and edited by Arne Tiselius and Sam Nilsson had the title *The Place of Value in a World of Facts*. This theme is certainly not new. Carl Linnaeus was only thirty-two years old when he became one of the five founding fathers of the Royal Swedish Academy of Sciences established in 1739 and he was its first president. When he left this honorary position after three months on October 3, he gave a lecture on the curious lives of insects. Apparently he was

The young Carl von Linné (1707–1778).

the leading world authority on plants and introduced his sexual system for classification of them, but he also had many other interests in the wide field of biology and even mineralogy. He ended his lecture highlighting "reason" as the most important quality of humans and finished by citing the Latin sentence "Vivitur ingenio, caetera mortis erunt" — the creative spirit lives, but all the rest is transient.

References

Chapter 1

1. Mukherjee, S. (2010) *The Emperor of All Maladies. A Biography of Cancer.* Scribner, New York.
2. Norrby, E. (2016) *Nobel Prizes and Notable Discoveries.* World Scientific, Singapore.
3. Norrby, E. (2013) *Nobel Prizes and Nature's Surprises.* World Scientific, Singapore.
4. Lewis, S. (1925) Arrowsmith, Jonathan Cape, London.
5. Fangerau, H.M. (2006) The novel Arrowsmith, Paul de Kruif (1890–1971) and Jacques Loeb (1859–1924): A literary portrait of "medical science". *J. Med. Ethics; Medical Humanities* 32:82–87.
6. Norrby, E. (2010) *Nobel Prizes and Life Sciences.* World Scientific, Singapore.
7. De Kruif, P. (1926) *Microbe Hunters.* Harcourt, New York.
8. Karlfeldt, E.A. (1931) *Nobelpriset I litteratur för år 1930.* Introductory speech in Swedish with an associated English translation. In *Les Prix Nobel en 1930.* Imprimerie Royale, P.A. Norstedt & Söner, Stockholm, pp. 42–45.
9. Liljestrand, G. and Bernhard, C.G. (1972) The Prize in Physiology or medicine. In Odelberg, W. (Coordinating Editor) *Nobel. The Man and His Prizes.* American Elsevier Publishing Co.
10. Andrewes, C.H. (1971) Francis Peyton Rous. 1879–1970. In *Biogr. Mems. Fell. R. Soc.* 17:643–662.
11. Stolt, C.-M., Klein, G. and Jansson, A.T.R. (2004) An analysis of a wrong Nobel Prize —Johannes Fibiger, 1926: A study in the Nobel Archives. *Cancer Research* 92:1–12.
12. Kildebaek-Nielsen, A., Thorling, E. (2001) Johannes Fibiger (1926) Backing the wrong horse? In *Neighbouring Nobel. The History of Thirteen Danish Nobel Prizes.* Aarhus University Press, pp. 461–493.
13. Bartholomew, J.R. (2002) Katsusaburo Yamagiwa's Nobel candidacy: physiology or medicine in the 1920s. In *Historical Studies in the Nobel Archives. The Prizes in Science and Medicine,* Universal Academy Press, Tokyo, pp. 107–131.
14. Sankaran, N. and van Helvoort, T (2016) Andrewes's Christmas fairy tale: Atypical thinking about cancer aetiology in 1935. *Royal Soc., Notes Rec.* 70, 175–201.

15. Murphy, F.A. (2016) Historical perspective: What constitutes discovery (of a new virus)? *Advances in Virus Research* 95: 197–220.
16. Van Regenmortel, M.H.V. (2010) Nature of viruses. In van Regenmortel, M.H.V. and Mahy, B.W.J. (Ed.) *Desk Encyclopedia of General Virology*, Academic Press, London, pp. 19–23.
17. Bookchin, D., Schumacher, J. (2004) *The Virus and the Vaccine: Contaminated Vaccine, Deadly Cancers and Government Neglect*. St. Martin's Press, New York.
18. Wadman, M. (2017) *The Vaccine Race. How scientists used human cells to combat killer viruses*. Transworld Publishers, London.
19. Murphy, F.A. (2012) *The Foundations of Virology*. Infinity Publishing, West Conshohocken, PA.

Chapter 2

1. Mukherjee, S. (2010) *The Emperor of All Maladies. A Biography of Cancer*. Scribner, New York.
2. Weinberg, R.A. (2014) *The Biology of Cancer*. 2nd ed. Garland Science, Taylor & Francis, New York.
3. Norrby, E. (2013) *Nobel Prizes and Nature's Surprises*. World Scientific, Singapore.
4. Norrby, E. (2016) *Nobel Prizes and Notable Discoveries*. World Scientific, Singapore.
5. Norrby, E. (2010) *Nobel Prizes and Life Sciences*. World Scientific, Singapore.
6. Stolt, C.-M. (2002) Why did Freud never receive the Nobel Prize? In *Historical Studies in the Nobel Archives* (Ed. Crawford, E.), Universal Academy Press, Tokyo, pp. 95–105.
7. Klein, G. (1967) Introductory speech to the Nobel Prize in physiology or medicine (English translation of a speech given in Swedish). In *Les Prix Nobel en 1966*, Almqvist & Wiksell International, Stockholm, pp. 49–52.
8. Rous, P. (1967). The challenge to man of the neoplastic cell. In *Les Prix Nobel en 1966*, Almqvist & Wiksell International, Stockholm, pp. 162–171.
9. Huggins, C. (1967). Endocrine-induced regression of cancers. In *Les Prix Nobel en 1966*, Almqvist & Wiksell International, Stockholm, pp. 172–182.
10. Dulbecco, R. (1976) Francis Peyton Rous, 1879–1970. A Biographical Memoir. National Academy of Sciences, Washington, D.C., pp. 275–306.
11. Ohlmarks, Å. (1969) *Nobelpristagarna* (in Swedish), Forsell, G.B. (ed.). F. Beck & Son, Stockholm.
12. Murphy, F.A. (2012) *The Foundations of Virology*. Infinity Publishing, West Conshohocken, PA.

Chapter 3

1. Rous, P. (1967) The challenge to man of the neoplastic cell. In *Les Prix Nobel en 1966*, Almqvist & Wiksell International, Stockholm, pp. 162–171.
2. Bishop, J.M. (2003) *How to Win the Nobel Prize (An Unexpected Life in Science)*. Harvard University Press, Cambridge, Mass.

3. Todaro, G.J. and Huebner, R.J. (1972) The viral oncogene hypothesis: New evidence. *Proc. Natl. Acad. Sci.* 69:1009–1015.
4. Norrby, E. (2016) *Nobel Prizes and Notable Discoveries*. World Scientific, Singapore.
5. Wadman, M. (2017) *The Vaccine Race. How scientists used human cells to combat killer viruses*. Transworld Publishers, London.
6. Norrby, E. (2010) *Nobel Prizes and Life Sciences*. World Scientific, Singapore.
7. Norrby, E. (2013) *Nobel Prizes and Nature's Surprises*. World Scientific, Singapore.
8. Temin, H.M. (1976) The DNA provirus hypothesis. In *Les Prix Nobel en 1975*. Imprimerie Royale, P.A. Norstedt & Söner, Stockholm, pp. 185–203.
9. Baltimore, D. (1976) Viruses, polymerases and cancer. In *Les Prix Nobel en 1975*. Imprimerie Royale, P.A. Norstedt & Söner, Stockholm, pp. 155–166.
10. Reichard, P. (1976) Introductory speech to the Nobel Prize in physiology or medicine. In *Les Prix Nobel en 1975*. Imprimerie Royale P.A. Norstedt & Söner, Stockholm, pp. 26–27.
11. Norrby, E. (1990) Introductory speech to the Nobel Prize in physiology or medicine. In *Les Prix Nobel en 1989*. Almqvist & Wiksell International, Stockholm, pp. 22–23.
12. Bishop, J.M. (1990) Retroviruses and oncogenes II. *In Les Prix Nobel en 1989*. Almqvist & Wiksell International, Stockholm, pp. 220–238.
13. Varmus, H.F. (1990) Retroviruses and oncogenes I. In *Les Prix Nobel en 1989*. Almqvist & Wiksell International, Stockholm, pp. 194–212.
14. Stehelin, D., Varmus, H.E., Bishop, J.M. and Vogt, P.K. (1976) DNA related to the transforming gene(s) of Rous sarcoma virus is present in normal avian DNA. *Nature* 260:70-73.
15. Fischer, E.H. (1993) Protein phosphorylation and cellular regulation, II. In *Les Prix Nobel en 1992*. Almqvist & Wiksell International, Stockholm, pp. 121–139.
16. Hall, K. (1987) Introductory speech to the Nobel Prize in physiology in 1986. In *Les Prix Nobel en 1986*, Almqvist & Wiksell International, Stockholm, pp. 25–27.
17. Weinberg, R.A. (2014) *The Biology of Cancer*, 2nd ed. Garland Science, Taylor & Francis.
18. Hartwell, L.H. (2002) Yeast and cancer. In *Les Prix Nobel en 2001*. Almqvist & Wiksell International, Stockholm, pp. 246–265.
19. Horvitz, H.R. (2003) Worms, life and death. In *Les Prix Nobel en 2002*. Almqvist & Wiksell International, Stockholm, pp. 320–351.
20. Larsson N.-G. (2017) Introductory speech to the Nobel Prize in physiology or medicine in 2016. In *The Nobel Prizes 2016*. Thomson-Shore, Dexter, MI.
21. Mitelman, F. and Heim, S. (2015) How it all began: cancer cytogenetics before sequencing. In *Cancer Cytogenetics: Chromosomal and Molecular Genetic Aberrations of Tumor Cells*, 4th ed. John Wiley & Sons, pp. 1–10.
22. Mertens, F., Johansson, B., Fioretos, T. and Mitelman, F. (2015) The emerging complexity of gene fusions in cancer. *Nat. Rev, Cancer* 15:371–381.
23. Uhlén, M., Zhang, C., Lee, S., Sjöstedt, E. Fagerberg, L. et al. (2017) A pathology atlas of the human cancer transcriptome. *Science* 357:660–671.
24. Garber, J.E. and Offit, K. (2017) Hereditary cancer predisposition syndromes. *J. Clin. Onc.* 23:1–40.
25. zur Hausen, H (2009) The search for infectious causes of human cancers: where and why. In *Les Prix Nobel en 2008*. Edita Norstedts Tryckeri, Stockholm, pp. 223–243.

26. Miller, J.F.A.P. and Sadelain, M. (2015) The journey from discoveries in fundamental immunology to cancer immunotherapy. *Cancer Cell* 27:1–11.
27. Chen, D.S. and Mellman, I. (2013) Oncology meets immunology: The cancer-immunity cycle. *Immunity* 39:1–10.
28. Brudno, J.N. and Kochenderfer, J.N. (2017) Chimeric antigen receptor T-cell therapies for lymphomas, *Nature Reviews, Clinical Oncology.* DOI 10,1038/nrclinonc 2017.128.
29. Rosenberg, S.A. and Restifo, N.P. (2015) Adoptive cell transfer as personalized immunotherapy for human cancer. *Science* 348:62–68.
30. Rajagopala, S.V., Vashee, S., Oldfield, L.M., Suzuki, Y., Venter, J.C. Telenti, A. and Nelson, K.E. (2017) The human microbiome and cancer. *Canc. Prev. Res.* 6, 2017.
31. Darnell, J. (2011) *RNA. Life's Indispensible Molecule.* Cold Spring Harbor Laboratory Press, New York.
32. Griffiths, D.J. (2001) Endogenous retroviruses in the human genome sequence. *Genome Biol.* 2(6): reviews pp. 1–5.
33. Kassiotis, G. (2014) Endogenous retroviruses and the development of cancer. *J. Immunol.* 192:1343–1349.
34. zur Hausen, H. (2009) Papillomaviruses in the causation of human cancers — a brief historical account. *Virology* 384:260–265.
35. Cattaneo, R. and Russell, S.J. (2017) How to develop viruses into anticancer weapons. *PLOS Pathogens.* DOI:10.1371/journal.ppat. 1006190. March.
36. Hanahan, D. and Weinberg, R.A. (2000) The hallmarks of cancer. *Cell* 100:57–70.
37. Hanahan, D. and Weinberg, R.A. (2011) Hallmarks of Cancer: The Next Generation. *Cell* 144:646–674.
38. Murphy, F.A. (2012) *The Foundations of Virology.* Infinity Publishing, West Conshohocken, PA.
39. Ohlmarks, Å. (1969) *Nobelpristagarna* (in Swedish), Forsell, G.B. (ed.). F. Beck & Son, Stockholm.

Chapter 4

1. Granit, R. (1941) *Ung mans väg till Minerva* (in Swedish). P. A. Norstedt & Söner, Stockholm.
2. Granit, R. (1983) *Hur det kom sig. Forskarminnen och motiveringar* (in Swedish). P.A. Norstedt & Söner, Stockholm.
3. Enroth-Cugell, C. (1994) Ragnar Granit (30 October 1900–12 March 1991). *Proceedings of the American Philosophical Society* 138: 328–333.
4. Grillner, S. (1995) Ragnar Granit. 30 October 1900–12 March 1991. *Biogr. Mems. Fell. R. Soc.* 41: 184–197.
5. Norrby, E. (2016) *Nobel Prizes and Notable Discoveries.* World Scientific, Singapore.
6. Norrby, E. (2013) *Nobel Prizes and Nature's Surprises.* World Scientific, Singapore.
7. Granit, R. and Svaetichin, G. (1939) Upsala Läkaref. *Handl.* 65:161–177.
8. Bernhard, C.G. (2000) Huset på höjden. (in Swedish). Atlantis, Stockholm.
9. Granit, R. (1960) *Neurofysiologi* (in Swedish). In *Karolinska Mediko-Kirurgiska Institutets historia* 1910–1960. Almqvist & Wiksell, Stockholm, pp. 170–199.

10. Norrby, E. (2010) *Nobel Prizes and Life Sciences*. World Scientific, Singapore.
11. Granit, R. (1947) *Sensory Mechanisms of the Retina*. Oxford Univ. Press, London.
12. Granit, R. (1955) *Receptors and Sensory Perception*. Yale University. Press, New Haven.
13. Granit, R. (1962) In *The Eye*. Editor H. Davson. Academic Press, New York and London, pp. 534–796.
14. Granit, R. (1966) *Charles Scott Sherrington. An appraisal*. Thomas Nelson & Son, London.
15. Granit, R. (1970) *The Basis of Motor Control*. Academic Press, New York and London.
16. Granit, R. (1977) *The Purposive Brain*. MIT Press, Cambridge, Mass.
17. Granit, R. (Ed.) *Utur Stubbotan Rot* (in Swedish). Essays commemorating 200 years after Linne's death. P.A. Norstedt & Söner, Stockholm.

Chapter 5

1. Norrby, E. (2013) *Nobel Prizes and Nature's Surprises*. World Scientific, Singapore.
2. Norrby, E. (2016) *Nobel Prizes and Notable Discoveries*, World Scientific, Singapore.
3. Ratliff, F. (1990) *Haldan Keffer Hartline. A Biographical Memoir*, National Academy of Sciences. pp. 197–213.
4. Granit, R., Ratliff, F. (1985) Haldan Keffer Hartline, 22 December 1903–18 March 1983. *Biogr. Mems. Fell. R. Soc.* 31, 262–292.
5. Granit, R. (1983) *Hur det kom sig. Forskarminen och motiveringar* (in Swedish). P.A. Norstedt & Söner, Stockholm.
6. Bernhard, C.D. ed. (1966) *The Functional Organization of the Compound Eye*. Pergamon Press, Oxford.
7. Medawar, J. Pyke, D. (2000) *Hitler's Gift. The True Story of the Scientists Expelled by the Nazi Regime*. Arcade Publishing, New York.
8. Dowling, J.H. (2000) George Wald, November 18, 1906–April 12, 1997. *Biogr. Mems. Fell. R. Soc.* 78, 297–316.
9. Bernhard, C.G. (2000) Huset på höjden (in Swedish). Atlantis, Stockholm.
10. Norrby, E. (2010) *Nobel Prizes and Life Sciences*. World Scientific, Singapore.
11. Eigen, M. (2013) *From Strange Simplicity to Complex Familiarity*. Oxford University Press.
12. Bernhard, C.G. (1968) The Nobel Prize in physiology or medicine. Introductory speech. In *Les Prix Nobel en 1967*. Imprimerie Royale, P.A. Norstedt & Söner, Stockholm, pp. 63–66.
13. Granit, R. (1968) The development of retinal neurophysiology. In *Les Prix Nobel en 1967*, Imprimerie Royale, P.A. Norstedt & Söner, pp. 232–241.
14. Granit, R. (1962) In *The Eye*. Editor H. Davson. Academic Press, New York and London, pp. 534–796.
15. Hartline, H.K. (1968) Visual receptors and retinal interaction. In *Les Prix Nobel en 1967*, Imprimerie Royale, P.A. Norstedt & Söner, Stockholm, pp. 242–259.
16. Wald, G. (1968) The molecular basis of visual excitation. In *Les Prix Nobel en 1967*, Imprimerie Royale, P.A. Norstedt & Söner, Stockholm, pp. 260–280.
17. Ohlmarks, Å. (1969) *Nobelpristagarna* (in Swedish), Forsell, G.B. (ed.). F. Beck & Son, Stockholm.

Chapter 6

1. Norrby, E. (2010) *Nobel Prizes and Life Sciences*. World Scientific, Singapore.
2. Norrby, E. (2013) *Nobel Prizes and Nature's Surprises*. World Scientific, Singapore.
3. Norrby, E. (2016) *Nobel Prizes and Notable Discoveries*, World Scientific, Singapore.
4. Exton, J.H. (2013) *Crucible of Science*. Oxford University Press, New York.
5. Watson, J.D. (1963) The involvement of RNA in the synthesis of proteins. In *Les Prix Nobel en 1962*. Imprimerie Royale, P.A. Norstedt & Söner, Stockholm, pp. 155–178.
6. Friedberg, E.C. (2014) *A Biography of Paul Berg. The Recombinant DNA Controversy revisited*. World Scientific, Singapore.
7. Crick, F. (1988) *What Mad Pursuit: A Personal View of Scientific Discovery*. Basic Books, New York.
8. Ochoa, S. (1960) Enzymatic synthesis of ribonucleic acid. In *Les Prix Nobel en 1959*. Imprimerie Royale, P.A. Norstedt & Söner, Stockholm, pp. 146–164.
9. Judson, H. (1996) *The Eighth Day of Creation: Makers of the Revolution in Biology*, expanded edition. Cold Spring Harbor Laboratory Press, New York.
10. Nirenberg, M. (2004) Historical review: Deciphering the genetic code — a personal account. *Trends in Biochemical Sciences* 29:46–54.
11. Wilkins, M.H.F. (1963) The molecular configuration of nucleic acids. In *Les Prix Nobel en 1962*. Imprimerie Royale, P.A. Norstedt & Söner, Stockholm, pp. 93–125.
12. Watson, J.D. (with Andrew Berry) (2004) *DNA. The Secret of Life*. Arrow Books, London.
13. Mukherjee, S. (2016) *The gene. An intimate story*. Scribner, New York.
14. Ridley, M. (2006) *Francis Crick: Discoverer of the genetic code*. Harper Collins, New York.
15. Crick, F.C.H. (1963) On the genetic code. In *Les Prix Nobel en 1962*. Imprimerie Royale, P.A. Norstedt & Söner, Stockholm, pp. 179–187.
16. Olby, R. (1974) *The Path to the Double Helix: The Discovery of DNA*. Macmillan, London; revised edition (1994), Dover, New York.
17. Portugal, F.H. and Cohen, J.S. (1977) *A Century of DNA. A History of the Discovery of the Structure and Function of the Genetic Substance*. MIT Press, Cambridge, Mass.
18. Portugal, F.H. (2015) *The Least Likely Man. Marshall Nirenberg and the Discovery of the Genetic Code*. MIT Press, Cambridge, Mass.
19. Sela, M. (2004) My world through science. *Comprehensive Biochemistry* 43:1–100.
20. Nirenberg, M.W. and Matthei, J.H. (1961) *Proc. Natl. Acad. Sci.* 47:1588–1592
21. Nirenberg, M. (1969) The genetic code. In *Les Prix Nobel en 1968*, Imprimerie Royale, P.A. Norstedt & Söner, Stockholm, pp. 221–241.
22. Goldstein, J.L. and Brown, M.S. (2012) *Science* 23:1033–34.
23. Murphy, F.A. (2012) *The Foundations of Virology*. Infinity Publishing, West Conshohocken, PA.

Chapter 7

1. Watson, J.D. (1963) The involvement of RNA in the synthesis of proteins. In *Les Prix Nobel en 1962*. Imprimerie Royale, P.A. Norstedt & Söner, Stockholm, pp. 155–178.
2. Norrby, E. (2016) *Nobel Prizes and Notable Discoveries*, World Scientific, Singapore.

3. Darnell, J. (2011) *RNA. Life's Indispensible Molecule*. Cold Spring Laboratory Press, New York.
4. Norrby, E. (2010) *Nobel Prizes and Life Sciences*. World Scientific, Singapore.
5. Todd, A. (1958) Synthesis in the study of nucleotides. In *Les Prix Nobel en 1957*. Imprimerie Royale, P.A. Norstedt & Söner, Stockholm, pp. 119–133.
6. Norrby, E. (2013) *Nobel Prizes and Nature's Surprises*. World Scientific, Singapore.
7. Holley, R.W. (1969) Alanine transfer RNA. In *Les Prix Nobel en 1968*, Imprimerie Royale, P.A. Norstedt & Söner, Stockholm, pp. 183–195.
8. Khorana, H.G. (1969) Nucleic acid synthesis in the study of the genetic code. In *Les Prix Nobel en 1968*, Imprimerie Royale, P.A. Norstedt & Söner, Stockholm, pp. 196–220.
9. Nirenberg, M. (1969) The genetic code. In *Les Prix Nobel en 1968*, Imprimerie Royale, P.A. Norstedt & Söner, Stockholm, pp. 221–241.
10. Crick, F.C.H. (1963) On the genetic code. In *Les Prix Nobel en 1962*. Imprimerie Royale, P.A. Norstedt & Söner, Stockholm, pp. 179–187.
11. Sakmar, T.P. (2017) *Biogr. Mems. Fell. R. Soc.* Har Gobind Khorana (1922–2011). *Proceedings of the American Philosophical Society* 161:268–275.
12. Khorana, H.G. (2000) World Scientific Series in 20th Century Biology. Vol. 5. *Chemical Biology. Selected Papers of H. Gobind Khorana*.
13. Nelson, P.G. (2011) Biographical memoirs. Marshall Warren Nirenberg (1927–2010). *Proceedings of the American Philosophical Society* 155:368–375.
14. Venter, J.C. (2007) *A Life Decoded. My Genome: My Life*. Penguin Group, New York.
15. Klug, A. (1983) From macromolecules to biological assemblies. In *Les Prix Nobel en 1982*, Almqvist & Wiksell International, Stockholm, pp. 93–125.
16. Ohlmarks, Å. (1969) *Nobelpristagarna* (in Swedish), Forsell, G.B. (ed.). F. Beck & Son, Stockholm.

Chapter 8

1. Ehrenberg, M. (2015) The Nobel Prize in chemistry. Laudatory speech. In *The Nobel Prizes 2014*. Science History Publications, Watson Publishing International, Mass. pp. xxv–xvi.
2. Ruska E. (1987) The development of the electron microscope and of electron microscopy. In *Les Prix Nobel en 1986*. Almqvist & Wiksell International, Stockholm, Sweden, pp. 58–83.
3. Norrby, E. (2010) *Nobel Prizes and Life Sciences*. World Scientific, Singapore.
4. Norrby, E. (2013) *Nobel Prizes and Nature's Surprises*. World Scientific, Singapore.
5. Norrby, E. (2016) *Nobel Prizes and Notable Discoveries*, World Scientific, Singapore.
6. Kruger, D.H., Schneck, P., Gelderblom, H.R. (2000) Helmut Ruska and the visualisation of viruses. *The Lancet* 355:1713–17.
7. Gelderblom, H.R., Krüger, D.H. (2014) Helmut Ruska (1908–1973): His role in the evolution of electron microscopy in the life of sciences, and especially virology. *Advances in Imaging and Electron Physics* 182:1–79.
8. Van Gerkom, J. van Delft, D., van Helvoort, T. (2018) The electron microscopes: A critical study. *Advances in Imaging and Electron Physics* 205: 1–137.

9. Claude, A. (1975) The coming of age of the cell. In *Les Prix Nobel en 1974*. Imprimerie Royale, P.A. Norstedt & Söner, Stockholm, pp. 133–137.
10. Palade, G.E. (1975) Intracellular aspects of the process of protein secretion. In *Les Prix Nobel en 1974*. Imprimerie Royale, P.A. Norstedt & Söner, Stockholm, pp. 167–196.
11. Noller, H.F. (1999) On the origin of the ribosome: coevolution of subdomains of tRNA and rRNA. *The RNA World*. Second Edition. Cold Spring Laboratory Press. pp. 197–219.
12. Koonin, E.V., Novozhilov, A.S. (2017) Origin and evolution of the universal genetic code. *Ann. Rev. Genetics* 51:45–62.
13. Sengupta, S., Higgs, P.G. (2015) Pathways of genetic code evolution in ancient and modern organisms. *J. Mol. Evol.* 80:229–243.
14. Liu, C.C., Schultz, P.G. (2010). Adding new chemistry to the genetic code. *Annu. Rev. Biochem.* 79:413–444.
15. Crick, F.H. (1968) The origin of the genetic code. *J.Mol.Biol.* 38:367–379.
16. Lagerkvist, U. (2012) *The Periodic Table and a Missed Nobel Prize*. World Scientific, Singapore.
17. Cech, T.R. (1990) Self-splicing and enzymatic activity of an intervening sequence RNA from *Tetrahymena*. In *Les Prix Nobel en 1989*, Almqvist & Wiksell International, Stockholm, pp. 165–188.
18. Altman, S. (1990) Enzymatic cleavage of RNA by RNA. In *Les Prix Nobel en 1989*, Almqvist & Wiksell International, Stockholm, pp. 140–160.
19. Samanta, B., Joyce, F.J. (2017) A reverse transcriptase ribozyme. DOI: https://doi.org/10.7554/eLIfe.31153.001, p1–10.
20. Cech, T.R. (2009) Crawling out of the RNA world. *Cell* 136: 599–603.
21. Diener, O.D. (2016) Viroids: "living fossils" of primordial RNAs? *Biol. Direct* 11:15 (doi 10.1186/s13062-016-0116-7)
22. Darnell, J (2011) *RNA. Life's Indispensible Molecule*. Cold Spring Harbor Laboratory Press, Cold Spring Harbor, New York.
23. Singer, M., Berg, P. (1991) *Genes and Genomes: A Changing Perspective*. Mill Valley: University Science Books.
24. Murphy, F.A. (2012) *The Foundations of Virology*. Infinity Publishing, West Conshohocken, PA.

Index

A

Abbe, Ernst, 328
actinomycin, 93
adenine, 225, 241
adenosine, 214, 226, 237, 285–286, 297, 302, 356
adenoviruses, 28, 45, 87, 357–359
adrenal glands, 53, 55, 56
Adrian, Edgar D., 147, 159, 162, 193, 195, 217, 217f, 220
Adrian, Marc, 345
Aenid (Vergil), 79–81
af Wirsén, C. David, 81, 82
age, of recipients, 318
aging, of cells, 106, 111–114
Agnon, Joseph, 67
Åkerman, Jules H., 51
Åland controversy, 168–169
alanine, 292, 297, 299f, 301–304, 303f, 316, 317, 351
Albertsson, Per-Åke, 298, 299
Alhazen, 327
Allfrey, Vincent G., 270–271
Allison, Anthony C., 45
Allison, James P., 125
Altman, Sidney, 360, 362, 362f
Alzheimer's disease, x, 87, 307, 322
amino acids: atypical, 350–351
  codons and, 295f, 348, 351 (*see also* codons)
  DNA and, 246 (*see also* deoxyribonucleic acid)
  enzymes and, 229–230, 351
  genetic code and (*see* genetic code)
  levo/dextro, 254
  life and, 112, 364–381, 367
  Nirenberg system and, 276
  non-canonical, 350–351
  nucleotides and, 247, 347 (*see also* nucleotides)
  polypeptides, 231 (*see also* polypeptides)
  protein synthesis and, 245, 276–277, 301–303 (*see also* protein synthesis)
  ribosomes and, 345
  RNA and, 258, 297, 298, 351 (*see also* ribonucleic acid)
  sequencing, 301–302, 319
  synthesis of, 246
  translation and, 231
  two forms of, 254
  wobble phenomenon, 347–348, 351
  See also specific types, topics
Anderson, William R., 169
Andrewes, Christopher H., 16, 25–29
androgen, 58
Ångström, Anders, 346, 346f
animal studies, 15, 23, 127, 163, 173, 207. *See also specfic types, topics*
antibiotics, 11, 292, 320
antibodies, x, 121, 124, 125, 376
antigens, 124
anti-metabolites, 50
apoptosis, 105, 110, 113–114, 135
Arber, Werner, 357, 357f
Arnold, Frances, 376
Arnon, Ruth, 257
Arrowsmith (Lewis), 11–17, 12f, 199
autoimmune diseases, 29, 117, 122, 125, 130, 133, 202, 256, 376
autophagy, 113, 114

Avery, Oswald, vii, 226, 242, 243f, 244, 245

B

B cells, 114, 118, 122
Bacon, Francis, 380
bacteria, 328
   atypical amino acids, 350
   bacilli, 328
   cancer and, 127, 130
   genetic material, 227
   infectious disease and, 328
   inflammation and, 127
   nucleotides and, 285
   protein metabolism, 91, 273, 349
   restriction enzymes, 357
   RNA and, 229–230, 252, 316, 344
   ultrafiltration and, 4, 7, 9, 25, 31, 329
   viruses and (see bacteriophages)
bacteriophages, viii, 18, 19, 276
   Arrowsmith and, 14
   Burnet and, 27–28
   genetics and, 75, 226–227
   H. Ruska and, 335
   labelling, 226
   latency, 371
   nucleic acid, 226
   replication and, 35
   RNA and, 254
   tadpole structure, 333
   ultrafiltration of, 335
Baltimore, David, 42, 94–96, 94f, 97
band-wagon effect, 315–316
Bang, Oluf, 8, 8f
Banting, Frederick, 51
Barnard, F. I., 18
Barr, Murray L., 61
Beijerinck, Martinus W., 31
Békésy, Georg von, 44, 176
Beljansky, Mirko, 259
Ben May Laboratory, 53, 57, 59
Benzer, Seymor, 248, 274, 276
Berg, Paul, 230, 236, 374f
Berger, Arieh, 257
Bergstrand, Hilding, 29–30, 31, 32
Bergström, Sune, 63
Bernhard, Carl Gustaf, 153, 153f, 170f
   autobiography, 217
   color and, 219
   flawed ceremony, 217
   Granit and, 158, 170, 175, 179, 197
   Hartline and, 197
Bertani, Giuseppe, 275, 276

Berzelius, Jöns Jacob, 286
Billroth, Theodor, 126
bird studies, 4, 9, 10, 24, 30, 101, 207
Bishop, J. Michael, 85, 98, 99f, 100, 104
Bittner, John, 38
Bittner factor, 35, 38, 43
Björklund, Bertil, 123–124
Black, James W., 50
Blackburn, Elizabeth H., 111, 112f
blood group system, 17, 64
Blumberg, Baruch, 132
Boman, Hans G., 11, 11f
Boveri, Theodor, 114
breast cancer, 53, 57, 73, 106, 120
Brenner, Sydney, 113, 245, 246f, 274, 276, 277
broad band elements, 173
Bronk, Detlev, 187, 187f, 188
Brown, Paul, 203
Buchtal, Fritz, 177
Buist, John, 30
Bunson-Roscoe law, 186
Burkitt, Dennis P., 72, 118
Burnet, Frank Macfarlane, 27, 122

C

c-AMP. See cyclic adenosine monophosphate
carcinomas, 8, 36
carotenoids, 208, 211
Carrel, Alexis, 24, 50–51
Caskey, Tom, 264
Caspersson, Torbjörn, 38, 41, 61, 242
   Bittner factor and, 38
   chromosomes and, 116, 117
   nucleic acid and, 35, 240, 287
   Rous and, 35–36, 41
   viruses and, 35
Castle, William E., 38
cat studies, 163, 180, 186
categories, of disciplines, 60–61
Cattaneo, Roberto, vi
Cech, Thomas R., 360, 360f, 361, 362
cecropines, 11
cells, 119, 321, 340
   aging of, 112–114
   central dogma, 35, 96–97, 231, 244
   communication and, 106, 109
   contact inhibition, 42, 112
   cycle, 3, 105, 109, 110, 113, 114, 135
   division of, 85, 109, 136
   electron microscopy, 341 (see also electron microscope)
   evolution of, 1–2

fusion, 107, 108
genetic material (*see* genetic code)
growth factors, v, 102, 103, 130–131
information and, 106
lateral transfer in, 371
LUCA and, 1–2
membrane, 107, 259, 329
mitochondria, 307, 349–350
multicellular organisms, 2
mutation, 84, 103, 120, 356
primitive forms of, 371
replication of, 3, 42, 102, 103, 130, 371
signaling in, 106
topography of, 341–342
transport in, 341
See also specific types, topics
cervical cancer, 131
Chain, Ernst B., 155
Chance, Britton, 63
Chanock, Stephen, vi
Chargaff, Erwin, vii, 226, 236–245, 236f, 271
base contents, 240
Chargaff rules, 237
chromatography, 237
nominations, 238, 240, 243
nucleotides and, 237, 239, 241
purine-pyrimidine, 242
species specificity, 242
Chase, Martha, 226
chemical weapons, 49
chemotherapy, 48–50, 59, 72–73, 106, 110, 137
children: Burkitt's disease, 72
congenital disorders, 307
EBV and, 118
leukemia, vi, 49, 50, 72, 89, 118
RS virus, 107
See also specific disorders
chromatography, 237, 287, 298, 301, 302
chromosomes: aneuploid lines, 321
banding technique, 117f
Barr body, 61
cancer and, 116–119
diploid lines, 88
DNA and, 225, 369
genetic disorders, 61–62, 116, 307
human, 115, 117, 117f
Ph chromosome, 116, 117
plants, 115
reverse transcriptase and, 96
telomers, 111–113, 369
translocation, 111

vision and, 148, 212
Chumakov, Mikhail, 354
cisplatin, 109–110
cis-trans isomerization, 203, 220, 223
Claude, Albert, 229, 340, 341
cloning, 101
clover leaf structure, 302, 303f, 317
codons: amino acids and (*see* amino acids)
DNA and (*see* deoxyribonucleic acid)
genetic code and, 248, 259, 261f, 262–267, 273, 292, 347–350, 350f (*see also* genetic code)
nucleotides and, 261, 313 (*see also* nucleotides)
RNA and, 295f (*see also* ribonucleic acid)
start/stop, 264, 274, 279, 296, 314, 348–350
triplets, 248, 261, 264–268, 295, 347
Cohen, Stanley, 102–103
Cold War, 236, 258
Coller, Fredrick A., 51
colon cancer, 21, 127
color: blindness, 148, 204, 211, 212
chromophoric sites, 223
cis-trans changes and, 203, 220, 223
early studies, 220
Granit and, 165–169, 177–180, 221
rhodopsin, 201–202, 209, 223, 319
rods and cones, 149–150, 150f, 162, 219
vision and, 151, 163–169, 177, 178, 200, 207, 211, 220 (*see also* eyes)
Wald and, 213
yellow spot, 211
Young–Helmholtz theory, 152, 164
Colton, Francis B., 62
computer-assisted tomography, 136, 345
continental drift, 140
cortisone, 58
Cournand, André F., 50
Craig, Lyman C., 301
Crick, Francis, 231, 258, 347, 364–365, 364f
double helix, 225–226, 237–248, 288, 355
hypothesis, 279
Nobel lecture, 245–248; RNA and, 297
triplets, 276
wobble hypothesis, 279, 347–348, 351
CRISP/Cas9 technology, 357
cryoelectron microscopy, 345, 360, 363
crystallography, 231, 343–346, 345
CTL-4 structure, 125
Curie, Marie, 48
Cutter episode, 40
cyclic adenosine monophosphate (c-AMP), 285

Index 393

cytosine, 226, 237, 241
cytotoxic drugs, 116, 117

D

dark field microscopy, 30
Darnell, James E., 356, 356f, 373
Darnell, James (Jim), viii, 230
Darwin, Charles, 148, 181
de Duve, Christian, 340, 341
de Kruif, Paul, 12–15, 13f
Dean, Henry R., 34
DeBakey, Michael, 91
deep sequencing, 137
Delbrück, Max, viii
Diener, Theodor O., 370, 370f
deoxyribonucleic acid (DNA), 96f, 100, 105, 258
  amino acids and, 246 (*see also* amino acids)
  base contents, 240
  base pairing, 244
  base sequences, 275
  central dogma and, 96, 231, 244, 246
  chromosomes and, 225, 368, 369 (*see also* chromosomes)
  double helix, 225–226, 237, 240–246, 288, 355, 367, 368
  evolution and, 1, 181 (*see also* evolution)
  four bases, 226
  genetic code and, 227, 373 (*see also* genetic code)
  genotoxins, 126
  information flow and, 97
  mitochondria and, 349–350
  mutagenesis, 49, 111, 130, 164, 319, 349 (*see also* mutations)
  nucleotides and, 88, 89, 226, 285, 286, 287f, 352, 354–355 (*see also* nucleotides)
  ORFs, 348, 349
  p53 gene, 110–111
  polymerases, 228–229
  proteins and, 225–226, 230, 349, 355 (*see also* protein synthesis)
  recombinant, 260
  redesign of, 357
  replication of, 93–94, 228, 229, 231, 368–369
  reverse transcription, 93, 94, 231–232
  RNA and, 94, 97, 229, 252, 291, 327–381, 359f, 368 (*see also* ribonucleic acid; specific topics)
  sequencing, 95, 119, 137, 275, 349
  species specificity, 242
  splicing, 360

synthesis of, 50, 350
transcription factors, 94, 231–232, 348, 372
viruses and, 86, 226, 252, 369, 371 (*see also* viruses)
diploid cell lines, 88
discovery, in science, 222, 325, 340, 375, 377, 380
  concept of, 66
  confirmation and, 20–21
  corroboration, 39
  defined, 156, 160, 244; luck and, 259
Djerassi, Carl, 62
DNA. *See* deoxyribonucleic acid
Doctor's plot (USSR), 354
Dodds, Edward C., 55, 60
dog studies, 52, 54, 57
Doisy, Edward A., 45, 56
Domagh, Gerhard, 71
double helix, 225–226, 237, 240–246, 288, 355, 367, 368
Dounce, Alexander, 244, 251
Dubochet, Jacques, 345
Duesberg, Peter, 99
Dulbecco, Renato, 42, 84, 97, 97f

E

E6 protein, 111
EBV. *See* Epstein-Barr virus
Eccles, John C., 146, 146f
economic sciences, 310–311
Eddy, Bernice, 39, 40, 40f
EGF. *See* epidermal growth factor
Eigen, Manfred, vii, 215–217, 216f, 376
Einstein, Albert, 187
electron microscope, 46, 329–345, 331f, 358–359, 381f
electrophoresis, 110
electroretinogram (ERG), 147, 172, 221
Elion, Gertrude, 50
Ellerman, Vilhelm, 8, 8f
Elliott, T. S., 319
Emperor of All Maladies, The (Mukherjee), 47
English language, 275–276
Engström, Arne, 337
Enroth-Cugell, Christina, 157
enzymes, 54, 376
  amino acids and, 229–230, 351 (*see also* amino acids)
  degrading, 253
  nucleic acid and, 228
  nucleotides and, 233, 295

proteins and, 367
restriction, 357
ribozymes and, 367–369
RNA and, 96, 253, 259, 298
viruses and, 96, 97
epidermal growth factor (EGF), 103
Epstein, Richard, 348
Epstein–Barr virus (EBV), 73, 118, 131
ERG. *See* electroretinogram
Eriksson, Nancy, 123
Erlander, Tage, 156, 162
estrogen, 52, 53, 55, 56, 57, 58, 73
etymology, of terms, 3, 72
evolution, 181, 202
    experimental, 376
    eyes and, 148–149 (*see also* eyes)
    genetic code and, 1, 348–351, 367, 371 (*see also* genetic code)
    in laboratory, 376
    natural selection, 148
    origin of life and, 365, 376
    survival of fittest, 3
    virocentric view, viii, 371
    See also specific topics
eyes, 162, 180, 186, 188, 193, 196
    amacrine cells, 164
    animals and, 163, 173, 207
    axons, 148, 149
    blind spot, 189
    camera and, 219
    cancer, 108–109
    carotenoids, 200, 208, 211
    chromoproteins, 211
    color and, 151–152, 200, 210 (*see also* color)
    Darwin on, 220
    ERG studies, 221
    evolution and, 148, 149
    Granit and, 158, 163, 164, 178 (*see also* Granit, R.)
    Hartline and, 195 (*see also* Hartline, H. K.)
    horseshoe crab and, 183–188, 221
    lateral inhibition, 147, 160, 188, 190, 195, 196, 198, 220
    natural selection and, 148
    ommatides, 196
    on-off phenomena, 162, 189, 221
    receptors, 147–150, 167, 170–174, 180, 184, 189, 193, 197, 203, 207, 211, 223
    retina, 148–149, 149f, 158, 163, 174, 189, 195, 203–210, 221, 357
    rods and cones, 149–150, 150f, 151, 162, 173, 203, 208, 222f, 223

threshold conditions and, 198
Wald and, 210

F

Fagraeus, Astrid, 124
fairy tale week, 68
fairy-story, of viruses, 25–27, 28
Farber, Sidney, 49, 49f
Feynman, Richard, 341
Fibiger, Johannes, 19–21, 34, 36, 60
fibromas, 23, 31, 33
finger-printing technique, 305
Fire, Andrew Z., 372, 373f
Fischer, Edmund H., 102
Fischer-Hjalmars, Inga, 70, 70f
fish studies, 174, 202, 207
Fleming, Alexander, 155
Flexner, Simon, 7, 7f, 9, 15, 16
Florey, Howard W., 33, 155
foot-and-mouth disease, 31
Forbes, Alexander, 192
Forssmann, Werner, T. O., 50
Frank, Joachim, 345
Franksson, Curt, 55
Fredga, Arne, 227, 290, 291
Freud, Sigmund, 62
Friberg, Sten, 65, 65f
Friend, Charlotte, 43
Friend virus, 43
Frisch, Karl von, 44
frog studies, 198, 201, 202
Frosch, Paul, 31

G

G proteins, 104, 319
Gahrton, Gösta, vi
Galileo, 327
gall bladder, 10
Gallo, Robert C., vi, vii, 102, 130, 262, 265
Gamow, George, 246, 247, 247f
Gard, Sven, 28, 29, 123, 334f
Gasser, Herbert S., 162, 166–167, 193
Geissendörfer, R., 55
Gelderblom, Hans, viii
genetic code, vii, 225–282, 310, 316f, 376
    alleles, 108–109, 110
    amino acids and, 347f (*see also* amino acids)
    bases in, 129
    central dogma, 96
    codons and (*see* codons)

Index 395

disorders, 307
DNA and (*see* deoxyribonucleic acid)
evolution and, 347–352, 367
genes, defined, 372, 373
genome analysis, 119, 129, 358, 360, 373
heredity and, 38, 311
Holley and, 313
Khorana and, 284, 290, 291, 314, 315
mammalian, 264
mutagenesis, 10, 28, 49, 111, 130, 164, 319, 349 (*see also* mutations)
mutation, 84, 103, 120, 356
Nirenberg and, 265, 266, 268, 269, 271, 273, 278, 280, 281
nucleic acids (*see* nucleic acids)
nucleotides and, 245, 262, 263, 282, 283, 284, 347
overlap in, 355
phenylalanine, 256
proteins, 304, 315 (*see also* protein synthesis)
restriction enzymes and, 358
RNA and (*see* ribonucleic acid)
sequencing, 119, 125, 136, 234, 252, 305, 319, 349
stuttering, 355–356
synthetic, 274, 284
triplets, 247, 266, 268, 273–276, 295, 313
tumors and, 48
universality, 264, 314
geological history, 139–142, 181–183
germ-free animals, 127
Gilbert, Walter, 352
Gilman, Alfred G., 104
Glass, John, viii
Golden Rice, 204
Goldstein, Joseph, vi, 265
Göthlin, Gustaf, 152
Gowans, James L., 64
Grand Canyon (U. S.), 192
Grandin, Karl, ix, x, 78, 78f
Granit, Ragnar, vi, vii, 62, 77, 143, 143f, 157f, 161f, 215f, 220f
   Bernhard and, 158, 170, 175, 179, 197
   color vision and, 165–169, 177–180, 221
   dominator-modulator theory, 163, 165, 173, 178, 221
   education, 144
   electroretinogram and, 172, 174
   eyes and, 158, 163–164, 174, 178
   gamma system, 174
   Hartline and, 171, 173–179, 190–197, 214
   international contacts, 145
   lateral inhibition, 147, 160, 221
   microelectrodes, 162
   muscle control, 159, 177
   neural studies, 147; Nobel lecture, 221
   nominations, 161–166, 170–171, 175–177
   on/off elements, 162, 166–167
   sailing and, 191–192
   Sherrington and, 164
   summer house, 176f
   vitamin A, 206
   Wald and, 205–207, 208, 214
   World War II and, 152, 154
   as writer, 160
   Zotterman and, 164, 171, 194
Greider, Carol W., 111, 112f
Grillner, Sten, vii
Gross, Ludwig, 36, 43, 39f
growth factors, v, 102, 103, 130–131
Grubb, Rune, 64–65
Grunberg-Manago, Marianne, vii, 234, 234f, 259, 277
Grunberg-Manago enzyme, 235
guanidine, 241, 281
guanine, 104, 225, 237, 244
Gullstrand, Allvar, 50
Gustafsson, Bengt, 63, 280
Gustav VI Adolf, 78, 153, 218, 218f
Gye, William E., 18, 25

H

Haeckel, Ernst, 144
Hall, Ben, 354
Halstead, William Steward, 5f, 6
Hammarsten, Einar, 20, 169, 240, 242, 243
hammer-head enzyme, 369–370
Hansemann, David von, 114
Hansson, Göran, 276
Harris, Henry, 107, 107f
Harrison, J. Hartwell, 58
Hartline, Haldane K., 62, 147, 184f, 215f, 220f
   Bernhard, and, 197
   Bronk and, 188
   as experimenter, 190
   family, 190
   Granit and, 167, 173, 175–179, 190–197, 214
   horseshoe crab and, 186–187, 188, 221
   lateral inhibition, 160, 188, 196–197, 198, 220, 221
   neural networks, 189
   Nobel lecture, 221–222

nominated, 166, 193–198
on-off phenomena, 189, 196, 221
Ratliff and, 197–198
retina and, 186, 195
sailing and, 191–192
Wald and, 214–215
Zotterman and, 194
Hartline-Ratliff equations, 197–198
Hartridge, H., 165
Hartwell, Leland H., 109
Haselkorn, Robert, vii
Hayflick, Leonard, 113
Hayflick limit, 113
Hecht, Selig, 152, 185–186, 185f
HeLa cells, 88, 113
Helicobacter pylori, 126
Hellström, John, 54, 55, 58
Helmholtz, Hermann von, 151
Henderson, Richard, 345
Henschen, Folke, 18, 20, 24, 33, 36
hepatitis viruses, 132, 369–371
Heppel, Leon, 259, 262
heredity, 38, 311. *See also* genetic code
herpes virus, 73, 133
Hershey, Alfred D., viii, 226
Hess, Walter, 51
Hilleman, Maurice, 41
Hitching, George H., 50
Hitler, Adolf, 200, 237
HIV/AIDS, 29, 122, 130, 133
Hoagland, Mahlon B., 229, 278
Hodges, Clarence, 53
Hodgkin, Alan, 7, 68–69, 78
Holley, Robert W., vii, 283, 296f, 312f
  alanine and, 304
  clover-leaf structure, 298
  genetic code and, 313
  growth factors, 324
  Nobel lecture, 313, 317, 318
  nominations, 296, 298–306
  Reichard and, 304
  at Salk Institute, 323–324
  transfer RNA, 298, 304, 323
Holmgren, Gunnar, 153
Honjo, Tasuku, 125
hormone therapy: adrenal glands, 53, 55
  androgen, 53
  ATP and, 285–286
  cortisone, 58
  estrogen, 52, 53, 55, 57, 58
  growth and, v, 35, 47–57
  Huggins and, 52–59, 67, 69, 70, 73

mammary cancer, 58, 73
prostate cancer, 52, 54, 73
receptors, 57
horseshoe crab, 183–188, 183f, 194, 221
Horvitz, H. Robert, 113
Houssay, Bernardo A., 166
How to Win a Nobel Prize (Bishop), 98
Hubbard, Ruth, 203, 207, 209–210, 211
Huebner, Robert J., 86, 87f, 129
Huggins, Charles, v, 50, 51f, 55, 58f, 65, 67f, 69f, 72, 72f
  funding, 53
  hormone therapy and, 52–59, 67, 69–73 (*see also* hormone therapy)
  Klein and, 70
  Nobel lecture, 73
  nomination of, 54–59
  orchiedectomy and, 54
  prostate cancer and, 57, 59
  Rous and, 71, 72
  surgery and, 56
  viruses and, 69
  Warburg and, 52
Huggins, Margaret, 53–54
human genome, 119, 129, 358, 360, 373. *See also* genetic code, *specific topics*
Hungerford, David A., 116
Hunt, Tim, 109
Hunter, Tony, viii
Huntington's disease, 307
Huxley, Andrew, 62
Huxley, Hugh, 62
hybridization (of nucleic acids), 89, 95–96, 352, 355

I

immunology, 10–11, 64, 121–125, 130, 132, 137, 349
inflammation, 121, 127, 130
influenza, 22, 28
information, 106
infrared, 202–203
Ingle, Dwight J., 60
innate immunology, 11
insects, 186, 196
International Agency for Cancer Research (France), 126
International Council for Scientific Unions, 71
Ivanovsky, Dimitry, 31

J

Jackson Laboratory, 38
Jacob, François, 46, 60, 230, 267

Jakoby, William, 251
Jangfeldt, Bengt, x
Javits, Jacob, 89
Jefferson, Thomas, 182–183
Jensen, Elwood, 57, 58f
Johns Hopkins University, 5–6, 5f
Jorpes, Erik, 168, 168f, 169, 175
Joyce, Gerald, viii, 366, 366f

K

Kahlson, Georg, 60, 165
Kandel, Eric, 321
Karlfeldt, Erik Axel, 14
Karolinska Institute, ix, 66, 76, 77, 81, 153, 166, 175, 232, 275, 283, 293, 306, 379
Kärre, Klas, vi
Kastler, Alfred, 67
Kawabata, Yasunari, 310
Kelly, Howard, 5f, 6
Kennedy, Ted, 89
Khorana, H. Gobind, vii, 289f, 312f, 318
  biochemical approach, 296
  education, 289
  genetic code and, 279–285, 290–293, 314, 315
  later research, 319
  membrane protein and, 319
  Nirenberg and, 291, 292, 293
  Nobel lecture, 313, 314
  nominations, 290–296
  nucleotides, 295
  protein synthesis and, 292
  Reichard and, 293
  Todd and, 290
kinases, 102–103
Klein, Georg, vi, 36–38, 37f, 41–44, 69–70, 76
Klug, Aaron, 323
Knaus, Herman, 62
Knoll, Max, 331f
Knudson, Alfred G., 108–109, 108f, 110, 120
Knudson hypothesis, 109, 110, 120
Koch, Robert, 31
Kocher, Emil T., 50
Kornberg, Arthur, 227, 227f, 228, 233, 344
Kornberg, Roger, 343–345, 343f
Kornberg enzyme, 233
Koshland, Daniel E., Jr., 91
Krebs, Edwin G., 102
Kretschmer, Ernst, 380
Krumbhaar, Edward, 49
Krumbhaar, Helen, 49

Kuffler, Stephen W., 62
Kugelberg, Erik, 173, 173f

L

Lacks, Henrietta, 88, 113
Laidlow, Patrick P., 28
Landsteiner, Karl, 17, 18, 24, 29, 64
Lasker, Mary, 89, 89f
Lasker Awards, 89–91
last universal cellular ancestor (LUCA), 1–2
lateral inhibition, 147, 160, 188, 190, 195, 196, 198, 220, 221
Laurent, Torvald C., 272f, 277–278, 294f
Leder, Philip, 262, 262f, 263
Lejeune, Jérome, 61
Leksell, Lars G. F., 172, 172f
leukemia, 53, 244
  abl gene, 118
  bird studies, 30, 35, 332
  childhood, vi, 49, 50, 72, 89, 118
  chromosomes, 116
  EBV and, 118
  Friend virus, 43
  Gross virus, 39, 43
  growth factors, 131
  immune system and, 122
  lymphoblastic, 49
  mice studies, 39, 43, 96, 118
  sarcomas and, 30
  T-cell, 89, 130
  types of, 49
  virotherapy, 134
  white blood cells, 8, 49
Levan, Albert, 115, 115f, 116, 117
Levi-Montalcini, Rita, 102–103
Levinson, Warren, 98
Lewis, Paul, 22
Lewis, Sinclair, 12f, 14, 199
Li, Choh Hao, 56
Library of Congress (U.S.), 183
Lichtenstein, Adolf, 154
life, origin of, 112, 364–381
lifestyles, 63, 138
Li-Fraumeni syndrome, 121
Liljestrand, Göran, 36, 155, 155f
Limulus. *See* horseshoe crab
Lindberg, Erik, 82
Lindsten, Jan, 61
Linnaeus, Carl, 380–381, 381f
Lipmann, Fritz, 45
Little, Clarence C., 38

liver cancer, 132
Loeb, Jacques, 15, 15f, 57, 185
Loeffler, Friedrich A. J., 31
Loewi, Otto, 208–209
Lowy, Douglas R., 132
LUCA. *See* last universal cellular ancestor
Lucia fest, 73–74
luck, in science, 259, 275
Luft, Rolf, 55, 55f, 56, 59, 60
lung cancer, 63, 125
Luria, Salvador E., viii
Lwoff, André, 28, 46, 60
lymphomas, 72, 105, 118, 128–132
lysogeny, 28, 84
lysosome, 341

M

macromolecules, 343–346
magnetic resonance imaging, 136
Maizel, Jacob, V., 110
Marshall, Barry J., 126, 127f
Matthaei, Johannes H., 253–256, 255f, 260, 280
Matthews, Bryan, 172, 174
Maxam, Allan, 352
May, Ben, 53
McClintock, Barbara, 111
measles, 106, 107, 134
Medawar, Peter B., 122
Meige, Joe V., 58
melanoma, 124, 125, 137
Mello, Craig C., 372, 373f
memes, social, 378
Mendeleev, Dmitri, 353
mepacrine, 117
Merkel cell carcinoma, 133–134
Meselson, Matthew, 258
metastasis, 10, 57, 125, 135, 136–137
methionine, 316, 348
Meyerhof, Otto F., 29, 200–201
mice studies, 21, 24, 38, 39, 43, 88
Michelsson, Jeremias, 142
Microbe Hunters (de Kruif), 13
microbiome, 125–128
microelectrode techniques, 162
microscopy, 327, 328
 cryoelectronic, 360, 363
 dark field, 30
 electron microscope, 46, 329–345, 331f, 358–359, 381f
 PALM method, 328

 scanning tunneling and, 329
Mitelman, Felix, vi
mitochondria, 307, 349–350
Mizutani, Satoshi, 94
Modest, Edward, 117
molecular biology, 214, 318–319
 cancer and, 98, 99
 chemistry and, 293, 296
 genetics and, 84, 119, 136, 246, 267, 288, 293, 311, 352, 373
 history of, 34–35, 47, 245, 246, 284–285
 NCI laboratory, 99
 virus and, 98
 See also specific persons, topics
Moniz, Antonio, 50
monkey studies, 40, 107, 180
monoclonal antibodies, 124, 137
Monod, Jacques, 46, 60, 267, 274
Montagnier, Lucy, 29
Morgan, Thomas H., 61
Morton, R. A., 169
Muller, Hermann J., 48
Mulliken, Robert S., 67, 70
Munthe, Axel, xi
Murphy, Frederick, ix, 28
Murphy, William P., 58
Murray, Joseph E., 50
muscular system, 159, 174
music, 76, 127, 192
mustard gas, 49
mutations: alleles and, 109
 genetic, 103, 120, 349, 352, 362, 373
 heredity and, 120–121, 252, 307
 HTLV-1 and, 130
 insertion, 130, 228, 367
 mustard gas, 49
 mutagenesis, 49, 111, 130, 164, 319, 349
 non-sense, 274
 oligonucleotide, 319
 one-step, 305
 p53 and, 110–111
 plus/minus, 248
 primitive, 367
 random, 124, 137
 Rous virus, 100
 site-directed, 319
 somatic, 73, 84, 85, 115
 transmissible mutagens, 10, 28
 viruses and, 10, 28, 30, 31, 100, 137–138
 x-rays and, 48, 164
Myrbäck, Karl, 232f
myxomas, 23, 30

Index 399

N

Nathans, Daniel, 357, 357f
National Cancer Institute (U.S.), vi, 90, 96, 99, 132
Nelson, Karen, vi, 126, 126f
nematodes, 19, 21
nervous system, 8, 114, 145, 167, 191, 196, 198, 321
Newton, Isaac, 151
Nirenberg, Marshall, vii, 63, 205, 248f, 255f, 308, 309f, 312f, 318
   amino acids and, 276
   cell biology and, 321
   early years, 248–250
   education of, 248–251
   genetic code and, 265, 266, 268, 271–274, 278, 280, 281, 293
   Khorana and, 291, 292, 293
   later research, 320, 322
   linking of tRNA to ribosomes, 263, 264
   Matthaei and, 254, 256, 280
   nerve cells, 320
   at NIH, 251, 260–261, 262
   Nobel lecture, 313
   nominations, 265–282
   nucleotides and, 279, 294
   Ochoa and, 271
   personality, 265, 271, 321–323
   phenylalanine and, 259
   poly-U studies, 254–259, 272, 275, 279, 313
   protein synthesis, 254–262, 273, 313
   Reichard and, 269, 271, 272, 276, 279
   solvent story, 257–258
   Theorell and, 281
Niven, Janet S. F., 60
Nixon, Richard, 90
Nobel, Alfred, 81f, 336, 374–375, 374f
Nobel, Ludvig, 15
Nobel, Robert, 15
Nobel Committee, 60, 66–67, 77, 92, 156, 165, 166, 271–272, 275, 282, 378–380
Nobel Foundation, 76, 77, 80–81, 311
Nobel medallions, 15, 79–83, 81f, 83f
Nordling, Carl O., 119
Norrby, E., x, 37, 46, 74, 76, 77, 107, 142f, 156, 243, 243f
nucleic acids, 44, 95, 286, 299
   Caspersson and, 35
   Chargaff and, 240
   discovery of, 311
   DNA (*see* deoxyribonucleic acid)
   enzymes and, 228
   genetic code and (*see* genetic code)
   G-proteins and, 104
   hybridization, 275
   nucleotides and, 252, 355 (*see also* nucleotides)
   proteins and, 104, 299, 302, 366 (*see also* protein synthesis)
   replication, 349
   RNA (*see* ribonucleic acid)
   sequencing, 125, 128, 252
   synthesis of, 232, 233
   viruses and, 226, 227, 252, 367, 369, 371 (*see also* viruses)
   See also specific persons, topics
nucleotides, 252, 285, 287, 301, 351
   amino acids and, 247, 347
   base pairing, 237
   block polymerization, 295
   Chargaff and, 237
   clover leaf structure and, 302, 303f, 317
   codons and, 261, 313 (*see also* codons)
   DNA and, 89, 226, 286, 287f, 352, 354–355 (*see also* deoxyribonucleic acid)
   double helix, 225–226, 237 (*see also* double helix)
   editing, 356
   energy and, 288
   enzymes and, 233, 295; finger-printing technique, 305
   genetic code and, 245, 262, 263, 284, 347 (*see also* genetic code)
   modifications, 351
   Nirenberg and, 294
   nucleic acids and, 252, 355 (*see also* nucleic acids)
   nucleosides, 287
   palindromic, 357
   polynucleotides, 295
   reductase, 271
   ribosomes and, 345
   RNA and, 89, 263, 286, 287f (*see also* ribonucleic acid)
   sequence, 89, 305, 352, 354–355, 357
   synthesized, 294, 295
   triplets, 248, 261–268, 274, 279, 313, 347
   wobble and, 351; See also specific persons, topics
Nurse, Paul M., 109

O

Ochoa, Severo, 59, 227, 227f, 232–235, 258, 271, 279
Ochoa enzyme, 231–236
Ohsumo, Yoshinori, 113–114
Okazaki fragments, 228
oncogenes, 72, 85, 89, 101–108, 118, 128–130, 137
oncolytic effects, vi, 134, 138
Onsager, Lars, 270
Orgel, Leslie, 364–365, 364f
origin of life, 112, 364–381
Origin of the Species (Darwin), 148, 181
Osler, William, 5f, 6
ovaries, 53, 57, 58, 109, 120

P

p53 gene, 110, 111
Palade, George E., 229, 340
pancreas, 138
papillomas, 23, 31–33, 39, 41, 45, 111, 132
papovavirus, 41, 45
Pavlov, Ivan P., 353
penicillin, 155
peptides, 11, 217, 264, 273–274, 292. *See also* polypeptides
Perlmann, Thomas, ix
Pestka, Sidney, 263, 265
Pettersson, Ulf, viii, 358
Phemister, Dallas, 51
phenylalanine, 271, 323
    alanine and, 316
    childhood disorder and, 307
    first codon, 255–257, 259
    Nirenberg and, 259
    poly-u and, 255–257, 259, 272, 279, 313
    ribosomes and, 263
    triplets, 292
    tRNA, 292, 315, 316, 323, 362
phenylketonuria, 307
Philadelphia chromosome, 116, 117
Philipson, Lennart, 356, 356f
Phua, Kok Khoo, ix
picornaviruses, 95
Pincus, Gregory, 62
plaques, 42
plate tectonics, 140
polio: Cutter episode, 40
    Eddy and, 40
    Flexner and, 7
    Gard and, 334
    picornaviruses, 95
    RNA and, 95
    Rous and, 17
    SV40, 41
    tumors and, 41
    two-phase system, 298
    ultrafiltration and, 31
    vaccines, 40–41, 64, 354
polymerases, 344
polyoma virus, 39, 40, 43, 133–134
polypeptides: amino acids and, 231, 246, 247, 254, 258, 295, 317, 348 (*see also* amino acids)
    antigens, 256
    genetic code and, 270, 274, 290, 292
    nucleotides and, 274
    ORFs and, 372
    phenylalanine, 255
    proteins and, 252, 254, 264, 290
    ribozymes and, 368
    RNA and, 258, 259, 268, 273, 372
    See also specific persons, topics
precision medicine, 136
Prelog, V., 314–315
priority, in science, 160
progression, concept of, 24
proline, 110
"Prostata" (song), 73–74
prostate cancer, 52–55, 57, 59, 67
protein synthesis, 110, 285, 291, 317
    amino acids and, 245, 276–277, 299, 301
    DNA and, 225–226, 230, 349, 355 (*see also* deoxyribonucleic acid)
    emergence of life, 367
    enzyme functions and, 367
    genetic code and, 304, 315
    G-proteins and, 104
    Khorana and, 292
    measurements of, 271, 272
    membrane protein, 319
    methionine, 305
    Nirenberg and, 313
    nucleic acids and, 302, 366 (*see also* nucleic acids)
    origin of life, 365–368, 376
    polypeptides (*see* polypeptides)
    ribozymes, 368
    RNA and, 225–226, 247, 252, 278, 349, 355, 368, 372 (*see also* ribonucleic acid)

R

rabbit studies, 21–25, 30, 32, 33
Ramakrishnan, Venkatraman, 344, 344f
Ramel, Povel, 73
Ratliff, Floyd, 189, 190, 220
RB1 gene, 120
receptors: cancer and, 103, 134
  discovery of, 57, 210, 211
  EGF and, 103, 106
  G protein-coupled, 319
  HER2 oncogene and, 106
  hormones and, 57
  immunotherapy and, 124
  lateral inhibition, 147, 160, 188, 190, 195, 198, 221
  light activated, 147–150, 167, 170–174, 180, 184, 189, 193, 197, 203, 207, 211, 223
  muscles and, 159, 188
  See also specifc types, topics
Regenmortel, Marc van, 31
Reichard, Peter, viii, 239, 268–276, 272f, 279, 293, 304
Reimers, Knud H., 154–155
respiratory syncytial (RS) virus, 107–108
restriction enzymes, 357, 358
retinoblastoma, 120
retinol, 203
retroviruses, 94, 98, 102, 104, 128, 129, 130
reverse transcriptase, 94, 96, 97, 368–369
Rexed, Bror, 191, 191f
Rh system, 200–202
rhodopsin, 200–202, 209, 223, 319
ribonucleic acid (RNA), 88, 96f, 105, 144, 235, 246, 285
  alanine and, 292, 297, 299f, 301–304, 303f, 316, 351
  amino acids and, 258, 297, 298, 351 (*see also* amino acids)
  base sequences, 275, 361f
  caps, 356
  catalytic properties, 360
  central dogma and, 96, 97, 231
  codons and, 295f, 304, 317 (*see also* codons)
  Crick and, 297 (*see also* Crick, F.)
  crystallography, 323
  DNA and, 94, 97, 229, 230, 254, 291, 327–381, 355, 359f, 368 (*see also* deoxyribonucleic acid)
  double helix and, 225–226, 237, 240–246, 288, 355
  editing, 356
  enzymes and, 96, 253, 259, 298, 362
  eukaryotes and, 356
  evolution and, 371 (*see also* evolution)
  evolution of, 351
  exons, 359–360
  folding of, 362
  forms of, 284
  gene silencing, 372–373
  genetic code and (*see* genetic code)
  genome and, 129
  hybridization, 359, 359f
  information and, 367, 371–372;
  interfering, 320–321
  introns, 359–360, 362
  lateral transfer and, 371
  ligation, 360
  mRNA, 95–96, 230, 259, 266–268, 277, 279, 285, 303, 341, 345, 355–358, 359f
  nucleotides and, 89, 263, 286, 287f, 351 (*see also* nucleotides)
  ORFs, 360
  origin of life and, 112, 364–381
  polymerases, 96, 370–371
  processing, 363
  protein synthesis and, 225–226, 247, 252, 349, 355, 368, 372 (*see also* protein synthesis)
  replication, 93–94, 229, 231, 368–369
  reverse transcriptase, 94, 231–232
  ribosomes and, 229, 285, 305, 314, 369
  ribozymes and, 362, 363, 363f, 367
  RNA world, 363
  role of, 245, 349, 366, 371–372, 373
  self-splicing, 362
  sequencing, 95, 299, 305
  shielding effects, 362
  splicing, 359, 360, 361, 363, 372
  synthesis of, 50, 93, 232, 233, 271, 272, 291, 295f
  TMV and, 252, 253
  transcription and, 94, 231, 358, 372
  transesterification, 362
  tRNA, 230, 263, 266, 276–279, 294, 297, 299f, 300–304, 303f, 315–317, 345, 351, 362, 363f, 368
  viroids, 370–371
  viruses and, 63, 86, 252, 369 (*see also* viruses)
  See also specific persons, topics
ribonucleotide reductase, 271, 368
ribosomes, 229, 230, 263, 294, 304, 317, 341, 344–345, 369

ribozymes, 285, 363, 363f, 367–370
Richards, Dickinson W., 50
Rivers, Thomas, 33f
RNA. See ribonucleic acid
RNA tie club, 247, 252
Roberts, Richard J., 358, 358f
Robinson, Robert, 52
Rock, John, 62
Rockefeller Foundation, 22
Rockefeller Institute, 15–17, 22, 34–36, 188, 245, 340
Rodbell, Martin, 104
Röntgen, W. Conrad, 48
roundworms, 19
Rous, Peyton, v, 7f, 28, 56, 67f, 69f, 70f, 72f
    Arrowsmith and, 12, 16
    chemotherapy and, 72
    correspondence, 78–79
    early life, 4–7
    education, 6
    experiments, 7–10
    filtrate, 32
    Flexner and, 9
    Hodgkin and, 78
    Huggins and, 71, 72–73
    Klein and, 41, 44, 69, 76
    Landsteiner and, 17
    Nobel lecture, 72–73, 85
    nominations of, 16–18, 29–36, 44–46
    polyoma and, 39
    progression and, 103, 244
    research pursuits, 10, 34–36
    Rockefeller Institute and, 16–17, 34–36
    Rous virus, 23, 42, 69, 72, 98–104, 116, 128, 130, 135, 256, 340
    sarcomas and, 4, 9, 10, 28, 32, 98, 116
    Shope and, 45
    somatic mutation theory, 73, 84, 85, 115
    in Stockholm, 68
    three musketeers, 25
    tuberculosis, 6–7
    tumor transfer and, 4, 7–10
    TV program and, 75
    ultra-filtrates and, 9, 25
    viruses and, 10, 21–25, 65, 67, 76, 84
    See also specific topics
RS virus. See respiratory syncytial virus
Rubin, Harry, 42, 42f
Rushton, William, 147, 158, 158f
Ruska, Ernst, viii, 329–338, 331f, 342–343
Ruska, Helmut, 330–335, 333f, 337, 339
Russia, 353–354

S

Sabin vaccine, 41, 64, 354
Sachs, Nelly, 67
Salk Institute, viii, 102, 323–325, 324f
Salk vaccine, 40, 41, 64
Sanarelli, Giuseppi, 24
Sanger, Frederick, 299–300, 305, 352
sarcomas, 8, 98, 116
    carcinomas and, 22–23
    forms of, 104
    heredity and, 121
    HIV and, 122, 130, 133
    immune suppression and, 122
    leukemia and, 30, 43
    ras gene, 104
    Rous and, 4, 9–10, 28, 32, 98, 116
    srs gene, 98
    ultrafiltration and, 8, 24
    viruses and, 1–10, 23, 28, 35, 43, 98, 116, 122, 130, 332
scanning tunneling microscopy, 329
Schiller, John T., 132
Schmitt, Francis, 321
scientific process, 17–22, 159, 222, 315, 325, 340, 377
Scolnick, Ed, 265
SDS gel technique, 110
Segerstedt, Torgny T., 272f
Sela, Michael, vii, 256, 256f, 257
selenocysteine, 350
self-consciousness, 205, 365, 377
sequencing, 119, 125, 128, 136, 234, 252, 301–302, 305, 319, 349
Sharp, Phillip A., 358, 358f
Sherrington, Charles S., 20, 145, 147, 159, 164
Shope, Richard E., 22, 22f, 28, 31, 32, 36, 39, 45, 65
Siegbahn, K. Manne G., 338, 339, 339f
simian virus (SV), 40, 41, 133
Singer, Maxine, 259, 260, 374f
Sjöstrand, Fritiof S., 338, 339f
skin, 2, 22, 32, 45, 111, 134
Skoglund, Carl R., 153f
Smith, George, 376
Smith, Hamilton O., vii, 323, 357, 357f
Smith, Wilson, 28
Snell, George D., 38
Snow, C. P., 160–161
somatic mutation theory, 73, 84, 85, 115
songs, 73, 74–75
Soviet Union, 152, 352–354
Speed, Kellogg, 54

Index 403

Spiegelman, Sol, 88, 352, 354, 354f
splicing, genetic, 359–363, 368, 372
Stanley, Wendell, 332
start/stop codons, 264, 274, 277, 279, 296, 314, 348–350
Stehelin, Dominique, 100, 101
Stein, William H., 44
Steinberg, Charles, 348
Steitz, Thomas A., 344, 344f
Stewart, Sarah E., 36, 39, 40f
Stewart, Timothy, 265
stilbestrol, 55
Sulston, John E., 113
suppressor genes, 106–111
surgery, 48, 50
SV. *See* simian virus
Svaetichin, Gunnar, 152, 158, 220
Svartz, Nanna, 165
Svedberg, The, 33, 34, 266f
Swedish coast, 139–142
Swedish language, 275–276
Sweet, Ben, 41
swine, 22
Sylvén, Christer, vi, 68
syncytia, 107, 129
Szent-Györgyi, Albert, 59, 238
Szilárd, Léo, 335–336
Szostak, Jack W., 111, 112f

T

T cells, 118, 122, 125, 128
tar-induced tumors, 21, 24, 30, 32, 33
telomeres, 111–112, 113, 369
Temin, Howard M., 42, 93, 93f, 96, 97
testosterone, 52
Theorell, Hugo, 63, 169, 209, 210, 212, 213, 232f, 271, 281
Thomas, E. Donnell, 50
thymidine, 226, 237, 241
Tigerstedt, Carl, 145
Tiselius, Arne, 266, 266f, 267, 333, 334f
TMV. *See* tobacco mosaic virus
tobacco mosaic virus (TMV), 30, 31, 226, 252, 253, 332
Todaro, George, 86, 87f, 129
Todd, Alexander, 227, 285, 285f, 286, 287, 288, 290
Tomkins, Gordon, 253, 253f
tomography, 345
Törnebladh, Ragnar, 81
transcription, 94–97, 105, 231, 368
triangulation technique, 51

triplets, codons and, 248, 261, 264–268, 295, 347. *See also* codons
tumors: BRCA genes, 120
chromosomes and, 116–119 (*see also* chromosomes)
emergence of, 3; fibromas, 23, 31, 33
lymphomas, 72, 105, 118, 128–132
oncogenes, 48, 72, 85, 89, 101–108, 118, 128–130
papillomas, 23, 31–33, 39, 41, 45, 111, 132
pathology atlas, 119
sarcomas (*see* sarcomas)
suppressor genes, 119, 121, 137
surgery and, 48, 50
ultra-filtrates and, 9, 23, 25, 30, 31
viruses and, 9, 25–29, 67, 84 (*see also* viruses)
See also specific types, topics
Turner, William, 183
Turpin, Raymond, 61
TV programs, 75–76
two-hit rule, 109, 110, 120
tyrosine, 102, 106, 118, 301, 304, 305, 317, 362, 363f
Tyrrell, David, 28, 29

U

ultracentrifugation, 229, 240, 258, 287, 340, 341
ultrafiltration, 4, 9, 23, 25, 30, 31, 362
papillomas, 32
phages and, 25
poxviruses, 30
Rous experiment, 7, 8
sarcomas and, 8
Shope and, 22–23
viruses and, 9, 30–31, 329, 332
ultramicrotome, 332
ultrastructural studies, 329, 330, 333, 335, 339–340, 341
UV light, 111, 237

V

vaccines, 47
cancer and, 123, 132, 137–138
cell line and, 88, 320
hepatitis and, 132–133
inactivated, 107
polio, 40–41, 64, 354
production of, 88, 132
smallpox, 30
testing of, 354

yellow fever, 156
    See also specific types
van Leeuwenhoek, Antonie, 327–328
Varmus, Harold E., 98, 99, 99f, 100
Venter, Craig, x–xi
Vietnam War, 260
Vigeland, Gustav, 82
Villa, Luigi, 329–330, 334
Virchow, Rudolf, 1
Virgil, 79–80
viroids, 370
virology: bacteriophages (*see* bacteriophages)
    cell lines and, 320
    electron microscopy and, 334
    history of, ix, 27
    International Congress, 353
    Karolinska Institute, 123, 156
    nucleic acids and, 44
    retroviruses, 94
    in Russia, 354
    tumor, 44, 65, 79
    See also specific persons
virotherapy, 134, 138
viruses, 44, 76, 95, 128–134
    bacteria and, viii, 14, 18, 27, 35, 63, 226, 335, 357, 371
    bacteriophages (*see* bacteriophages)
    Bittner factor and, 35
    cancer and, 25–29, 35, 67, 84, 128–138, 340–341
    Caspersson and, 35
    cells and, v, 42, 107, 108
    classification structure, 335
    contagious, 31
    diversity of, 369
    DNA and, 86, 226, 252, 369, 371 (*see also* deoxyribonucleic acid)
    early studies, 328–329
    electron microscope and, 333f, 381f (*see also* electron microscope)
    endogenous, 35
    enzyme and, 96, 97 (*see also* enzymes)
    evolution and, viii
    fairy-story, 25–27
    genetics of, 100, 305–306, 367, 371 (*see also* genetic code)
    history of, 25–27
    lysogeny, 28, 84
    microscopy and, 331–339 (*see also* microscopy)
    morphology of, 334
    negative contrast, 333
    nucleic acids and, 28, 226, 252, 367, 369, 371 (*see also* nucleic acids)
    oncogenes and, 128–129
    oncolytic effects, vi, 134, 138
    particulate nature of, 30
    polyoma and, 39
    pre-cellular, 371
    quasi-species, 349
    recombination and, 260
    replication, 35, 93, 355, 371
    restriction enzymes, 357
    retroviruses, 94, 98, 102, 104, 128, 129, 130
    ribozymes and, 369
    RNA and, 63, 86, 252, 369 (*see also* ribonucleic acid)
    Rous and, 10, 21–25, 67, 84 (*see also* Rous, P.)
    sarcomas and, 23
    splicing and, 359
    src and, 100
    structure of, 332, 334
    subviral agents, 370
    tumors and, 9, 25–29, 67, 84 (*see also* tumors)
    types of, 131
    ultrafiltration and, 9, 23, 30, 328, 329 (*see also* ultrafiltration)
    virocentric view, viii
    viroids, 370
    virology (*see* virology)
    virome, 125
    virotherapy, 134, 138
    virus theory, 28
    warts and, 21–25, 45, 111
    See also specific types, topics
vitamin A, 200–201
vitamin C, 59
vitamin D, 56
Vogelstein, Bert, 119, 120f
Vogt, Marguerite, 42
Vogt, Peter, 100
von Borries, Bodo, 330, 332, 334, 337
von Braun, W., 335
von Euler, Hans, 206
von Euler, Ulf S., 153, 162, 162f, 166, 173, 218, 309

W

Wagner-Jauregg, Julius, 20
Waksman, Byron, 201–202
Waksman, Selman A., 36
Wald, George, 62, 177, 178, 197, 199f, 215f, 220f

cis-trans formation, 203, 220, 223
color and, 204, 212, 213
education of, 199
as educator, 204
Granit and, 205–207, 208, 214
Hartline and, 214
Hubbard and, 209–210
Nobel lecture, 222–224
nomination of, 205–214
Theorell and, 213
vision and, 204, 210, 212–213
vitamin A and, 200, 210–211
Warburg and, 200
Woods Hole, 202
Wallenberg Foundation, 153
Warburg, Otto, 20, 52, 54, 55, 58, 200
Warren, J. Robin, 126
warts, 21–25, 45–46, 111
Watson, Harry, ix
Watson, James, 226, 238, 242, 244, 246
Wegener, Alfred, 140
Weinberg, Robert A., 105f
Weiss, Samuel D., 235
Welch, William Henry, 5f, 6
Westman, Axel, 56, 57–58, 59
Westman, Sven, 74
white blood cells, 8, 49
Wilkins, Maurice, 238, 246
Winter, Gregory P., 125, 376

wobble phenomenon, 279, 347–348, 351
women, 70, 74, 82, 111, 234–236, 260, 323.
  See also specific topics
Woods Hole Laboratory, 201
World Health Organization, 41

X

X-rays, 48

Y

Yamagiwa, Katsusaburo, 18, 18f, 19, 20, 21
Yonath, Ada E., 344, 344f
Young, Thomas, 151
Young–Helmholtz theory, 152, 164

Z

Zamecnik, Paul, 278
Zamenhof, Stephan, 268
Zech, Lore, 117
Zhang, Shuguang, vi
Zhdanov, Viktor, 352, 353, 353f
Zotterman, Yngve, 62, 154, 154f, 164, 171, 178, 195–196, 211, 212, 220
zur Hausen, Harald, vi, 122, 131, 131f